Natural Language Processing
and
Information Retrieval

TANVEER SIDDIQUI
Assistant Professor
Indian Institute of Information Technology
Allahabad

U.S. TIWARY
Professor
Indian Institute of Information Technology
Allahabad

OXFORD
UNIVERSITY PRESS

OXFORD
UNIVERSITY PRESS

Oxford University Press is a department of the University of Oxford.
It furthers the University's objective of excellence in research, scholarship,
and education by publishing worldwide. Oxford is a registered trademark of
Oxford University Press in the UK and in certain other countries

Published in India by
Oxford University Press
22 Workspace, 2nd Floor, 1/22 Asaf Ali Road, New Delhi 110 002

First published 2008
Seventh impression 2019

ISBN-13: 978-0-19-569232-7
ISBN-10: 0-19-569232-2

Typeset in Baskerville
by The Composers, New Delhi 110 063
Printed in India by Rakmo Press, New Delhi 110 020

For product information and current price, please visit www.india.oup.com

PREFACE

Natural language processing (NLP) is among the most heavily researched areas in computer science today. Historically, NLP has been concerned with developing machine translation (MT) and natural language generation (NLG) systems. Textbooks on NLP mainly focus on development of grammar-based language models and various types of processing involved in the development of these systems such as morphological processing and parsing. But nowadays, information retrieval (IR) has emerged as one of the most important applications of NLP. Natural language processing has to deal with IR and related applications such as information extraction and text summarization. Further, as the Web is becoming multilingual, the need for tools and techniques for automatic processing of languages, besides English and other major European languages, is evident. This book covers these recent applications apart from the traditional ones. Multilingual issues are dealt with by giving examples from Indian languages, wherever possible. We have tried to give a balanced mix of theory and practice.

Natural Language Processing and Information Retrieval aims to give the readers a sound understanding of NLP and prepare them for taking up challenging tasks in the area. We have tried to acquaint the readers with the tools, techniques, applications, and challenges existing in the area. We have written this book as much for ourselves as for our students. Some of the matter in this book is the outcome of the hands-on research experience of the authors. This book will be useful for the students involved in project and research work in text processing and related applications.

Who should Read this Book?

This book is suitable for one-semester undergraduate and graduate courses. The applications of NLP covered in the book make it useful for the students doing project work. It can also be used as a reference and resource to young researchers involved with NLP; they will find it to be a helpful guide to the newly established techniques of rapidly growing research field of NLP and IR.

Content and Structure

The book is organized into twelve chapters that are almost evenly distributed between theory and application.

Chapter 1 offers an insight into NLP and IR and sheds light on the factors that make NLP challenging. It also discusses various levels of analysis involved in NLP and the knowledge used by these levels of analysis.

Chapter 2 deals with language modelling. The models covered in this chapter include lexical functional grammar, government and binding, Paninian grammar, and *n*-gram based model.

Chapters 3 to 6 are devoted to various levels of NLP. *Chapter 3* focuses on word level analysis, whereas *Chapter 4* deals with syntactic analysis. Besides grammar framework to describe phrase structure of English language and various parsing approaches, Chapter 4 also includes a discussion on Paninian grammar-based framework, which is suitable for parsing Indian languages. *Chapter 5* is devoted to semantic analysis of text. It discusses meaning representation of a sentence, a general approach to semantic analysis based on semantic compositionality, and word sense disambiguation approaches. *Chapter 6* is on discourse processing. It includes a discussion on cohesive devices in languages, reference resolution methods, and discourse coherence and structure. This chapter also presents a theoretical framework to discourse analysis.

Chapters 7 to 11 cover various NLP applications. *Chapter 7* is concerned with automated natural language generation (NLG). It introduces the issues involved in NLG and describes the architecture of a generation system and various approaches for generating text automatically. *Chapter 8* is on machine translation (MT). This chapter discusses various types of MT approaches such as direct, rule-based (transfer and interlingua), corpus-based, and knowledge-based. It also highlights characteristics of Indian languages and translation strategies.

Chapters 9 and 10 both focus on information retrieval (IR). *Chapter 9* is concerned with the design of an IR system. It introduces various IR models and offers a detailed discussion of vector space retrieval model and its evaluation. This chapter also offers a brief introduction to some of the freely available resources and tools useful for research work in the area. *Chapter 10* introduces some of the difficulties associated with keyword-based retrieval systems and discusses the use of NLP in IR that goes beyond what has been used in vector-based approaches.

Chapter 11 emphasizes different approaches to design and evaluation of information extraction, text summarization, and question-answering systems.

Chapter 12 presents a ready reference of freely available tools and other useful lexical resources. It offers a discussion on tools such as stemmers, taggers, lexical resources such as WordNet, FrameNet, and text corpora/tests document collections useful for research work such as EMILLE, CACM, and TREC collections. This chapter also has information on major journals and conferences related to NLP and IR.

Acknowledgements

This book would not have been possible without the aid and collaboration of many people whom we wish to thank and remember.

Tanveer Siddiqui would like to acknowledge her brother for encouraging her to write a book. She would also like to thank her parents for giving her support.

U.S. Tiwary would like to thank his wife Padma, daughter Zoya, and son Kislay for giving him freedom for pursuing this task.

We would also like to thank our students Alkesh Patel, Pradeep Varma Dantuluri, Fahad Mahmood, and Vaibhav Rastogi for helping us in giving useful information on lexical resources given in Chapter 12. This chapter would not have been possible without the indirect support of open source providers of the resources covered in it. So, we extend our thanks to them. Fahad also helped us in editing Urdu and Hindi examples included in this book. Pradeep helped us in providing description on some of the taggers covered in Section 12.5.

We thank the reviewers of this book for their useful suggestions.

Finally, we would like to thank the editorial team of Oxford University Press involved with this book for their encouragement and support.

Tanveer Siddiqui
U.S. Tiwary

CONTENTS

CHAPTER 1

INTRODUCTION

CHAPTER OVERVIEW

This chapter gives an idea of natural language processing (NLP) and information retrieval (IR). Various levels of analysis involved in NLP along with the knowledge used by these levels of analysis are discussed. Some of the difficulties in analysing text and specific factors that make automatic processing of languages difficult are also touched upon. The chapter underlines the role of grammar in language processing and introduces transformational grammar. Indian languages differ a lot from English. These differences are clearly pointed out. Further, a number of NLP applications are introduced along with some of the early NLP systems. Towards the end, information retrieval is discussed.

1.1 WHAT IS NATURAL LANGUAGE PROCESSING (NLP)

Language is the primary means of communication used by humans. It is the tool we use to express the greater part of our ideas and emotions. It shapes thought, has a structure, and carries meaning. Learning new concepts and expressing ideas through them is so natural that we hardly realize how we process natural language. But there must be some kind of representation in our mind, of the content of language. When we want to express a thought, this content helps represent language in real time. As children, we never learn a computational model of language, yet this is the first step in the automatic processing of languages. *Natural language processing* (NLP) is concerned with the development of computational models of aspects of human language processing. There are two main reasons for such development:

1. To develop automated tools for language processing
2. To gain a better understanding of human communication

Building computational models with human language-processing abilities requires a knowledge of how humans acquire, store, and process language. It also requires a knowledge of the world and of language.

Historically, there have been two major approaches to NLP—the *rationalist* approach and the *empiricist* approach. Early NLP research took a rationalist approach, which assumes the existence of some language faculty in the human brain. Supporters of this approach argue that it is not possible for children to learn a complex thing like natural language from limited sensory inputs. Empiricists do not believe in existence of a language faculty. Instead, they believe in the existence of some general organization principles such as pattern recognition, generalization, and association. Learning of detailed structures can, therefore, take place through the application of these principles on sensory inputs available to the child.

1.2   ORIGINS OF NLP

Natural language processing sometimes mistakenly termed natural language understanding—originated from machine translation research. While natural language understanding involves only the interpretation of language, natural language processing includes both understanding (interpretation) and generation (production). The NLP also includes speech processing. However, in this book, we are concerned with text processing only, covering work in the area of computational linguistics, and the tasks in which NLP has found useful application.

Computational linguistics is similar to theoretical- and psycho-linguistics, but uses different tools. Theoretical linguists mainly provide structural description of natural language and its semantics. They are not concerned with the actual processing of sentences or generation of sentences from structural description. They are in a quest for principles that remain common across languages and identify rules that capture linguistic generalization. For example, most languages have constructs like noun and verb phrases. Theoretical linguists identify rules that describe and restrict the structure of languages (grammar). Psycholinguists explain how humans produce and comprehend natural language. Unlike theoretical linguists, they are interested in the representation of linguistic structures as well as in the process by which these structures are produced. They rely primarily on empirical investigations to back up their theories.

Computational linguistics is concerned with the study of language using computational models of linguistic phenomena. It deals with the application of linguistic theories and computational techniques for NLP. In computational linguistics, representing a language is a major problem; most knowledge representations tackle only a small part of knowledge.

Representing the whole body of knowledge is almost impossible. The words knowledge and language should not be confused. This is discussed in detail in Section 1.3.

Computational models may be broadly classified under knowledge-driven and data-driven categories. Knowledge-driven systems rely on explicitly coded linguistic knowledge, often expressed as a set of handcrafted grammar rules. Acquiring and encoding such knowledge is difficult and is the main bottleneck in the development of such systems. They are, therefore, often constrained by the lack of sufficient coverage of domain knowledge. Data-driven approaches presume the existence of a large amount of data and usually employ some machine learning technique to learn syntactic patterns. The amount of human effort is less and the performance of these systems is dependent on the quantity of the data. These systems are usually adaptive to noisy data.

As mentioned earlier, this book is mainly concerned with computational linguistics approaches. We try to achieve a balance between semantic (knowledge-driven) and data-driven approaches on one hand, and between theory and practice on the other. It is at this point that the book differs significantly from other textbooks in this area. The tools and techniques have been covered to the extent that is needed to build sufficient understanding of the domain and to provide a base for application.

The NLP is no longer confined to classroom teaching and a few traditional applications. With the unprecedented amount of information now available on the web, NLP has become one of the leading techniques for processing and retrieving information. In order to cope with these developments, this book brings together information retrieval with NLP. The term *information retrieval* is used here in a broad manner to include a number of information processing applications such as information extraction, text summarization, question answering, and so forth. The distinction between these applications is made in terms of the level of detail or amount of information retrieved. We consider retrieval of information as part of processing. The word 'information' itself has a much broader sense. It includes multiple modes of information, including speech, images, and text. However, it is not possible to cover all these modes due to space constraints. Hence, this book focuses on textual information only.

1.3   LANGUAGE AND KNOWLEDGE

Language is the medium of expression in which knowledge is deciphered. We are not competent enough to define language and knowledge and its implications. We are here considering the text from of the language and the content of it as knowledge.

Language, being a medium of expression, is the outer form of the content it expresses. The same content can be expressed in different languages. But can language be separated from its content? If so, how can the content itself be represented? Generally, the meaning of one language is written in the same language (but with a different set of words). It may also be written in some other, formal, language. Hence, to process a language means to process the content of it. As computers are not able to understand natural language, methods are developed to map its content in a formal language. Sometimes, formal language content may have to be expressed in a natural language as well. Thus, in this book, language is taken up as a knowledge representation tool that has historically represented the whole body of knowledge and that has been modified, maybe through generation of new words, to include new ideas and situations. The language and speech community, on the other hand, considers a language as a set of sounds that, through combinations, conveys meaning to a listener. However, we are concerned with representing and processing text only. Language (text) processing has different levels, each involving different types of knowledge. We now discuss various levels of processing and the types of knowledge it involves.

The simplest level of analysis is *lexical analysis*, which involves analysis of words. Words are the most fundamental unit (syntactic as well as semantic) of any natural language text. Word-level processing requires morphological knowledge, i.e., knowledge about the structure and formation of words from basic units (morphemes). The rules for forming words from morphemes are language specific.

The next level of analysis is *syntactic analysis*, which considers a sequence of words as a unit, usually a sentence, and finds its structure. Syntactic analysis decomposes a sentence into its constituents (or words) and identifies how they relate to each other. It captures grammaticality or non-grammaticality of sentences by looking at constraints like word order, number, and case agreement. This level of processing requires syntactic knowledge, i.e., knowledge about how words are combined to form larger units such as phrases and sentences, and what constraints are imposed on them. Not every sequence of words results in a sentence. For example, 'I went to the market' is a valid sentence whereas 'went the I market to' is not. Similarly, 'She is going to the market' is valid, but 'She are going to the market' is not. Thus, this level of analysis requires detailed knowledge about rules of grammar.

Yet another level of analysis is *semantic analysis*. Semantics is associated with the meaning of the language. Semantic analysis is concerned with creating meaningful representation of linguistic inputs. The general idea of semantic interpretation is to take natural language sentences or utterances and map them onto some representation of meaning. Defining meaning components is difficult as grammatically valid sentences can be meaningless. One of the famous examples is, 'Colorless green ideas sleep furiously' (Chomsky 1957). The sentence is well-formed, i.e., syntactically correct, but semantically anomalous. However, this does not mean that syntax has no role to play in meaning. Bach (2002) considers:

> *'... semantics to be a projection of its syntax. That is semantic structure is interpreted syntactic structure.'*

But definitely, syntax is not the only component to contribute meaning. Our conception of meaning is quite broad. We feel that humans apply all sorts of knowledge (i.e., lexical, syntactic, semantic, discourse, pragmatic, and world knowledge) to arrive at the meaning of a sentence. The starting point in semantic analysis, however, has been lexical semantics (meaning of words). A word can have a number of possible meanings associated with it. But in a given context, only one of these meanings participates. Finding out the correct meaning of a particular use of word is necessary to find meaning of larger units. However, the meaning of a sentence cannot be composed solely on the basis of the meaning of its words. Consider the following sentences:

Kabir and Ayan are married. (1.1a)

Kabir and Suha are married. (1.1b)

Both sentences have identical structures, and the meanings of individual words are clear. But most of us end up with two different interpretations. We may interpret the second sentence to mean that Kabir and Suha are married to each other, but this interpretation does not occur for the first sentence. Syntactic structure and compositional semantics fail to explain these interpretations. We make use of pragmatic information. This means that semantic analysis requires pragmatic knowledge besides semantic and syntactic knowledge.

A still higher level of analysis is *discourse analysis*. Discourse-level processing attempts to interpret the structure and meaning of even larger units, e.g., at the paragraph and document level, in terms of words, phrases, clusters, and sentences. It requires the resolution of anaphoric references and identification of discourse structure. It also requires discourse knowledge, that is, knowledge of how the meaning of a sentence is determined by preceding sentences—e.g., how a pronoun refers to the preceding noun—and how to determine the function of a sentence in the text. In fact, pragmatic knowledge may be needed for resolving anaphoric references. For example, in the following sentences, resolving the anaphoric reference 'they' requires pragmatic knowledge:

> *The district administration refused to give the trade union permission for the meeting because they feared violence.*   (1.2a)
> *The district administration refused to give the trade union permission for the meeting because they oppose government.*
>   (1.2b)

The highest level of processing is *pragmatic analysis*, which deals with the purposeful use of sentences in situations. It requires knowledge of the world, i.e., knowledge that extends beyond the contents of the text. The Cyc project (Lenat 1986) at University of Austin is an attempt to utilize world knowledge in NLP. However, its usefulness in a general-domain NLP system is yet to be demonstrated. Furthermore, whether or not semantics can be associated with a symbol manipulator and whether humans use logic in the same way as the Cyc project, are both issues of debate.

1.4 THE CHALLENGES OF NLP

There are a number of factors that make NLP difficult. These relate to the problems of representation and interpretation. Language computing requires precise representation of content. Given that natural languages are highly ambiguous and vague, achieving such representation can be difficult. The inability to capture all the required knowledge is another source of difficulty. It is almost impossible to embody all sources of knowledge that humans use to process language. Even if this were done, it is not possible to write procedures that imitate language processing as done by humans. In this section, we detail some of the problems associated with NLP.

Perhaps the greatest source of difficulty in natural language is identifying its semantics. The principle of compositional semantics considers the meaning of a sentence to be a composition of the meaning of words appearing in it. In the earlier section, we saw a number of examples where this principle failed to work. Our viewpoint is that words alone do not make a sentence. Instead, it is the words as well as their syntactic and semantic relation that give meaning to a sentence. As pointed out by Wittgenstein (1953): 'The meaning of a word is its use in the language.' A language keeps on evolving. New words are added continually and existing words are introduced in new context. For example, most newspapers and TV channels use 9/11 to refer to the terrorist act on the World Trade Centre in the USA in 2004. When we process written text or spoken utterances, we have access to underlying mental representation. The only way a machine can learn the meaning of a specific word in a message is by considering its context, unless some explicitly coded general world or domain knowledge is available. The context of a word is defined by co-occurring words. It includes everything that occurs before or after a word. The frequency of a word being used in a particular sense also affects its meaning. The English word 'while' was initially used to mean 'a short interval of time'. But now it is more in use as a conjunction. None of the usages of 'while' discussed in this chapter correspond to this meaning.

Idioms, metaphor, and ellipses add more complexity to identify the meaning of the written text. As an example, consider the sentence:

The old man finally kicked the bucket. (1.3)

The meaning of this sentence has nothing to do with the words 'kick' and 'bucket' appearing in it.

Quantifier-scoping is another problem. The scope of quantifiers (the, each, etc.) is often not clear and poses problem in automatic processing.

The ambiguity of natural languages is another difficulty. These go unnoticed most of the times, yet are correctly interpreted. This is possible because we use explicit as well as implicit sources of knowledge. Communication via language involves two brains not just one—the brain of the speaker/writer and that of the hearer/reader. Anything that is assumed to be known to the receiver is not explicitly encoded. The receiver possesses the necessary knowledge and fills in the gaps while making an interpretation. As humans, we are aware of the context and current cultural knowledge, and also of the language and traditions, and utilize these to process the meaning. However, incorporating contextual and world knowledge poses the greatest difficulty in language computing. An example of cultural impact on language is the representation of different shades of white in the Eskimo world. It may be hard for a person living in plain to distinguish among various shades. Similarly, to an Indian, the word 'Taj' may mean a monument, a brand of tea, or a hotel, which may not be so for a non-Indian. Let us now take a look at the various sources of ambiguities in natural languages.

The first level of ambiguity arises at the word level. Without much effort, we can identify words that have multiple meanings associated with them, e.g., bank, can, bat, and still. A word may be ambiguous in its part-of-speech or it may be ambiguous in its meaning. The word 'can' is ambiguous in its part-of-speech whereas the word 'bat' is ambiguous in its meaning. We hardly consider all possible meanings of a word to get the correct one. A program on the other hand, must be explicitly coded to resolve each meaning. Hence, we need to develop various models and algorithms to resolve them. Deciding whether 'can' is a noun or a verb is solved by 'part-of-speech tagging' whereas identifying whether a particular use of 'bank' corresponds to 'financial institution' sense or 'river bank' sense is solved by 'word sense disambiguation'. 'Part-of-speech tagging' and 'word sense disambiguation' algorithms are discussed in Chapters 3 and 5 respectively.

A sentence may be ambiguous even if the words are not, for example, the sentence: '*Stolen rifle found by tree.*' None of the words in this sentence is ambiguous but the sentence is. This is an example of structural ambiguity. Verb sub-categorization may help to resolve this type of ambiguity but not always. Probabilistic parsing, which is discussed in Chapter 4, is another solution. At a still higher level are pragmatic and discourse ambiguities. Ambiguities are discussed in Chapter 5.

A number of grammars have been proposed to describe the structure of sentences. However, there are an infinite number of ways to generate them, which makes writing grammar rules, and grammar itself, extremely complex. On top of it, we often make correct semantic interpretations of non-grammatical sentences. This fact makes it almost impossible for grammar to capture the structure of all and only meaningful text.

1.5   LANGUAGE AND GRAMMAR

Automatic processing of language requires the rules and exceptions of a language to be explained to the computer. Grammar defines language. It consists of a set of rules that allows us to parse and generate sentences in a language. Thus, it provides the means to specify natural language. These rules relate information to coding devices at the language level—not at the world-knowledge level (Bharati et al. 1995). However, since world knowledge affects both the coding (i.e., words) and the coding convention (structure), this blurs the boundary between syntax and semantics. Nevertheless such a separation is made because of the ease of processing and grammar writing.

The main hurdle in language specification comes from the constantly changing nature of natural languages and the presence of a large number of hard-to-specify exceptions. Several efforts have been made to provide such specifications, which has led to the development of a number of grammars. Main among them are transformational grammar (Chomsky 1957), lexical functional grammar (Kaplan and Bresnan 1982), government and binding (Chomsky 1981), generalized phrase structure grammar, transformational grammar (Chomsky 1957), dependency grammar, Paninian grammar, and tree-adjoining grammar (Joshi 1985). Some of these grammars focus on derivation (e.g., phrase structure grammar) while others focus on relationships (e.g., dependency grammar, lexical functional grammar, Paninian grammar, and link grammar). We discuss some of these in Chapter 2. The greatest contribution to grammar comes from Noam Chomsky, who proposed a hierarchy of formal grammar based on level of complexity. These grammars use phrase structure rules (or rewrite rules). The term 'generative grammar' is often used to refer to the general framework introduced by Chomsky. Generative grammar basically refers to any grammar that uses a set of rules to specify or generate all and only grammatical (well-formed) sentences in a language. Chomsky argued that phrase structure grammars are not adequate to specify natural language. He proposed a complex system of transformational grammar in his book on *Syntactic Structures* (1957), in which he suggested that each sentence in a language has two levels of representation, namely, a deep structure and a surface structure (See Figure 1.1). The mapping from deep structure to surface structure is carried out by transformations. In the following paragraphs, we introduce transformational grammar.

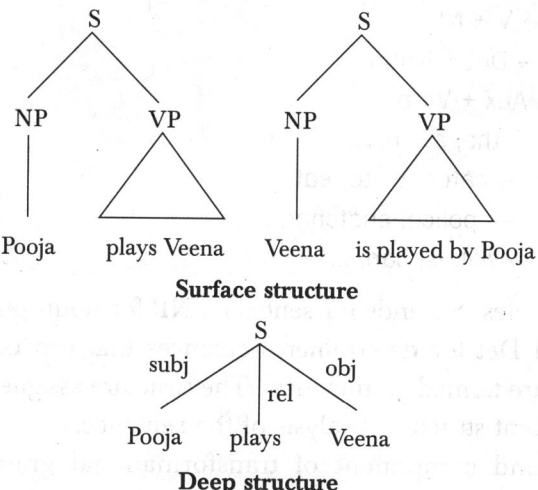

Figure 1.1 Surface and deep structures of sentence

Transformational grammar was introduced by Chomsky in 1957. Chomsky argued that an utterance is the surface representation of a 'deeper structure' representing its meaning. The deep structure can be transformed in a number of ways to yield many different surface-level representations. Sentences with different surface-level representations having the same meaning, share a common deep-level representation. Chomsky's theory was able to explain why sentences like

Pooja plays veena.   (1.4a)

Veena is played by Pooja.   (1.4b)

have the same meaning, despite having different surface structures (roles of subject and object are inverted). Both the sentences are being generated from the same 'deep structure' in which the deep subject is Pooja and the deep object is the veena.

Transformational grammar has three components:

1. Phrase structure grammar
2. Transformational rules
3. Morphophonemic rules—These rules match each sentence representation to a string of phonemes.

Each of these components consists of a set of rules. Phrase structure grammar consists of rules that generate natural language sentences and assign a structural description to them. As an example, consider the following set of rules:

```
S → NP + VP
VP → V + NP
NP → Det + Noun
V → Aux + Verb
Det → the, a, an, ...
Verb → catch, write, eat, ...
Noun → police, snatcher, ...
Aux → will, is, can, ...
```

In these rules, S stands for sentence, NP for noun phrase, VP for verb phrase, and Det for determiner. Sentences that can be generated using these rules are termed grammatical. The structure assigned by the grammar is a constituent structure analysis of the sentence.

The second component of transformational grammar is a set of transformation rules, which transform one phrase-maker (underlying) into another phrase-marker (derived). These rules are applied on the terminal string generated by phrase structure rules. Unlike phrase structure rules, transformational rules are heterogeneous and may have more than one symbol on their left hand side. These rules are used to transform one surface representation into another, e.g., an active sentence into passive one. The rule relating active and passive sentences (as given by Chomsky) is

$$NP_1 - Aux - V - NP_2 \rightarrow NP_2 - Aux + be + en - V - by + NP_1$$

This rule says that an underlying input having the structure NP–Aux–V–NP can be transformed to NP – Aux + be + en – V – by + NP. This transformation involves addition of strings 'be' and 'en' and certain re-arrangements of the constituents of a sentence. Transformational rules can be obligatory or optional. An obligatory transformation is one that ensures agreement in number of subject and verb, etc., whereas an optional transformation is one that modifies the structure of a sentence while preserving its meaning. Morphophonemic rules match each sentence representation to a string of phonemes.

Consider the active sentence:

The police will catch the snatcher. (1.5)

The application of phrase structure rules will assign the structure shown in Figure 1.2 to this sentence.

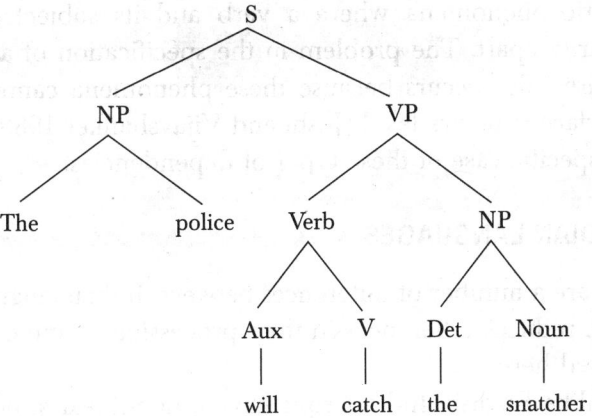

Figure 1.2 Parse structure of sentence (1.5)

The passive transformation rules will convert the sentence into:
The + culprit + will + be + en + catch + by + police (Figure 1.3).

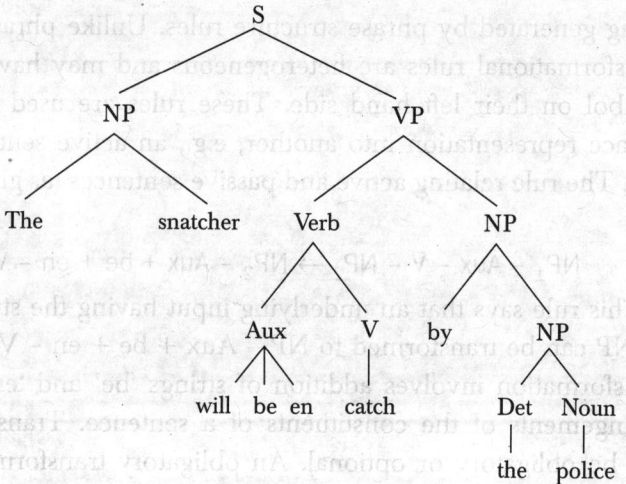

Figure 1.3   Structure of sentence (1.5) after applying passive transformations

Another transformational rule will then reorder 'en + catch' to 'catch + en' and subsequently one of the morphophonemic rules will convert 'catch + en' to 'caught'. In general, the noun phrase is not always as simple as in sentence (1.5). It may contain other embedded structures, such as adjectives, modifiers, relative clause, etc. Long distance dependencies are other language phenomena that cannot be adequately handled by phrase structure rules. Long distance dependency refers to syntactic phenomena where a verb and its subject or object can be arbitrarily apart. The problem in the specification of appropriate phrase structure rules occurs because these phenomena cannot be localized at the surface structure level (Joshi and Vijayshanker 1989). Wh-movement[1] are a specific case of these types of dependencies.

1.6   PROCESSING INDIAN LANGUAGES

There are a number of differences between Indian languages and English. This introduces differences in their processing. Some of these differences are listed here.

- Unlike English, Indic scripts have a non-linear structure.
- Unlike English, Indian languages have SOV (Subject-Object-Verb) as the default sentence structure.

---

[1]It refers to a syntactic phenomenon in which interrogative words, called wh-words, appear at the beginning of the sentence. For example, when the direct object of the verb 'read' in the sentence 'She is reading a book' is replaced with a wh-word, the sentence becomes 'What is she reading?' instead of 'She is reading what?'.

- Indian languages have a free word order, i.e., words can be moved freely within a sentence without changing the meaning of the sentence.
- Spelling standardization is more subtle in Hindi than in English.
- Indian languages have a relatively rich set of morphological variants.
- Indian languages make extensive and productive use of complex predicates (CPs).
- Indian languages use post-position (*Karakas*) case markers instead of prepositions.
- Indian languages use verb complexes consisting of sequences of verbs, e.g., गा रहा है (*ga raha hai—*singing) and खेल रही है (*khel rahi hai—*playing). The auxiliary verbs in this sequence provide information about tense, aspect, modality, etc.

Except for the direction in which its script is written, Urdu is closely related to Hindi. Both share similar phonology, morphology, and syntax. Both are free-word-order languages and use post-positions. They also share a large amount of their vocabulary. Differences in the vocabulary arise mainly because a significant portion of Urdu vocabulary comes from Persian and Arabic, while Hindi borrows much of its vocabulary from Sanskrit.

Paninian grammar provides a framework for Indian language models. These can be used for computation of Indian languages. The grammar focuses on extraction of Karaka relations from a sentence. We talk about the details of modelling in Chapter 2. A parsing framework based on Paninian grammar is introduced in Chapter 4 and issues involved in Indian language translation (using Paninian grammar theory) are discussed in Chapter 8.

1.7 NLP APPLICATIONS

Machine translation is the first application area of NLP. It involves the complete linguistic analysis of a natural language sentence, and linguistic generation of an output sentence. It is one of the most comprehensive and most challenging tasks in the area (AI-complete). However, the recent dramatic progress in the field of NLP has found interesting applications in information retrieval, information extraction, text summarization, etc. This book offers an extensive coverage of these recent applications, and also of traditional ones like machine translation and natural language generation. The focus has been on bridging the gap between theory and practice rather than on offering a gamut of linguistic, psychological, and computational theories.

The applications utilizing NLP include the following.

Machine Translation

This refers to automatic translation of text from one human language to another. In order to carry out this translation, it is necessary to have an understanding of words and phrases, grammars of the two languages involved, semantics of the languages, and world knowledge.

Speech Recognition

This is the process of mapping acoustic speech signals to a set of words. The difficulties arise due to wide variations in the pronunciation of words, homonym (e.g. dear and deer) and acoustic ambiguities (e.g., in the rest and interest).

Speech Synthesis

Speech synthesis refers to automatic production of speech (utterance of natural language sentences). Such systems can read out your mails on telephone, or even read out a storybook for you. In order to generate utterances, text has to be processed. So, NLP remains an important component of any speech synthesis system.

Natural Language Interfaces to Databases

Natural language interfaces allow querying a structured database using natural language sentences.

Information Retrieval

This is concerned with identifying documents relevant to a user's query. NLP techniques have found useful applications in information retrieval. Indexing (stop word elimination, stemming, phrase extraction, etc.), word sense disambiguation, query modification, and knowledge bases have also been used in IR system to enhance performance, e.g., by providing methods for query expansion. WordNet, LDOCE (*Longman Dictionary of Contemporary English*) and *Roget's Thesaurus* are some of the useful lexical resources for IR research.

Information Extraction

An information extraction system captures and outputs factual information contained within a document. Similar to an information retrieval system, it responds to a user's information need. However, unlike in an information retrieval system, the information need is not expressed as a keyword query. Instead it is specified as pre-defined database schemas or templates. Whereas an IR system identifies a subset of documents in a large repository of text database—e.g., in a library scenario, a subset of resources in a library—an information extraction system identifies a subset of information within a document that fits the pre-defined template.

Question Answering

Given a question and a set of documents, a question answering system attempts to find the precise answer, or at least the precise portion of text in which the answer appears. This is unlike an IR system, which returns whole document that seems relevant to the user's query. A question answering system is different from an information extraction system in that the content that is to be extracted is unknown. In general, a question answering system benefits from having an information extraction system to identify entities in the text. A question answering system requires more NLP than an information retrieval system or an information extraction system. It requires not only precise analysis of questions and portions of texts but also semantic as well as background knowledge to answer certain type of questions.

Text Summarization

This deals with the creation of summaries of documents and involves syntactic, semantic, and discourse level processing of text.

1.8 SOME SUCCESSFUL EARLY NLP SYSTEMS

ELIZA (Weizenbaum 1966)

ELIZA is one of the earliest natural language understanding programs. It uses syntactic patterns to mimic human conversation with the user. Here is a sample conversation.

> *Eliza:* Hello. I am ELIZA. How may I help you?
> *User:* I am feeling a little bit sleepy.
> *Eliza:* How long have you been feeling a little bit sleepy?
> *User:* For almost half an hour.
> *Eliza:* Please go on.

SysTran (System Translation)

The first SysTran machine translation system was developed in 1969 for Russian–English translation. SysTran also provided the first on-line machine translation service called Babel Fish, which is used by AltaVista search engines for handling translation requests from users.

TAUM METEO

This is a natural language generation system used in Canada to generate weather reports. It accepts daily weather data and generates weather reports in English and French.

SHRDLU (Winogard 1972)

This is a natural language understanding system that simulates actions of a robot in a block world domain. It uses syntactic parsing and semantic reasoning to understand instructions. The user can ask the robot to manipulate the blocks, to tell the blocks configurations, and to explain its reasoning.

LUNAR (Woods 1977)

This was an early question answering system that answered questions about moon rock.

1.9   INFORMATION RETRIEVAL

The availability of a large amount of text in electronic form has made it extremely difficult to get relevant information. Information retrieval systems aim at providing a solution to this.

The term 'information' should not be confused with the term 'entropy' (numerical measure of the uncertainty of an outcome) as it is used in communication theory. Information is being used here to reflect 'subject matter' or the 'content' of some text. We are not interested in 'digital communication', where bits and bytes are the information carriers. Instead our focus in on the communication taking place between human beings as expressed through natural languages. Information is always associated with some data (text, number, image, and so on): we are concerned with text only. Hence, we consider words as the carriers of information and written text as the message encoded in natural language.

As a cognitive activity, the word 'retrieval' refers to operation of accessing information from memory. We use the word 'retrieval' to refer to the operation of accessing information from some computer-based representation. Retrieval of information thus requires information to be processed and stored. Not all the information represented in computable form is retrieved. Instead, only the information relevant to the needs expressed in the form of query is located. In order to get this relevance, the stored and processed information needs to be compared against query representation. Information retrieval (IR) deals with all these facets. It is concerned with the organization, storage, retrieval, and evaluation of information relevant to the query.

Information retrieval deals with unstructured data. The retrieval is performed based on the content of the document rather than on its structure. The IR systems usually return a ranked list of documents. The IR components have been traditionally incorporated into different types of information systems including database management systems, bibliographic text retrieval systems, question answering systems, and more recently in search engines.

Current approaches for accessing large text collections can be broadly classified into two categories. The first category consists of approaches that construct topic hierarchy, e.g., Yahoo. This helps the user locate documents of interest manually by traversing the hierarchy. However, it requires manual classification of new documents within the existing taxonomy. This makes it cost ineffective and inapplicable due to rapid growth of documents on the Web. The second category consists of approaches that rank the retrieved documents according to relevance. We discuss various IR models that support ranked retrieval in Chapter 9.

Major Issues in Information Retrieval (Siddiqui 2006)

There are a number of issues involved in the design and evaluation of IR systems, which are briefly discussed in this section. The first important point is to choose a representation of the document. Most human knowledge is coded in natural language, which is difficult to use as knowledge representation language for computer systems. Most of the current retrieval models are based on keyword representation. This representation creates problems during retrieval due to polysemy, homonymy, and synonymy. Polysemy involves the phenomenon of a lexeme with multiple meaning. Homonymy is an ambiguity in which words that appear the same have unrelated meanings. Ambiguity makes it difficult for a computer to automatically determine the conceptual content of documents. Synonymy creates problem when a document is indexed with one term and the query contains a different term, and the two terms share a common meaning. Another problem associated with keyword-based retrieval is that it ignores semantic and contextual information in the retrieval process. This information is lost in the extraction of keywords from the text and cannot be recovered by the retrieval algorithms.

A related issue is that of inappropriate characterization of queries by the user. There can be many reasons for the vagueness and inaccuracy of the user's queries, say for instance, her lack of knowledge of the subject or even the inherent vagueness of the natural language. The user may fail to include relevant terms in the query or may include irrelevant terms. Inappropriate or inaccurate queries lead to poor retrieval performance. The problem of ill-specified query can be dealt with by modifying or expanding queries. An effective technique based on user-interaction is relevance feedback which modifies queries based on the feedback provided by the user on initial retrieval.

In order to satisfy the user's request, an IR system matches document representation with query representation. Matching query representation with that of the document is another issue. A number of measures have been proposed to quantify the similarity between a query and the document to produce a ranked list of results. Selection of the appropriate similarity measure is a crucial issue in the design of IR systems.

Evaluating the performance of IR systems is also a major issue. There are many aspects of evaluation, the most important being the effectiveness of the system. Recall and precision are the most widely used measures of effectiveness.

As the major goal of IR is to search a document in a manner relevant to the query, understanding what constitutes relevance is also an important issue. Relevance is subjective in nature (Saracevic 1991). Only the user can tell the true relevance; it is not possible to measure this 'true relevance'. One may however, define the degree of relevance. Relevance has been considered as a binary concept, whereas it is in fact a continuous function (a document may be exactly what the user wants or it may be closely related). Current evaluation techniques do not support this continuity as it is quite difficult to put into practice. A number of relevance frameworks have been proposed (Saracevic 1996). These include the system, communication, psychological, and situational frameworks. The most inclusive is the situational framework, which is based on the cognitive view of the information seeking process and considers the importance of situation, context, multi-dimensionality, and time. A survey of relevance studies can be found in Mizzaro (1997). Most of the evaluations of IR systems have so far been done on document test collections with known relevance judgments.

The size of document collections and the varying needs of users also complicate text retrieval. Some users require answers of limited scope, while others require documents with a wider scope. These differing needs can require different and specialized retrieval methods. However, these are research issues and have not been dealt with in this book.

SUMMARY

- Language is the primary means of communication used by humans.
- Natural language processing is concerned with the development of computational models of aspects of human language processing.
- Theoretical linguists are mainly interested in providing a description of the structure and semantics of natural language, whereas

computational linguists deal with the study of language from a computational point of view.

- Historically, there have been two major approaches to natural language processing, namely rationalist approach and empiricist approach.
- The highly ambiguous and vague nature of natural language makes it difficult to create a representation amenable to computing.

REFERENCES

Bach, Kent, 2002, *Meaning and Truth*, J. Keim Campbell, M. O'Rourke, and D. Shei (Eds.), Seven Bridges Press, New York, pp. 284–92.

Chomsky, Noam, 1957, *Syntactic Structures*, Mouton, The Hague.

———1981, *Lectures on Government and Binding*, Foris Publications, Dordreeht, The Netherlands.

Joshi, Aravind K., 1985, 'Tree adjoining grammar: How much sensitivity is required to provide reasonable structural description,' *Natural Language Parsing*, D. Dowty, L. Karttunen, and A. Zwicky (Eds.), Cambridge University Press, Cambridge.

Joshi, Aravind K. and K. Vijayshanker, 1989, 'Treatment of long distance dependencies in LFG and TAG: functional uncertainty in LFG is a corollary in TAG,' *Proceedings of the 27th Annual Meeting on Association for Computational Linguistics*, Vancouver, British Columbia, pp. 220–27.

Kaplan, R.M. and Joan Bresnan, 1982, 'Lexical functional grammar: A formal system for grammatical representation,' *The Mental Representation of Grammatical Relations*, Joan Bresnan (Ed.), MIT Press, Cambridge.

Lenat, D.B., M. Prakash, and M. Shepherd, 1986, 'Cyc: using common sense knowledge to overcome brittleness and knowledge acquisition bottlenecks,' *AI Magazine*, 6(4).

Mizzaro, S., 1997, 'Relevance: the whole history,' *Journal of the American Society for Information Science*, 48(9), pp. 810–32.

Saracevic, T., 1991, 'Individual differences in organizing, searching and retrieving information,' *Proceedings of the 54th Annual Meeting of the American Society for Information Science (ASIS)*, pp. 82–86.

———1996, 'Relevance reconsidered,' *Proceedings of CoLIS 2, Second International Conference on Conceptions of Library and Information Science: Integration in Perspective*, P. Ingwersen and N.O. Pors (Eds.), The Royal School of Librarianship, Copenhagen, pp. 201–18.

Siddiqui, Tanveer, 2006, 'Intelligent techniques for effective information retrieval: a conceptual graph-based approach,' *Ph.D. Thesis*, J.K. Institute of Applied Physics, Deptt. of Electronics and Communication, University of Allahabad.

Weizenbaum, R., 1966, 'ELIZA—A computer program for the study of natural language communication between man and machine,' *Communications of the ACM*, 9(1).

Winogard, Terry, 1972, *Understanding Natural Language*, Academic Press, New York.

Woods, William, 1977, 'Lunar Rocks in Natural English: Explorations in Natural Language Question-answering,' *Linguistic Structures Processing*, A. Zampolli (Ed.), Elsevier, North Holland.

EXERCISES

1. Differentiate between the rationalist and empiricist approaches to natural language processing.
2. List the motivation behind the development of computational models of languages.
3. Briefly discuss the meaning components of a language.
4. What makes natural language processing difficult?
5. What is the role of transformational rules in transformational grammar? Explain with the help of examples.

CHAPTER 2

LANGUAGE MODELLING

CHAPTER OVERVIEW

The domain of language is quite vast. It presents an almost infinite number of sentences to the reader (or computer). To handle such a large number of sentences, we have to create a model of the domain, which can subsequently be simplified and handled computationally. A number of language models have been proposed. We introduce some of these models in this chapter. To create a general model of any language is a difficult task. There are two approaches for language modelling. One is to define a grammar that can handle the language. The other is to capture the patterns in a grammar language statistically. This chapter has a mixed approach. It gives a glimpse of both grammar-based model and statistical language model. These include lexical functional grammar, government and binding, Paninian grammar, and *n*-gram based model.

2.1 INTRODUCTION

Why and how do we model a language? This question has been discussed by linguists since 500 BC. Computational linguists also have to confront this question. It is obvious that our purpose is to understand and generate natural languages from a computational viewpoint. One approach can be to just take a language, try to understand every word and sentence of it, and then come to a conclusion. This approach has not succeeded as there are difficulties at each stage, which we will understand as we go through this book. An alternative approach is to study the grammar of various languages, compare them, and if possible, arrive at reasonable models that facilitate our understanding of the problem and designing of natural-language tools.

A model is a description of some complex entity or process. A language model is thus a description of language. Indeed, natural language is a complex entity and in order to process it through a computer-based program, we need to build a representation (model) of it. This is known as *language modelling*. Language modelling can be viewed either as a problem of grammar inference or a problem of probability estimation. A grammar-based language model attempts to distinguish a grammatical sentence from a non-grammatical (ill-formed) one, whereas a probabilistic language model attempts to identify a sentence based on a probability measure, usually a maximum likelihood estimate. These two viewpoints have led to the following categorization of language modelling approaches.

Grammar-based language model

A grammar-based approach uses the grammar of a language to create its model. It attempts to represent the syntactic structure of language. Grammar consists of hand-coded rules defining the structure and ordering of various constituents appearing in a linguistic unit (phrase, sentence, etc.). For example, a sentence usually consists of noun phrase and a verb phrase. The grammar-based approach attempts to utilize this structure and also the relationships between these structures.

Statistical language modelling

The statistical approach creates a language model by training it from a corpus. In order to capture regularities of a language, the training corpus needs to be sufficiently large. Rosenfeld (1994) pointed out:

> *Statistical language modelling (SLM) is the attempt to capture regularities of natural language for the purpose of improving the performance of various natural language applications.*

Statistical language modelling is one of the fundamental tasks in many NLP applications, including speech recognition, spelling correction, handwriting recognition, and machine translation. It has now found applications in information retrieval, text summarization, and question answering also. A number of statistical language models have been proposed in literature. The most popular of these are the *n*-gram models. We discuss this model in Section 2.3. The following section discusses various grammar-based models.

2.2 VARIOUS GRAMMAR-BASED LANGUAGE MODELS

Various computational grammars have been proposed and studied, e.g., transformational grammar (Chomsky 1957), lexical functional grammar (Kaplan and Bresnan 1982), government and binding (Chomsky 1981), generalized phrase structure grammar (Gazdar et al. 1985), dependency grammar, paninian grammar, and tree-adjoining grammar (Joshi 1985).

This section focuses on lexical functional grammar (LFG), generalized phrase structure grammar (GPSG), government and binding (GB), and Paninian grammar (PG) and introduces various approaches to understand a language in a grammatical and rule-based format. It also introduces the dominant approaches to create statistical models of language and grammar.

2.2.1 Generative Grammars

In 1957, in his book on *Syntactic Structures*, Noam Chomsky wrote that we can generate sentences in a language if we know a collection of words and rules in that language. Only those sentences that can be generated as per the rules are grammatical. This point of view has dominated computational linguistics and is appropriately termed generative grammar. The same idea can be used to model a language. If we have a complete set of rules that can generate all possible sentences in a language, those rules provide a model of that language. Of course, we are talking only about the syntactical structure of language here.

Language is a relation between the sound (or the written text) and its meaning. Thus, any model of a language should also deal with the meaning of its sentences. As seen earlier, we can have a perfectly grammatical but meaningless sentence.

In this chapter, we will assume that grammars are a type of language models.

2.2.2 Hierarchical Grammar

Chomsky (1956) described classes of grammars in a hierarchical manner, where the top layer contained the grammars represented by its sub classes. Hence, Type 0 (or unrestricted) grammar contains Type 1 (or context-sensitive grammar), which in turn contains Type 2 (context-free grammar) and that again contains Type 3 grammar (regular grammar). Although this relationship has been given for classes of formal grammars, it can be extended to describe grammars at various levels, such as in a class-sub class (embedded) relationship.

2.2.3 Government and Binding (GB)

As discussed in Chapter 1, a common viewpoint taken by linguists (not computational linguists, however) is that the structure of a language (or how well its sentences are formed) can be understood at the level of its meaning, particularly while resolving structural ambiguity. However, the sentences are given at the syntactical level and the transformation from meaning to syntax or vice versa is not well understood.

Transformational grammars assume two levels of existence of sentences–one at the surface level and the other at the deep root level (this should not be confused with the meaning level). Government and binding (GB) theories have renamed them as s-level and d-level, and identified two more levels of representation (parallel to each other) called *phonetic form* and *logical form*. According to GB theories, language can be considered for analysis at the levels shown in Figure 2.1.

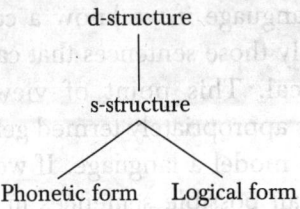

Figure 2.1   Different levels of representation in GB

If we describe language as the representation of some 'meaning' in a 'sound' form, then according to Figure 2.1, these two ends are the logical form (LF) and phonetic form (PF) respectively. The GB is concerned with LF, rather than PF. Chomsky was the first to put forward a GB theory (Peter Sells 1985).

Transformational grammars have hundreds of rewriting rules, which are generally language-specific and also construct-specific (say, different rules for assertive and interrogative sentences in English, or for active and passive voice sentences). Generation of a complete set of coherent rules may not be possible. The GB envisages that if we define rules for structural units at the deep level, it will be possible to generate any language with fewer rules. These deep-level structures are abstractions of noun-phrase, verb-phrase, etc., and common to all languages. It is possible to do if, as GB theory states, a child learns its mother tongue because the human mind is 'hard-wired' with some universal grammar rules. Thus, the data enters the mind and its abstract structure gives rise to actual phonetic structures. The existence of deep level, language-independent, abstract structures, and the expression of these in surface level, language-specific structures with the help of simple rules is the main concern of GB theories. Let us take an example to explain d- and s-structures.

Example 2.1

Mukesh was killed.   (2.1)

(i) In transformational grammar, this can be represented as S-NP Aux VP as given below:

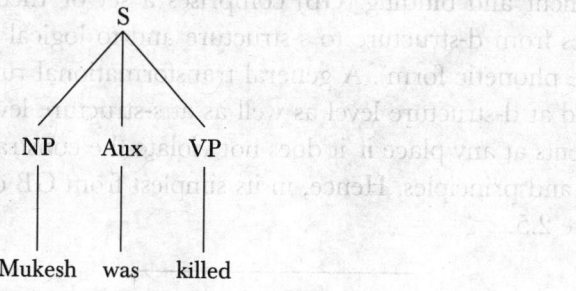

Figure 2.2 TG representation of sentence (2.1)

(ii) In GB, the s-structure and d-structure are as follows:

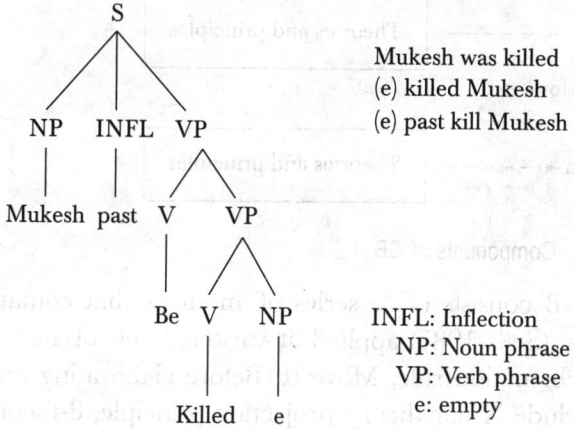

Mukesh was killed
(e) killed Mukesh
(e) past kill Mukesh

INFL: Inflection
NP: Noun phrase
VP: Verb phrase
e: empty

Figure 2.3 Surface structure of sentence (2.1) in GB

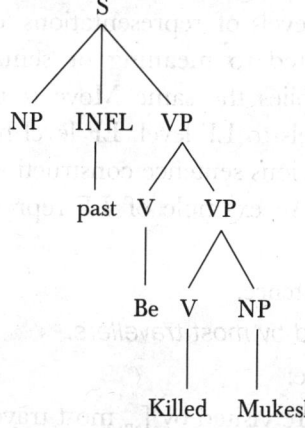

Figure 2.4 Deep structure of sentence (2.1) in GB

Components of GB

Government and binding (GB) comprises a set of theories that map the structures from d-structure to s-structure and to logical form (LF) (leaving aside the phonetic form). A general transformational rule called 'Move α' is applied at d-structure level as well as at s-structure level. This can move constituents at any place if it does not violate the constraints put by several theories and principles. Hence, in its simplest from GB can be represented by Figure 2.5.

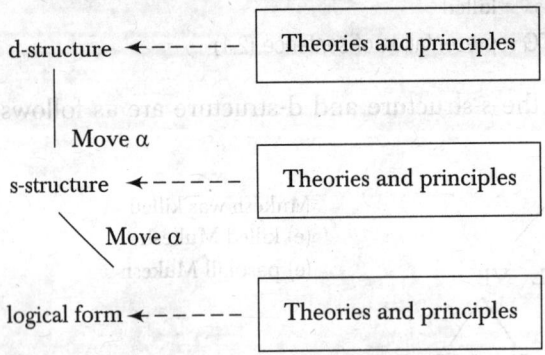

Figure 2.5 Components of GB

Hence, GB consists of 'a series of modules that contain constraints and principles' (Sells 1985) applied at various levels of its representations and the transformation rule, Move α. Before elaborating on these modules—which include X-bar theory, projection principle, θ-theory and θ-criterion, C-command and government, case theory, empty category principle (ECP), and binding theory—we discuss the general characteristics of GB.

The GB considers all three levels of representations (d-, s-, and LF) as syntactic, and LF is also related to meaning or semantic-interpretive mechanisms. However, GB applies the same Move α transformation to map d-levels to s-levels or s-levels to LF level. LF level helps in quantifier scoping and also in handling various sentence constructions such as passive or interrogative constructions. An example of LF representation may be helpful.

Example 2.2 Consider the sentence:

> *Two countries are visited by most travellers.* (2.2)

Its two possible logical forms are:

LF1: [$_S$ Two countries are visited by [$_{NP}$ most travellers]]
LF2: Applying Move α
     [$_{NP}$ Most travellers$_i$] [$_S$ two countries are visited by e$_i$]

In LF1, the interpretation is that most travellers visit the same two countries (say, India and China). In LF2, when we move [most travellers] outside the scope of the sentence, the interpretation can be that most travellers visit two countries, which may be different for different travellers.

One of the important concepts in GB is that of constraints. It is the part of the grammar which prohibits certain combinations and movements; otherwise Move α can move anything to any possible position. Thus, GB is basically the formulation of theories or principles which create constraints to disallow the construction of ill-formed sentences. To account for cross-lingual constraints of similar type, GB can specify that 'a constituent cannot be moved from position X' (where X can have value X_1 in one language, X_2 in another, and so on). These rules are so general and language-independent that 'language-particular details of description typically go uncharted in GB' (Sells 1985).

Figure 2.5 showed the application of various theories and principles at three different levels of representations in GB. Figure 2.6 mentions these explicitly to understand the organization of GB.

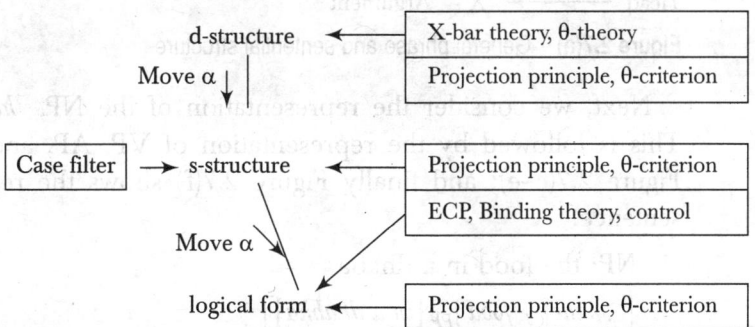

Figure 2.6 Organization of GB (adapted from Peter Sells 1985)

$\overline{\text{X}}$ *Theory*

The $\overline{\text{X}}$ theory (pronounced X-bar theory) is one of the central concepts in GB. Instead of defining several phrase structures and the sentence structure with separate sets of rules, $\overline{\text{X}}$ theory defines them both as maximal projections of some head. In this manner, the entities defined become language independent. Thus, noun phrase (NP), verb phrase (VP), adjective phrase (AP), and prepositional phrase (PP) are maximal projections of noun (N), verb (V), adjective (A), and preposition (P) respectively, and can be represented as head X of their corresponding phrases (where X = {N, V, A, P}). Not only that, even the sentence structure (S′, which is projection of sentence) can be regarded as the maximal projection of inflection (INFL). The GB envisages projections at two levels—first the projection of head at semi-phrasal level, denoted by $\bar{X}$, and then the second maximal projection at the phrasal level, denoted by $\bar{\bar{X}}$.

For sentences, the first level projection is denoted as S and the second level maximal projection is denoted by S′. We now illustrate phrase and sentence representations with the help of examples.

Example 2.3 Figure 2.7 depicts the general and particular structures with examples. We see the general structure in Figure 2.7(a).

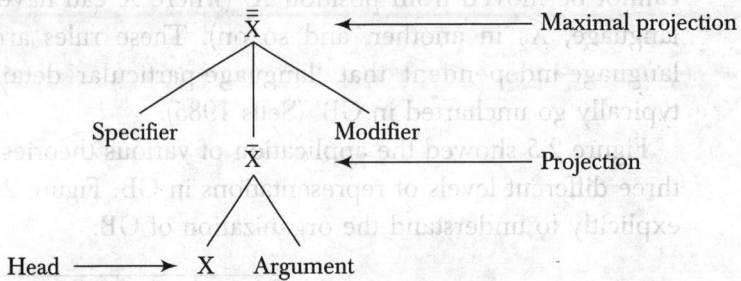

Figure 2.7(a)   General phrase and sentential structure

Next, we consider the representation of the NP, *the food in a dhaba*. This is followed by the representation of VP, AP, and PP structure in Figure 2.7(c–e); and finally Figure 2.7(f) shows the representation of a sentence.

1. NP: the food in a dhaba

$$\left[_{NP}\, the \left[_{N}\, food \,\right]_{PP} \left[in\ a\ dhabha \right] \right]$$

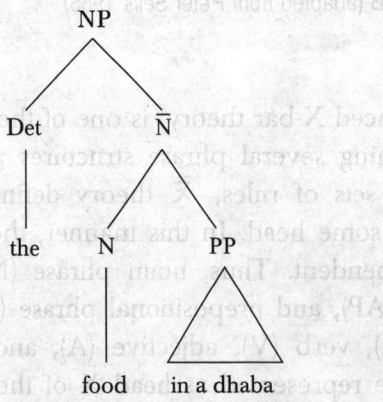

Figure 2.7(b)   NP structure

2. VP: ate the food in a dhaba

$$\left[_{VP} \left[_{\bar{V}} [_V \ ate \ ][_{NP} \ the \ food \ ] \right] [_{PP} \ in \ a \ dhaba \ ] \right]$$

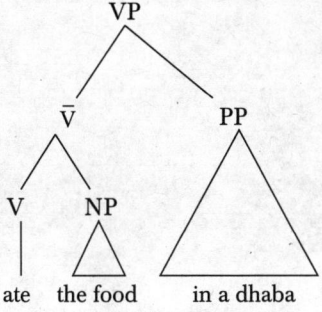

Figure 2.7(c) VP structure

3. AP: very proud of his country

$$\left[_{AP} [_{Deg} \ very \ ] [_{\bar{A}} [_A \ proud \ ] [_{PP} \ of \ his \ country \ ]] \right]$$

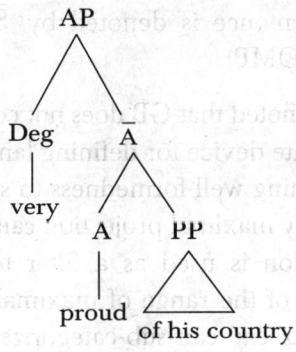

Figure 2.7(d) AP structure

4. PP: in a dhaba

$$\left[_{PP} \left[_{\bar{P}} [_P \ in \ ] [_{NP} [_{Det} \ a \ ] [_N \ dhabha \ ]] \right] \right]$$

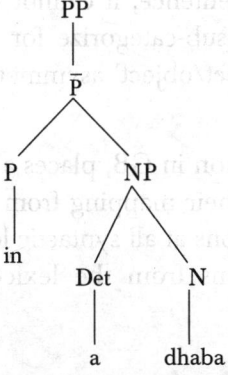

Figure 2.7(e) PP structure

5. S: that she ate the food in a dhaba

$$\left[\,_{\bar{S}}\left[_{COMP}\ that\,\right]\left[_{S}\left[_{Det}\ she\,\right]\left[_{INFL}\ past\,\right]\left[_{VP}\ ate\ the\ food\ in\ a\ dhaba\,\right]\right]\right]$$

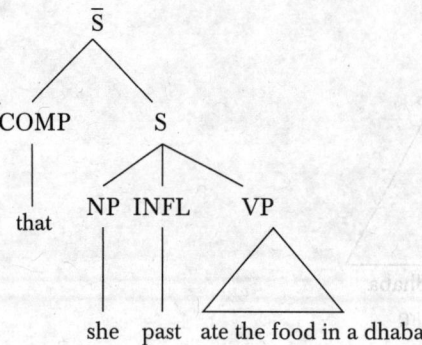

Figure 2.7(f)   Maximal projection of sentence structure

As shown in Figure 2.7(f), the sentence is considered to be the head of INFL and the projection of sentence is denoted by S̄, which has the specifier as complementizer (COMP).

Sub-categorization   It is to be noted that GB does not consider traditional phrase structure as an appropriate device for defining language constructs. It places the burden of ascertaining well-formedness to sub-categorization frames of heads. In principle, any maximal projection can be the argument of a head, but sub-categorization is used as a filter to permit various heads to select a certain subset of the range of maximal projections. For example, we know that the verb 'eat' can sub-categorize for NP, whereas the verb 'sleep' cannot. Hence, 'ate food' is well-formed, but the sentence '*slept the bed' is not. GB claims that defining phrase structures as head projections and sub-categorization helps ensure well-formed structures, even at the sentence level.

As 'verb' is not the head of the sentence, it cannot sub-categorize for 'subject NP', while it can perfectly sub-categorize for 'object NP'. This explains, to certain extent, the 'subject/object' asymmetry (Sells 1985).

Projection Principle

The projection principle, a basic notion in GB, places a constraint on the three syntactic representations and their mapping from one to the other. The principle states that representations at all syntactic levels (i.e., d-level, s-level, and LF level) are projections from the lexicon. Thus, lexical

---

*will precede the incorrect sentences.

properties of categorical structure (sub-categorization) must be observed at each level.

This can be understood with an example. Suppose 'the object' is not present at d-level, then another NP cannot take this position at s-level. This principle, in conjunction with the possibility of presence of empty category and other theories (like Binding Theory) ensures correct movement and well-formed structure.

Theta Theory (θ-Theory) or The Theory of Thematic Relations

As discussed earlier, 'sub-categorization' only places a restriction on syntactic categories which a head can accept. GB puts another restriction on the lexical heads through which it assigns certain roles to its arguments. These roles are pre-assigned and cannot be violated at any syntactical level as per the projection principle. These role assignments are called theta-roles and are related to 'semantic-selection'.

Theta-role and Theta-criterion   There are certain thematic roles from which a head can select. These are called θ-roles and they are mentioned in the lexicon, say for example the verb 'eat' can take arguments with θ-roles '(Agent, Theme)'. Agent is a special type of role which can be assigned by a head to outside arguments (external arguments) whereas other roles are assigned within its domain (internal arguments). Hence in 'Mukesh ate food', the verb 'eat' assigns the 'Agent' role to 'Mukesh' (outside VP) and 'Theme' (or 'patient') role to 'food'. As roles are assigned based on the syntactic positions of the arguments, it is important that there should be a match between the number of roles and number of arguments as depicted by θ-criterion.

Theta-Criterion states that 'each argument bears one and only one θ-role, and each θ-role is assigned to one and only one argument' (Sells 1985). Thus, each argument will have a unique θ-role and cannot be moved to a position where it may acquire another θ-role.

In GB, d-structure is conceived as some kind of 'pure' representation of arguments and hence, θ-roles are assigned at d-level only, whereas theta-criterion is applied at all the three levels, as shown in Figure 2.6.

C-command and Government

As 'Government' is a special case of 'C-command', we will first define C-command.

C-command   C-command defines the scope of maximal projection. It is a basic mechanism through which many constraints are defined on Move α. If any word or phrase (say α or β) falls within the scope of and is determined by a maximal projection, we say that it is dominated by the maximal projection. Now, if there are two structures α and β related in such a way that 'every maximal projection dominating α dominates β', we say that α C-commands β, and this is the necessary and sufficient condition (iff) for C-command.

The definition of C-command does not include all maximal projections dominating β, only those dominating α. If we put this extra constraint, it becomes a kind of mutual C-command (Sells 1985), called government.

Government

α governs β iff:

α C-commands β

α is an X (head, e.g., noun, verb, preposition, adjective, and inflection), and every maximal projection dominating β dominates α.

Thus no maximal projection can intervene between the governor and governee. In GB literature, this has been stated as: 'Maximal projections are barriers to government.'

Movement, Empty Category, and Co-indexing

Briefly let us discuss Move α. In GB, Move α is described as 'move anything anywhere', though it provides restrictions for valid movements.

In GB, the active to passive transformation is the result of NP movement as shown in sentence (2.3). Another well-known movement is the wh-movement, where wh-phrase is moved as follows.

What did Mukesh eat ?   (2.3)
[Mukesh INFL eat what]

As discussed in the projection principle, lexical categories must exist at all the three levels. This principle, when applied to some cases of movement leads to the existence of an abstract entity called empty category. In GB, there are four types of empty categories, two being empty NP positions called wh-trace and NP trace, and the remaining two being pronouns called small 'pro' and big 'PRO'. This division is based on two properties—anaphoric (+a or –a) and pronominal (+p or –p).

Wh-trace   –a, –p
NP-trace   +a, –p
small 'pro' –a, +p
big 'PRO'   +a, +p

Co-indexing is the indexing of the subject NP and AGR (agreement) at d-structure which are preserved by Move α operations at s-structure.

When an NP-movement takes place, a trace of the movement is created by having an indexed empty category (e_i) from the position at which the movement began to the corresponding indexed NP, i.e. NP_i. All A-positions (argument positions) at s-level are also freely indexed. These categories and indices are used to define Binding Theory.

It is interesting to note that for defining constraints to movement, the theory identifies two positions in a sentence. Positions assigned θ-roles are called θ-positions, while others are called $\bar{θ}$-positions.

In a similar way, core grammatical positions (where subject, object, indirect object, etc., are positioned) are called A-positions (arguments positions), and the rest are called $\bar{A}$-positions.

Binding Theory

Binding is defined by Sells (1985) as follows:

α binds β iff

α C-commands β, and

α and β are co-indexed

As we noticed in sentence (2.1),

[e_i INFL kill Mukesh]

[Mukesh$_i$ was killed (by e_i)]

Mukesh was killed.

Empty clause (e_i) and Mukesh (NP_i) are bound. This theory gives a relationship between NPs (including pronouns and reflexive pronouns).

Now, binding theory can be given as follows:

(a) An anaphor (+a) is bound in its governing category.

(b) A pronominal (+p) is free in its governing category.

(c) An R-expression (−a, −p) is free.

This theory applies to binding at A-positions. Governing category is the local domain (the smallest only) NP or S containing it (G or p or R-expression) and its governor.

Example 2.4

A: Mukesh$_i$ knows himself$_i$

B: Mukesh$_i$ believes that Amrita knows him$_i$

C: Mukesh believes that Amrita$_j$ knows Nupur$_k$

Similar rules apply on empty categories also:

NP-trace: +a, −p: Mukesh$_i$ was killed e_i

wh-trace: −a, −p: Who$_i$ does he$_i$ like e_i

Empty Category Principle (ECP)

We have already defined 'government'. Now, let us define 'proper government':

α properly governs β iff:

α governs β and α is lexical (i.e. N, V, A, or P) or

α locally $\bar{\text{A}}$-binds β

The ECP says 'A trace must be properly governed'.

This principle justifies the creation of empty categories during NP-trace and wh-trace and also explains the subject/object asymmetries to some extent. As in the following sentences:

(a) What$_i$ do you think that Mukesh ate e$_i$?

(b) What$_i$ do you think Mukesh ate e$_i$?

Bounding and Control Theory

There are many other types of constraints on Move α. It is not possible to explain all of them here, for details, see Peter Sells (1985).

In English, the long distance movement for complement clause can be explained by bounding theory if NP and S are taken to be bounding nodes. The theory says that the application of Move α may not cross more than one bounding node. The theory of control involves syntax, semantics, and pragmatics. As stated previously, the empty category 'PRO' (+a, +p) behaves as an anaphor sometimes, when it is the subject of the clausal complement to verbs such as decide and try. However, it behaves as pronoun with some other verbs.

Case Theory and Case Filter

In GB, case theory deals with the distribution of NPs and mentions that each NP (with the possible exception of a few empty categories) must be assigned a case. In English, we have the nominative, objective, genitive, etc., cases, which are assigned to NPs at particular positions. Indian languages are rich in case-markers, which are carried even during movements.

Case Filter   An NP is ungrammatical if it has phonetic content or if it is an argument (with the exception of big 'PRO') and is not case-marked. Phonetic content here, refers to some physical realization, as opposed to empty categories. Thus, case filters restrict the movement of NP at a position which has no case assignment. It works in a manner similar to that of the θ-criterion.

In short, GB presents a model of the language which has three levels of syntactic representation. It assumes phrase structures to be the maximal projection of some lexical head and in a similar fashion, explains the structure of a sentence or a clause. It assigns various types of roles to these structures and allows them a broad kind of movement called Move α. It then defines various types of constraints which restrict certain movements and justifies others. GB gives a new insight for the modelling of languages, although the Chomskian Minimalist Programme has superseded GB.

2.2.4 Lexical Functional Grammar (LFG) Model

This section presents those features of LFG that throw a light on language modelling. For the details of lexical functional grammar, readers are encouraged to seek Darlymple et al. (1995).

Unlike GB, LFG represents sentences at two syntactic levels—constituent structure (c-structure) and functional structure (f-structure). Based on Woods' *Augmented Transition Networks* 1970, which used phrase structure trees to represent the surface structure of sentences and the underlying predicate–argument structure, Kaplan (1975a, b) proposed a concrete form for the register names and values (used in ATN implementation), which became the functional structures in LFG. On the other hand, Bresnan (1976a, 1977) was more concerned with the problem of explaining some linguistic issues, such as active/passive and dative alternations, in transformational approach. She proposed that such issues can be dealt with by using lexical redundancy rules. The unification of these two diverse approaches (with a common concern) led to the development of the LFG theory, which was presented as *Lexical Functional Grammar: A Formal System for Grammatical Representation* in 1982.

The LFG is a formalism that is both computationally and linguistically motivated and provides precise algorithms for linguistic issues it can handle. The term 'lexical functional' is composed of two terms: the 'functional' part is derived from 'grammatical functions', such as subject and object, or roles played by various arguments in a sentence. The 'lexical' part is derived from the fact that the lexical rules can be formulated to help define the given structure of a sentence and some of the long distance dependencies, which is difficult in transformational grammars.

C-structure and f-structure in LFG

As LFG is aimed at providing exact computational algorithms, it provides well-defined objects called constituent structure (c-structure) and functional structure (f-structure). The c-structure is derived from the usual phrase and sentence structure syntax, as in CFG (discussed in Chapter 4). However, as the grammatical-functional role cannot be derived directly from phrase and sentence structure, functional specifications are annotated on the nodes of c-structure, which when applied on sentences, results in f-structure. Hence, f-structure is the final product which encodes the information obtained from phrase and sentence structure rules and functional specifications.

Let us consider an example.

Example 2.5

    She saw stars in the sky.

CFG rules to handle this sentence are:

    S → NP VP

    VP → V {NP} {NP} PP* {S′}

    PP → P NP

    NP → Det N {PP}

    S′ → Comp S

where

S:	sentence	V:	verb
P:	preposition	N:	noun
S′:	clause	Comp:	complement

{ } optional

*: Phrase can appear any number of times including blank.

When annotated with functional specifications, the rules become:

Rule 1:   *S →*   *NP*   *VP*
                              ↑ subj = ↓   ↑ = ↓

Rule 2:   *VP → V*   *{NP}*   *{NP}*   *PP**   *{S′}*
                        ↑ obj = ↓ ↑ obj 2 = ↓ ↑ (↓ case) = ↓ ↑ comp = ↓

Rule 3:   *PP → P*   *NP*
                   ↑ obj = ↓

Rule 4:   *NP →* {Det} *N*   *{PP}*
                            ↑ Adjunct = ↓

Rule 5:   *S′ →*   Comp   *S*
                       ↑ = ↓

Here, ↑ (up arrow) refers to the f-structure of the mother node that is on the left hand side of the rule. The ↓ (down arrow) symbol refers to the f-structure of the node under which it is denoted.

Hence, in Rule 1, (↑ subj = ↓) indicates that the f-structure of the first NP goes to the f-structure of the subject of the sentence, while (↑ = ↓) indicates that the f-structure of the VP node goes directly to the f-structure of the sentence VP. Similarly, in Rule 2, the f-structure of VP is defined by the lexical item V, the two optional NPs, any number of PPs, and the optional clause(S'). The f-sructure of V can be obtained from the lexicon itself. All terminals in LFG can be thought of as annotated with $\uparrow = \downarrow$. The NPs can function as object and object 2 of the sentence, and their f-structures are obained using f-structure of Obj and Obj$_2$. '$\uparrow (\downarrow case) = \downarrow$' in rule 2 indicaes that the f-sructure of the PP and the case of PP (some literature refers it as P case) determines the f-structure of VP. 'Comp' refers to the compliment in a sentence, e.g., 'He said *that* she is powerful'.

Let us see first the lexical entries of various words in the sentence.

She saw stars. (2.4)

She N ($\uparrow$ Pred) = 'PRO'
      ($\uparrow$ Pers) = 3
      ($\uparrow$ Num) = SG
      ($\uparrow$ Gen) = FEM
      ($\uparrow$ Case) = NOM
Saw V $\uparrow$ Pred = 'see < ($\uparrow$ Subj) ($\uparrow$ Obj) >'
      ($\uparrow$ Tense = PAST)
Stars N $\uparrow$ Pred = 'Star'
      $\uparrow$ Pers = 3
      $\uparrow$ Num = PL

This will lead to the c-structure shown in Figure 2.8.

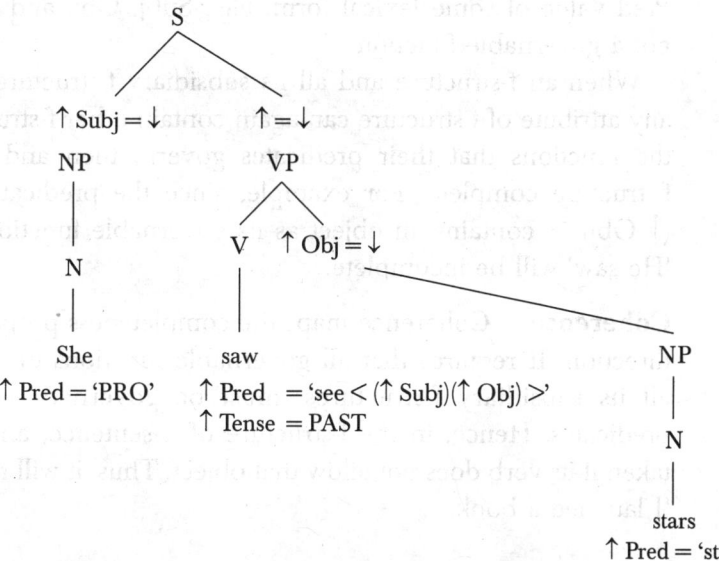

Figure 2.8 C-structure of sentence (2.4)

Finally, the f-structure is the set of attribute–value pairs, represented as

$$
\begin{bmatrix}
\text{subj} & \begin{bmatrix} \text{Pers} & 3 \\ \text{Num} & \text{SG} \\ \text{Gen} & \text{FEM} \\ \text{Case} & \text{NOM} \\ \text{Pred} & \text{‘PRO’} \end{bmatrix} \\
\text{obj} & \begin{bmatrix} \text{Pers} & 3 \\ \text{Num} & \text{PL} \\ \text{Pred} & \text{‘Star’} \end{bmatrix} \\
\text{Pred} & \text{‘see’} <(\uparrow \text{ subj}) \ (\uparrow \text{ obj})>
\end{bmatrix}
$$

It is interesting to note that the final f-structure is obtained through the unification of various f-structures for subject, object, verb, complement, etc. This unification is based on the functional specifications of the verb, which predicts the overall sentence structure.

LFG requires that all possible structures corresponding to passive constructs, dative constructs, etc., must be specified. If the given sentence does not match the specifications, it is said to be ill-formed. LFG imposes three conditions on f-structure (Sells 1985).

Consistency   In a given f-structure, a particular attribute may have at the most one value. Hence, while unifying two f-structures, if the attribute Num has value SG in one and PL in the other, it will be rejected.

Completeness   A function is called governable if it appears within the Pred value of some lexical form, e.g., Subj, Obj, and Obj 2. Adjunct is not a governable function.

When an f-structure and all its subsidiary f-structures (as the value of any attribute of f-structure can again contain other f-structures) contain all the functions that their predicates govern, then and only then is the f-structure complete. For example, since the predicate ‘see < ($\uparrow$ Subj) ($\uparrow$ Obj) >’ contains an object as its governable function, a sentence like ‘He saw’ will be incomplete.

Coherence   Coherence maps the completeness property in the reverse direction. It requires that all governable functions of an f-structure, and all its subsidiary f-structures, must be governed by their respective predicates. Hence, in the f-structure of a sentence, an object cannot be taken if its verb does not allow that object. Thus, it will reject the sentence, ‘I laughed a book.’

The completeness and coherence conditions are counterparts of θ-criterion in GB theory.

Lexical Rules in LFG

Different theories have different kinds of lexical rules and constraints for handling various sentence-constructs (active, passive, dative, causative, etc.). In GB, to express a sentence in its passive form, the verb is changed to its participial form and the ability of the verb to assign case and external (Agent) θ-role is taken away. In LFG, the verb is converted to the participial form, but the sub-categorization is changed directly. Consider the following example:

> Active: Tara ate the food.
> Passive: The food was eaten by Tara.
> Active: $\uparrow$ Pred = 'eat $< (\uparrow \text{Subj}) (\uparrow \text{Obj}) >$'
> Passive: $\uparrow$ Pred = 'eat $< (\uparrow \text{Obl}_{ag}) (\uparrow \text{Subj}) >$'

Here, Obl_{ag} represents oblique agent phrase. Similar rules can be applied in active and dative constructs for the verbs that accept two objects.

> Active: Tara gave a pen to Monika.
> Passive: Tara gave Monika a pen.
> Active: $\uparrow$ Pred = 'give $< (\uparrow \text{Subj}) (\uparrow \text{Obj}_2) (\uparrow \text{Obj}) >$'
> Passive: $\uparrow$ Pred = 'give $< (\uparrow \text{Subj}) (\uparrow \text{Obj}) (\uparrow \text{Obl}_{go}) >$'

Here, Obl_{go} stands for oblique goal phrase. Similar rules are also applicable to the process of causativization. This can be seen in Hindi, where the verb form is changed as follows:

हँसना $\xrightarrow{\text{causativization}}$ हँसाना
Laugh   Laugh-cause-past
                                      made to laugh

Example 2.6

Active: तारा हँसी
                 Taaraa hansii
                 Tara laughed

Causative: मोनिका ने तारा को हँसाया
                 Monika ne Tara ko hansaayaa
                 Monika Subj Tara Obj laugh-cause-past
                 Monika made Tara to laugh.

Active: $\uparrow$ Pred = 'Laugh $< \uparrow \text{Subj} >$'
Causative: $\uparrow$ Pred = 'cause $< (\uparrow \text{Subj}) (\uparrow \text{Obj}) (\text{Comp}) >$'

Here, a new predicate is formed which causes the action and requires a new subject, while the old subject becomes the object of the new predicate and the old verb becomes the X-complement (complement to infinital VPs).

Long Distance Dependencies and Coordination

In GB, when a category moved, it creates an empty category. In LFG, unbounded movement and coordination is handled by the functional identity and by correlation with the corresponding f-structure. An example will better explain these ideas.

Example 2.7 Consider the wh-movement in the following sentence.

Which picture does Tara like—most?

The f-structure can be represented as follows:

$$
\begin{bmatrix}
\text{Focus} & \begin{bmatrix} \text{Obl}_{th} & \begin{bmatrix} \text{pred} & \text{`PRO'} \\ \text{Refl} & + \end{bmatrix} \\ \text{Pred} & \text{`picture } \langle(\text{Obl}_{th})\rangle\text{'} \end{bmatrix} \\
\text{Subj} & [\text{pred} \quad \text{`Tara'}] \\
\text{Obj} & [\quad] \\
\text{Pred} & \text{`like } \langle(\uparrow\text{Subj}) (\uparrow\text{Obj})\rangle\text{'}
\end{bmatrix}
$$

The mechanism of handling these movements and coordination are not detailed here. The only aim is to highlight efforts and issues involved in modelling language.

2.2.5 Paninian Framework

Another very important model which has drawn much attention is the Paninian Grammar-based model (Kiparsky 1982, Bharti et al. 1995). Although *Paninian grammar* (PG) was written by Panini in 500 BC in Sanskrit (the original text being titled *Asthadhyayi*), the framework can be used for other Indian languages and possibly some Asian languages as well.

Unlike English, Asian languages are SOV (Subject-Object-Verb) ordered and inflectionally rich. The inflections provide important syntactic and semantic cues for language analysis and understanding. The Paninian framework takes advantage of these features. However, it should be noted that the research on this framework is still in progress and there are many complexities of Indian languages which are yet to be explained through this or other models. In this section, we briefly discuss some unique features of PG, to provide a glimpse of another potential model.

Some Important Features of Indian Languages

Indian languages have traditionally used oral communication for knowledge propagation. The purpose of these languages is to communicate ideas from the speaker's mind to the listener's mind. Such oral traditions have given rise to a morphologically rich language. Also, they are relatively word-order free. Some languages, like Sanskrit, have the flexibility to allow word groups representing subject, object, and verb to occur in any order. In others, like Hindi, we can change the position of subject and object. For example:

(a) माँ बच्चे को खाना देती है ।

Maan Bachche ko khanaa detii hai

Mother child to food give–(s)

Mother gives food to the child.

(b) बच्चे को माँ खाना देती है ।

Bachche ko Maan khanaa detii hai

Child to mother food give–(s)

Mother gives food to the child.

The auxilary verbs follow the main verb. In Hindi, they remain as separate words, whereas in south Indian (Dravidian) languages, they combine with the main verb. For example:

खा रहा है	करता रहा है
khaa raha hai	*kartaa rahaa hai*
eat-ing	doing been has
eating	has been doing

In Hindi, some verbs (main), e.g., give (देना), take (लेना), also combine with other verbs (main) to change the aspect and modality of the verbs.

Example 2.8

उसने खाना खाया ।	उसने खाना खा लिया ।
Usne khanaa khaayaa	*Usne khaanaa kha liyaa*
He (Subj) food ate	He (Subj) food eat taken
He ate food	He ate food (completed the action)
वह चला	वह चल दिया
	He move given
He moved	He moved (started the action)

In Indian languages, the nouns are followed by post-positions instead of prepositions. They generally remain as separate words in Hindi, except in the case of pronouns, for example

रेखा के पिता   उसके पिता
Rekha ke pita   *Uske pita*
Rekha of father
Father of Rekha   Her (His) father

In view of such differences between English (and English-like languages) and Indian languages, it is imperative that we find a new framework for handling Indian languages. Even among Indian languages, all features are not the same. As noted earlier, verb groups are formed differently in Indo-Aryan and Dravidian languages. Sanskrit is very different from the other Indian languages as it has five tenses and three numbers, and only one time aspect in each tense. Hence, the translation of 'He goes' and 'He is going' is the same in Sanskrit. Hindi is unique in the sense that it has no neuter gender. All nouns are categorized as feminine or masculine, and the verb form must have a gender agreement with the subject (sometimes with the object).

Thus, we have

ताला खो गया   चाभी खो गयी
Taalaa kho gayaa   *Chaabhii kho gayeee*
Lock lose (past)   key lose (past)
The lock was lost.   The key was lost.

Layered Representation in PG

The GB theory represents three syntactic levels: deep structure, surface structure, and logical form (LF), where the LF is nearer to semantics. This theory tries to resolve all language issues at syntactic levels only. Unlike GB, Paninian grammar framework is said to be syntactico-semantic, that is, one can go from surface layer to deep semantics by passing through intermediate layers. Although all these layers are not named, as per Bharti, et al. (1995), the language can be represented as follows:

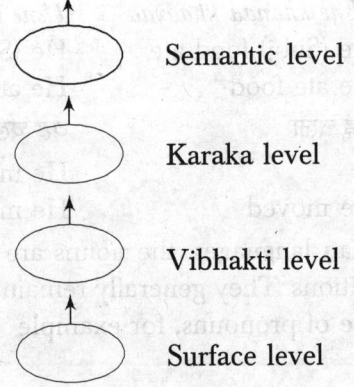
Semantic level

Karaka level

Vibhakti level

Surface level

The surface and the semantic levels are obvious. The other two levels should not be confused with the levels of GB. *Vibhakti* literally means inflection, but here, it refers to word (noun, verb, or other) groups based either on case endings, or post-positions, or compound verbs, or main and auxiliary verbs, etc. Instead of talking about NP, VP, AP, PP, etc., word groups are formed based on various kinds of markers (including the absence of it or θ). These markers are language-specific, but all Indian languages (and possibly Asian languages as well) can be represented at the Vibhakti level.

Karaka (pronounced *Kaaraka*) literally means Case, and in GB, we have already discussed case theory, θ-theory, and sub-categorization, etc. Paninian Grammar has its own way of defining Karaka relations, which we discuss in the next section. These relations are based on the way the word groups participate in the activity denoted by the verb group. In this sense, it is semantic as well as syntactic. However, things are not so straightforward. Complexities arise because of the absence of inflections, multiple categories of the words, multiple meanings, and above all, the presence of a large number of exceptions. These exceptions are not only applicable on stated rules but also on future rules. Such forward and backward chaining makes actual implementation difficult.

As the purpose of these languages is to communicate, generally between one human and another, the resolution of ambiguities is a contentious issue, often left to the listener. Hence, there may not be any particular number of semantic levels. Multiple-meaning texts are abundant in Indian literature as seen in the hundreds of interpretations of the epics.

Karaka Theory

Karaka theory is the central theme of PG framework. Karaka relations are assigned based on the roles played by various participants in the main activity. These roles are reflected in the case markers and post-position markers (parsargs). These relations are similar to case relations in English, but the types of relations are defined in a different manner and the richness of the case endings found in Indian languages has been used to its advantage.

We will discuss the various Karakas, such as Karta (subject), Karma (object), Karana (instrument), Sampradana (beneficiary), Apadan (separation), and Adhikaran (locus). These descriptions are just examples and not a complete discussion of PG or Karaka theory. For details of Karaka theory, see Shastri (1973) and Vasu (1977).

To explain various Karaka relations, let us consider an example. Consider the following Hindi sentence:

माँ बच्ची को आँगन में हाथ से रोटी खिलाती है ।   (2.5)

Maan bachchi ko aangan mein haath se rotii khilaatii hei

Mother child-to courtyard-in hand-by bread feed (s).

The mother feeds bread to the child by hand in the courtyard.

The first important Karak is subject, called 'Karta' in PG. Karta is defined as the noun group which is most independent (*swatantra* in Hindi). Karta has generally 'ne' or 'φ' case marker. It is an independent entity in the activity denoted by the main verb. As seen in GB also, verbs do not sub-categorize for subjects, although they assign a θ role to it. In sentence 2.5, 'maan' (mother) is the Karta. The concept of Karta is different from the 'agent' concept in the sense that Karta can also take up the role of experiencer, e.g.,

मुझसे रहा न गया ।

Mujhse rahaa na gayaa

Me hold -not -passive

I could not hold myself.

'Karma' is similar to object and is the locus of the result of the activity. In sentence (2.5), *rotii* (bread) is the Karma. As explained earlier, when the Karta is the experiencer, it (she) is also the locus of the result. Thus, the locus of the result is only when it is different from Karta termed Karma. Karma generally has 'φ' or 'KO' case marker. Another Karaka relation is 'Karan' (instrument), which is a noun group through which the goal is achieved. In sentence (2.5), *haath* (hand) is the Karan. It has the marker *dwara* (by) or *se*. 'Sampradan' is the beneficiary of the activity, e.g., *bachchi* (child) in sentence (2.5). It takes the marker *ko* (to) or *ke liye* (for). 'Apaadaan' denotes separation and the marker is attached to the part that serves as a reference point (being stationary), for example

माँ ने थाली से खाना उठाकर बच्चे को दिया ।

Maan ne thaali se khana uthakar bachche ko diyaa

Mother-Karta plate from Apaadan food taking-up child-to gave.

The mother gave food to the child taking it up from the plate.

Here *thaali* is the Apaadan. 'Adhikaran' is the locus (support in space or time) of Karta or Karma. In sentence (2.5), *aangan* (courtyard) is the Adhikaran. As these six relations are not sufficient to capture all possible relations, various others such as 'Sambandh' (relation) and 'Tadarthya' (purpose) have also been tried.

Issues in Paninian Grammar

The two problems challenging linguists are:

(i) Computational implementation of PG, and

(ii) Adaptation of PG to Indian, and other similar languages.

An approach to implementing PG has been discussed in Bharati, et al. (1995). This is a multilayered implementation. The approach is named 'Utsarga-Apvada' (default-exception), where rules are arranged in multiple layers in such a way that each layer consists of rules which are in exception to rules in the higher layer. Thus, as we go down the layer, more particular information is derived. Rules may be represented in the form of charts (such as Karaka chart and Lakshan chart).

However, many issues remain unresolved, specially in cases of shared Karak relations. Another difficulty arises when mapping between the Vibhakti (case markers and post-positions) and the semantic relation (with respect to verb) is not one to one. Two different Vibhakti can represent the same relation, or the same Vibhakti can represent different relations in different contexts. The strategy to disambiguate the various senses of words, or word groupings, are still the challenging issues.

As the system of rules is different in different languages, the framework requires adaptations to tackle various applications in various languages. Only some general features of PG framework has been described here.

2.3 STATISTICAL LANGUAGE MODEL

A statistical language model is a probability distribution $P(s)$ over all possible word sequences (or any other linguistic unit like words, sentences, paragraphs, documents, or spoken utterances). A number of statistical language models have been proposed in literature. The dominant approach in statistical language modelling is the *n*-gram model.

2.3.1 *n*-gram Model

As discussed earlier, the goal of a statistical language model is to estimate the probability (likelihood) of a sentence. This is achieved by decomposing sentence probability into a product of conditional probabilities using the chain rule as follows:

$$P(s) = P(w_1, w, w_3, \ldots, w_n)$$
$$= P(w_1)\, P(w_2/w_1)\, P(w_3/w_1\, w_2)\, P(w_4/w_1\, w_2\, w_3) \ldots$$
$$P(w_n/w_1\, w_2 \ldots w_{n-1}))$$

$$= \prod_{i=1}^{n} P(w_i/h_i)$$

where h_i is history of word w_i defined as

$$w_1\ w_2\ ..w_{i-1}$$

So, in order to calculate sentence probability, we need to calculate the probability of a word, given the sequence of words preceding it. This is not a simple task. An n-gram model simplifies the task by approximating the probability of a word given all the previous words by the conditional probability given previous $n-1$ words only.

$$P(w_i/h_i) \approx P(w_i/w_{i-n+1}.w_{i-1})$$

Thus, an n-gram model calculates $P(w_i/h_i)$ by modelling language as Markov model of order $n-1$, i.e., by looking at previous $n-1$ words only. A model that limits the history to the previous one word only, is termed a bi-gram $(n = 1)$ model. Likewise, a model that conditions the probability of a word to the previous two words, is called a tri-gram $(n = 2)$ model. Using bi-gram and tri-gram estimate, the probability of a sentence can be calculated as:

$$P(s) \approx \prod_{i=1}^{n} P(w_i/w_{i-1})$$

and

$$P(s) \approx \prod_{i=1}^{n} P(w_i/w_{i-2}.w_{i-1})$$

As an example, the bi-gram approximation of P(east/The Arabian knights are fairy tales of the) is

$$P(\text{east/the}),$$

whereas a tri-gram approximation is

$$P(\text{east/of the}).$$

A special word (pseudo word) $<s>$ is introduced to mark the beginning of the sentence in bi-gram estimation. The probability of the first word in a sentence is conditioned on $<s>$. Similarly, in tri-gram estimation, we introduce two pseudo-words $<s1>$ and $<s2>$.

Now, we discuss how to estimate these probabilities. This is done by training the n-gram model on the training corpus. We estimate n-gram parameters using the maximum likelihood estimation (MLE) technique, i.e., using relative frequencies. We count a particular n-gram in the training corpus and divide it by the sum of all n-grams that share the same prefix.

$$P(w_i/w_{i-n+1},...,w_{i-1}) = \frac{C(w_{i-n+1},...,w_{i-1},w_i)}{\sum_{w} C(w_{i-n+1},...,w_{i-1},w)}$$

The sum of all *n*-grams that share first *n*–1 words is equal to the count of the common prefix $w_{i-n+1}, ..., w_{i-1}$. So, we rewrite the previous expression as follows:

$$P(w_i/w_{i-n+1}, ..., w_{i-1}) = \frac{C(w_{i-n+1}, ..., w_{i-1}, w_i)}{C(w_{i-n+1}, ..., w_{i-1})}$$

The model parameter we get using these estimates, maximizes the probability of the training set *T* given the model *M*, i.e., *P*(*T*/*M*). The frequency with which a word occurs in a text may not be the same as in the training set; this model only provides the most likely solution.

A number of improvements have been suggested for the standard *n*-gram model. Before we discuss them, let us illustrate these ideas with the help of an example.

Example 2.9

Training set:

| The Arabian Knights |
| These are the fairy tales of the east |
| The stories of the Arabian knights are translated in many languages |

Bi-gram model:

P(the/<s>) = 0.67 P(Arabian/the) = 0.4 P(knights /Arabian) = 1.0

P(are/these) = 1.0 P(the/are) = 0.5 P(fairy/the) = 0.2

P(tales/fairy) = 1.0 P(of/tales) = 1.0 P(the/of) = 1.0

P(east/the) = 0.2 P(stories/the) = 0.2 P(of/stories) = 1.0

P(are/knights) = 1.0 P(translated/are) = 0.5 P(in /translated) = 1.0

P(many/in) = 1.0

P(languages/many) = 1.0

Test sentence(s): The Arabian knights are the fairy tales of the east.

P(The/<s>) × P(Arabian/the) × P(Knights/Arabian) × P(are/knights)
× P(the/are) × P(fairy/the) × P(tales/fairy) × P(of/tales) × P(the/of)
× P(east/the)

= 0.67 × 0.5 × 1.0 × 1.0 × 0.5 × 0.2 × 1.0 × 1.0 × 1.0 × 0.2

= 0.0067

As each probability is necessarily less than 1, multiplying the probabilities might cause a numerical underflow, particularly in long sentences. To avoid this, calculations are made in log space, where a calculation corresponds to adding log of individual probabilities and taking antilog of the sum.

The *n*-gram model suffers from data sparseness problem. An *n*-gram that does not occur in the training data is assigned zero probability, so that even a large corpus has several zero entries in its bi-gram matrix. This is because of the assumption that the probability of occurrence of a word depends only on the preceding word (or preceding *n*–1 words), which is not true in general. There are several long distance dependencies in natural language sentences, which this model fails to capture. Goodman (2003) pointed out that 'there is rarely enough data to accurately estimate the parameters of a language model.'

A number of smoothing techniques have been developed to handle the data sparseness problem, the simplest of these being add-one smoothing. In the words of Jurafsky and Martin (2000):

> *Smoothing in general refers to the task of re-evaluating zero-probability or low-probability n-grams and assigning them non-zero values.*

The word 'smoothing' is used to denote these techniques because they tend to make distributions more uniform by moving the extreme probabilities towards the average.

2.3.2 Add-one Smoothing

This is the simplest smoothing technique. It adds a value of one to each *n*-gram frequency before normalizing them into probabilities. Thus, the conditional probability becomes:

$$P(w_i / w_{i-n+1}, ..., w_{i-1}) = \frac{C(w_{i-n+1}, ..., w_{i-1}, w_i)}{C(w_{i-n+1}, ..., w_{i-1}) + V}$$

where V is the vocabulary size, i.e., size of the set of all the words being considered.

In general, add-one smoothing is not considered a good smoothing technique. It assigns the same probability to all missing *n*-grams, even though some of them could be more intuitively appealing than others. Gale and Church (1994) reported that variance of the counts produced by the add-one smoothing is worse than the unsmoothed MLE method. Another problem with this technique is that it shifts too much of the probability mass towards the unseen *n*-grams (*n*-grams with 0 probabilities) as there number is usually quite large. Good–Turing smoothing (Good 1953) attempts to improve the situation by looking at the number of *n*-grams with a high frequency in order to estimate the probability mass that needs to be assigned to missing or low-frequency *n*-grams.

2.3.3 Good–Turing Smoothing

Good–Turing smoothing (Good 1953) adjusts the frequency f of an n-gram using the count of n-grams having a frequency of occurrence $f+1$. It converts the frequency of an n-gram from f to f^* using the following expression:

$$f^* = (f + 1)\frac{n_{f+1}}{n_f}$$

where n_f is the number of n-grams that occur exactly f times in the training corpus. As an example, consider that the number of n-grams that occur 4 times is 25,108 and the number of n-grams that occur 5 times is 20,542. Then, the smoothed count for 5 will be

$$\frac{20542}{25108} \times 5 = 4.09$$

2.3.4 Caching Technique

Another improvement over basic n-gram model is caching. The frequency of n-gram is not uniform across the text segments or corpus. Certain words occur more frequently in certain segments (or documents) and rarely in others. For example, in this section, the frequency of the word 'n-gram' is high, whereas it occurs rarely in earlier sections. The basic n-gram model ignores this sort of variation of n-gram frequency. The cache model combines the most recent n-gram frequency with the standard n-gram model to improve its performance locally. The underlying assumption here is that the recently discovered words are more likely to be repeated.

A number of other smoothing techniques appear in literature. For these, readers are referred to Chen and Goodman (1998).

SUMMARY

The main topics covered in this chapter are as follows.

- Language modelling deals with providing a description of natural languages amenable to processing.
- There are two main approaches to language modelling: grammar-based language model and statistical language model.
- A grammar-based language model uses grammar to model language. A number of computational grammars have been proposed in literature. This chapter provides a discussion of models based on

government and binding, lexical functional grammar, and Paninian grammar.

- GB theory talks about representations at four different levels: s-structure, d-structure, logical form, and phonetic form.
- An important concept of GB is a general transformational rule called Move α which moves constituents freely, subject to certain constraints (defined by several theories and principles).
- LFG represents sentences at two different levels—constituent structure (c-structure) and functional structure (f-structure).
- Paninian grammar provides a framework for modelling Indian languages.
- The Paninian framework is syntactico-semantic. The central theme of this framework is Karaka relations.
- A statistical language model estimates the probability (likelihood) of a sentence. The most widely used statistical model is n-gram model.
- The *n*-gram model suffers from sparseness of data. Smoothing techniques such as add-one and Good–Turing can be used to handle this problem.
- Caching is another improvement over the standard *n*-gram model.

REFERENCES

Bharati, A., V. Chaitanya, and R. Sangal, 1995, 'Natural Language Processing: A Paninian Perspective,' Prentice-Hall of India, New Delhi.

Bresnan, Joan, 1976, 'Evidence for a theory of unbounded transformation,' *Linguistic Analysis.*

_____1977, 'Variables in theory of transformations,' *Formal Syntax*, Peter W. Culicover, Thoman Wasow, and Adrian Akmajian (Eds.), Academic Press, New York, pp. 157–96.

Chen, Stanley F. and Joshua T. Goodman, 1998, 'An Empirical Study of Smoothing Techniques for Language Modelling,' *Technical Report TR-10–98*, Computer Science Group, Harvard University.

Chomsky, Noam, 1957, *Syntactic Structures*, Mouton, The Hague.

_____1986, *Some Concepts and Consequences of the Theory of Government and Binding*, MIT Press, Cambridge, MA.

Gale, William A. and Kenneth W. Church, 1994, 'What's wrong with adding one?,' Corpus-based Research into Language, Oostdijk and P. de Haan (Eds.), Rodolpi, Amsterdam.

Gazdar, G., E. Klein, G. Pullum, and I. Sag, 1985, *Generalized Phrase Structure Grammar*, Blackwell.

Good, I.J., 1953, 'The population frequencies of species and the estimation of population parameters,' *Biometrika*, 40(3 and 4), pp. 237–64.

Goodman, Joshua, 2003, 'The state of the art in language modelling,' *Proceedings of the 2003 Conference of the North American Chapter of the Association for Computational Linguistics on Human Language Technology: Tutorials*, 5, Edmonton, Canada.

Jurafsky, Daniel and James H. Martin, 2000, 'Speech and Language Processing: An Introduction to Natural Language Processing,' *Computational Linguistics and Speech Recognition*, Prentice Hall, New Jersey.

Kaplan, Ronald M., 1975a, 'On Process Models for Sentence Comprehension,' *Explorations in Cognition*, Donald A. Norman and David E. Rumelhart (Eds.), W.H. Freeman, San Francisco.

————1975b, 'Transient Processing Load in Relative Clauses,' *Doctoral Dissertation*, Harvard University.

————1995, 'The Formal Architecture of Lexical-Functional Grammar,' *Formal Issues in Lexical-Functional Grammar*, M. Dalrymple, R.M. Kaplan, J.T. Maxwell III, and A. Zaenen (Eds.), CSLI Publications, Stanford.

Kaplan, Ronald M. and Joan Bresnan, 1982, *The Mental Representation of Grammatical Relations*, Joan Bresnan (Ed.), The MIT Press, Cambridge, MA.

Kiprasky, P., 1982, 'Some theoretical problem in Panini's grammar,' Bhandarkar Oriental Research Institute, Poona.

Rosenfeld, Ronald, 1994, 'Adaptive Statistical Language Modelling: A Maximum Entropy Approach,' *D.Phil. Thesis*, Carnegie Mellon University, Pittsburgh.

Sells, Peter, 1985, 'Lectures on Contemporary Syntactic Theories,' Center for the Study of Language and Information (CSLI), *Lecture Notes*, Number 3, Stanford.

Shastri, Charudev, 1973, *Vyakarana Chandrodaya*, (Vol. I–V), Motilal Banarsidass, in Hindi, New Delhi

Vasu, S.C. (Tr.), 1977, *The Ashtadhyayi of Panini* (2 volumes), Motilal Banarsidass, New Delhi.

Woods, William A., 1970, 'Transition Network Grammars for Natural Language Analysis,' *Communications of the ACM*, 13(10), pp. 591–606.

EXERCISES

1. Give the representation of a sentence in d-structure and s-structure in GB.
2. How is the structure of a sentence different from the structure of a phrase?

3. Discuss empty-category principle and give two examples of the creation of empty categories.
4. What is f-structure in LFG? What inputs to an algorithm create an f-structure?
5. Using the annotated rules in Example 2.5, find f-structures for the various phrases and the complete sentence, 'I saw Aparna in the market at night'.
6. What are Karaka relations? Explain the difference between Karta and Agent.
7. What are lexical rules? Give the complete entry for a verb in the lexicon, to be used in LFG.
8. Compare GB and PG. Why is PG called syntactico-semantic theory?
9. What are the problems associated with n-gram model? How are these problems handled?
10. How does caching improve the *n*-gram model?

LAB EXERCISES

1. Create a collection of 10 documents. Write a program to find the frequency of bi-grams in this collection.
2. Find the number of bi-grams that occur more than three times in this collection.

CHAPTER 3

WORD LEVEL ANALYSIS

CHAPTER OVERVIEW

This chapter focuses on processing carried out at word level, including methods for characterizing word sequences, identifying morphological variants, detecting and correcting misspelled words, and identifying correct part-of-speech of a word. The part-of-speech tagging methods covered in this chapter are: rule-based (linguistic), Stochastic (data-driven), and hybrid.

3.1 INTRODUCTION

As discussed in Chapter 1, natural language processing (NLP) involves different levels and complexities of processing. One way to analyse natural language text is by breaking it down into constituent units (words, phrases, sentences, and paragraphs) and then analyse these units. In Chapter 2, we discussed various language models that are used for analysing the syntax of natural language sentences. Before analysing syntax, we need to understand words, as words are the fundamental unit (syntactic as well as semantic) of any natural language text. This chapter focuses on NLP carried out at word level, including characterizing word sequences, identifying morphological variants, detecting and correcting misspelled words, and identifying correct part-of-speech of a word.

Regular expressions are a beautiful means for describing words. In many text applications, we wish to work with string patterns. Suppose you have just come across the word 'supernova'. It catches your interest and you jump to a search engine to find out more on 'supernovas'. But you do not know whether to type in 'supernova', 'Supernova', or 'supernovas'. Obviously, you need a system, which will retrieve relevant articles using any one of these word forms. Regular expressions are used for describing text strings in situations like this and in other information retrieval applications. In this chapter, we introduce regular expressions and discuss standard notations for describing text patterns.

After defining regular expressions, we discuss their implementation using finite-state automaton (FSA). Readers who have gone through a course in formal language theory will be familiar with FSA. The FSAs and their variants, such as finite state transducers, have found useful applications in speech recognition and synthesis, spell checking, and information extraction. As we will be using FSA throughout this book, we formally define FSAs in this chapter.

Errors in typing and spelling are quite common in text processing applications. Detecting and correcting these errors are the next topics of discussion in this chapter. Numerous web pages also contain misspelled words and often, query terms entered into the search engines are misspelled. An interactive spelling facility that informs users of such errors and presents appropriate corrections to them, is useful in these applications.

There are different classes of words. A word has many forms and the same word may have many different meanings depending on the context. Identifying the class to which a word belongs, its basic form, and its meaning are crucial to analysing text. This is the last topic covered in this chapter.

3.2   REGULAR EXPRESSIONS

Regular expressions, or regexes for short, are a pattern-matching standard for string parsing and replacement. They are a powerful way to find and replace strings that take a defined format. For example, regular expressions can be used to parse dates, urls and email addresses, log files, configuration files, command line switches, or programming scripts. They are useful tools for the design of language compilers and have been used in NLP for tokenization, describing lexicons, morphological analysis, etc. We have all used simplified forms of regular expressions, such as the file search patterns used by MS DOS, e.g., dir*.txt.

The use of regular expressions in computer science was made popular by a Unix-based editor, 'ed'. Perl was the first language that provided integrated support for regular expressions. It used a slash around each regular expression; we will follow the same notation in this book. However, slashes are not a part of regular expressions.

Regular expressions were originally studied as a part of theory of computation. They were first introduced by Kleene (1956). A regular expression is an algebraic formula whose value is a pattern consisting of a set of strings, called the language of the expression. The simplest kind of regular expression contains a single symbol. For example, the expression

/a/

denotes the set containing the string 'a'. A regular expression may specify a sequence of characters also. For example, the expression

/supernova/

denotes the set that contains the string 'supernova' and nothing else. In a search application, the first instance of each match to regular expression is underlined in Table 3.1.

Table 3.1 Some simple regular expressions

Regular expression	Example patterns
/book/	The world is a <u>book</u>, and those who do not travel read only one page.
/book/	Reporters, who do not read the style<u>book</u>, should not criticize their editors.
/face/	Not everything that is <u>face</u>d can be changed. But nothing can be changed until it is faced.
/a/	Re<u>a</u>son, Observation, and Experience—the Holy Trinity of Science.

3.2.1 Character Classes

Characters are grouped by putting them between square brackets. This way, any character in the class will match one character in the input. For example, the pattern /[abcd]/ will match (any of) a, b, c, and d. The use of brackets specifies a disjunction of characters. The regular expression / [0123456789]/ specifies any single digit. The character classes are important building blocks in expressions. They sometimes lead to cumbersome notation. For example, it is inconvenient to write the regular expression

/[abcdefghijklmnopqrstuvwxyz]/

to specify 'any lowercase letter'. In these cases, a dash is used to specify a range. The regular expression /[5–9]/ specifies any one of the characters 5, 6, 7, 8, or 9. The pattern /[m–p]/ specifies any one of the letter m, n, o, or p.

Regular expressions can also specify what a single character cannot be, by the use of a caret at the beginning. For example, the pattern /[^x]/ matches any single character except x. This interpretation is true only when a caret appears as the first symbol. If it occurs at any other place, it refers to the caret symbol itself. Table 3.2 shows a few examples explaining these concepts.

Table 3.2 Use of square brackets

RE	Match	Example patterns matched
[abc]	Match any of a, b, and c	'Refresher <u>c</u>ourse will start tomorrow'
[A–Z]	Match any character between A and Z (ASCII order)	'the course will end on <u>J</u>an. 10, 2006'
[^A–Z]	Match any character other than an uppercase letter	'TREC C<u>o</u>nference'
[^abc]	Match anything other than a, b, and c	'<u>T</u>REC Conference'
[+*?.]	Match any of +, *, ?, or the dot.	'3 <u>±</u> 2 = 5'
[a^]	Match a or ^	'<u>^</u> has three different uses.'

Regular expressions are *case sensitive*. The pattern /s/ matches a lower case 's' but not an uppercase 'S'. This means that the pattern /sana/ will not match the string /Sana/. This problem can be solved by using the disjunction of character s and S. The pattern /[sS]/ will match the strings containing either 's' or 'S'. The pattern /[sS]ana/ matches with the string 'sana' or 'Sana'.

While the use of square brackets solves the capitalization problem, we still need a solution for how to specify both 'supernova' and 'supernovas'. The pattern /[sS]upernova[sS]/ matches with any of the strings 'supernovas', 'supernovas', 'Supernovas', and 'SupernovaS', but not with the string 'supernova'. This is achieved with the use of a question mark /?/. A question mark makes the preceding character optional, i.e., zero or one occurrence of the previous character. The regular expression

/supernovas?/

specifies both 'supernova' and 'supernovas'. Often, we need to specify repeated occurrences of a character. The * operator, called the *Kleene ** (pronounced 'cleany star'), allow us to do this. The * operator specifies zero or more occurrences of a preceding character or regular expression. The regular expression /b*/ will match any string containing zero or more occurrences of 'b', i.e., it will match 'b', 'bb', or 'bbbb'. It will also match 'aaa', since that string contains zero occurrences of 'b'. To match a string containing one or more 'b's, the regular expression is /bb*/. The regular expression /bb*/ means a 'b' followed by zero or more 'b's. This will match with any of the strings 'b', 'bb', 'bbb', 'bbbb'. Similarly, the regular expression /[ab]*/ specifies 'zero or more "a"s or "b"s. This will match strings like 'aa', 'bb', or 'abab'. The kleene+ provides a shorter notation to specify one or more occurrences of a character. The regular expression /a+/ means one or more occurrences of 'a'. Using *Kleene+*, we can specify a sequence of digits by the regular expression /[0–9]+/. Complex regular expressions can be built up from simpler ones by means of regular expression operators.

The caret (^) is also used as an anchor to specify a match at the beginning of a line. The dollar sign, $, is an anchor that is used to specify a match at the end of the line. ^ and $ are important to regexes. If you wish to search for a line containing only the phrase 'The nature.' and nothing else, you need to specify a regular expression for this search. The anchors ^ and $ are of great help in this case. The regular expression / ^The nature\.$/ will search exactly for this line.

A number of special characters are also used to build regular expressions. One such character is the dot (.), called wildcard character, which matches any single character. The wildcard expression /./ matches any single character. The regular expression /.at/ matches with any of the string cat, bat, rat, gat, kat, mat, etc. It will also match the meaningless words such as nat, 4at, etc. Table 3.3 shows some of the special characters and their likely use.

Table 3.3 Some special characters

RE	Description
.	The dot matches any single character.
\n	Matches a new line character (or CR+LF combination).
\t	Matches a tab (ASCII 9).
\d	Matches a digit [0–9].
\D	Matches a non-digit.
\w	Matches an alphanumeric character.
\W	Matches a non-alphanumberic character.
\s	Matches a whitespace character.
\S	Matches a non-whitespace character.
\	Use \ to escape special characters. For example, \. matches a dot, * matches a * and \\ matches a backslash.

We can also use the wildcard symbol for counting characters. For instance /.....berry/ matches ten-letter strings that end in berry. This finds patterns like strawberry, sugarberry, and blackberry but fails to match with blueberry or hackberry.

Suppose you are searching a text for the presence of 'Tanveer' or 'Siddiqui'. You cannot use square brackets for this. You need a disjunction operator, shown by the pipe symbol(|). The pattern blackberry|blackberries matches either 'blackberry' or 'blackberries'. You might prefer to do this matching by merely writing blackberry|ies. Unfortunately, this does not work. The pattern blackberry|ies matches with either 'blackberry' or 'ies'. This is because sequences take precedence over disjunction. In order to apply the disjunction operator to a specific pattern, we need to enclose it within parentheses. The parenthesis operator makes it possible to treat the enclosed pattern as a single character for the purposes of neighboring operators. Now, we will consider an example from real application.

Example 3.1   Suppose we need to check if a string is an email address or not. An email address consist of a non-empty sequence of characters followed by the 'at' symbol, @, followed by another non-empty sequence of characters ending with pattern like .xx, .xxx, .xxxx, etc. The regular expression for an email address is

$$\text{^[A-Za-z0-9\_\backslash.-]}^{+}\text{+@[A-Za-z0-9\_\backslash.-]}^{+}\text{+[A-Za-z0-9\_][A-Za-z0-9\_]\$}$$

Table 3.4 shows the various parts of this 'rgex'.

Table 3.4   Parts of regular expression of Example 3.1

Pattern	Description
^[A-Za-z0-9_\.-]+	Match a positive number of acceptable characters at the start of the string.
@	Match the @ sign.
[A-Za-z0-9_\.-]+	Match any domain name, including a dot.
[A-Za-z0-9_][A-Za-z0-9_]$	Match two acceptable characters but not a dot. This ensures that the email address ends with .xx, .xxx, .xxxx, etc.

This example works for most cases. However, the specification is not based on any standard and may not be accurate enough to match all correct email addresses. It may accept non-working email addresses and reject working ones. Fine-tuning is required for accurate characterization.

A regular expression characterizes a particular kind of formal language, called a regular language. The language of regular expressions is similar to formulas of Boolean logic. Like logic formulas, regular expressions represent sets. Regular language is set of strings described by the regular expression. Regular languages may be encoded as finite state networks.

A regular expression may contain symbol pairs. For example, the regular expression /a:b/ represents a pair of string. The regular expression /a:b/ actually denotes a regular relation. A regular relation may be viewed as a mapping between two regular languages. The a:b relation is simply the cross product of the languages denoted by the expressions /a/ and /b/. To differentiate the two languages that are involved in a regular relation, we call the first one, the upper, and the second one, the lower language, of the relation. Similarly, in the pair /a:b/, the first symbol, a, can be called the upper symbol and the second symbol, b, the lower symbol. The two components of a symbol pair are separated in our notation by a colon (:) without any whitespace before or after. To make the notation less cumbersome, we ignore the distinction between the language A and the identity relation that maps every string of A to itself. Therefore, we also write /a:a/ simply as /a/.

Regular expressions have clean, declarative semantics (Voutilainen 1996). Mathematically, they are equivalent to finite automata, both having the same expressive power. This makes it possible to encode regular languages using finite-state automata, leading to easier manipulation of context free and other complex languages. Similarly, regular relations can be represented using finite-state transducers. With this representation, it is possible to derive new regular languages and relations by applying regular operators, instead of re-writing the grammars.

3.3 FINITE-STATE AUTOMATA

In our childhood, each of us must have played some game that fits the following description:

Pieces are set up on a playing board; dice are thrown or a wheel is spun, and a number is generated at random. Based on the number appearing on the dice, the pieces on the board are rearranged specified by the rules of the game. Then, it is your opponent's turn; she also does the same thing, resulting in rearrangement of the pieces based on the number generated. There is no skill or choice involved. The entire game is based on the values of the random numbers.

Consider all possible positions of the pieces on the board and call them *states*. The state in which the game begins is termed the *initial state*, and the state corresponding to the winning positions is termed the *final state*. We begin with the initial state of the starting positions of the pieces on the board. The game then changes from one state to another based on the value of the random number. For each possible number, there is one and only one resulting state given the input of the number and the current state. This continues until one player wins and the game is over. This is called a final state.

Now consider a very simple machine with an input device, a processor, some memory, and an output device. The machine starts in the initial state. It checks the input and goes to next state, which is completely determined by the prior state and the input. If all goes well, the machine reaches final state and terminates. If the machine gets an input for which the next state is not specified, and it gets stuck at a non-final state, we say the machine has failed or rejected the input.

A general model of this type of machine is called a finite automaton; 'finite' because the number of states and the alphabet of input symbols is finite; 'automaton' because the machine moves automatically, i.e., the change of state is completely governed by the input. This type of machine is more commonly called deterministic.

A finite automaton has the following properties:

1. A finite set of states, one of which is designated the *initial* or *start state*, and one or more of which are designated as the *final states*.
2. A finite alphabet set, Σ, consisting of input symbols.
3. A finite set of *transitions* that specify for *each* state and *each* symbol of the input alphabet, the state to which it next goes.

A finite automaton can be deterministic or non-deterministic. In a non-deterministic automaton, more than one transition out of a state is possible for the same input symbol.

Example 3.2   Suppose $\Sigma = \{a, b\}$, the set of states $= \{q_0, q_1, q_2, q_3, q_4\}$ with q_0 being the start state and q_4 the final state, we have the following rules of transition:

1. From state q_0 and with input a, go to state q_1.
2. From state q_1 and with input b, go to state q_2.
3. From state q_1 and with input c, go to state q_3.
4. From state q_2 and with input b, go to state q_4.
5. From state q_3 and with input b, go to state q_4.

This finite-state automaton is shown as a directed graph, called transition diagram, in Figure 3.1. The nodes in this diagram correspond to the states, and the arcs to transitions. The arcs are labelled with inputs. The final state is represented by a double circle. As seen in the figure, there is exactly one transition leading out of each state. Hence, this automaton is deterministic.

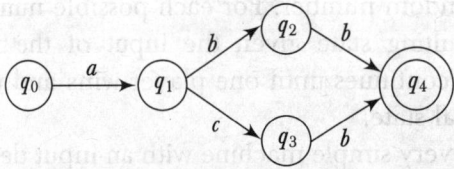

Figure 3.1   A deterministic finite-state automaton (DFA)

Finite-state automata have been used in a wide variety of areas, including linguistics, electrical engineering, computer science, mathematics, and logic. These are an important tool in computational linguistics and have been used as a mathematical device to implement regular expressions. Any regular expression can be represented by a finite automaton and the language of any finite automaton can be described by a regular expression. Both have the same expressive power. The following formal definitions of the two types of finite state automaton, namely, deterministic and non-deterministic finite automaton, are taken from Hopcroft and Ullman (1979).

A *deterministic finite-state automaton* (DFA) is defined as a 5-tuple $(Q, \Sigma, \delta, S, F)$, where Q is a set of *states*, Σ is an alphabet, S is the *start* state, $F \subseteq Q$ is a set of *final* states, and δ is a *transition function*. The transition function δ defines mapping from $Q \times \Sigma$ to Q. That is, for each state q and symbol a, there is at most one transition possible as shown in Figure 3.1.

Unlike DFA, the transition function of a *non-deterministic finite-state automaton* (NFA) maps $Q \times (\Sigma \cup \{\varepsilon\})$ to a subset of the power set of Q. That is, for each state, there can be more than one transition on a given symbol, each leading to a different state.

This is shown in Figure 3.2, where there are two possible transitions from state q_0 on input symbol a.

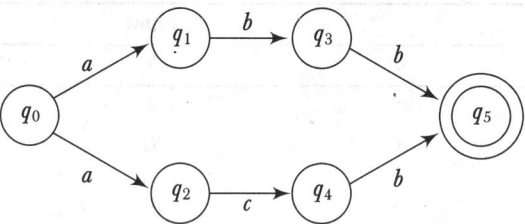

Figure 3.2 Non-deterministic finite-state automaton (NFA)

A *path* is a sequence of transitions beginning with the start state. A path leading to one of the final states is a *successful path*. The FSAs encode *regular* languages. The language that an FSA encodes is the set of strings that can be formed by concatenating the symbols along each successful path. Clearly, for automata with cycles, these sets are not finite.

We now examine what happens to various input strings that are presented to finite state automata. Consider the deterministic automaton described in Example 3.2 and the input, *ac*. We start with state q_0 and go to state q_1. The next input symbol is *c*, so we go to state q_3. No more input is left and we have not reached the final state, i.e., we have an unsuccessful end. Hence, the string *ac* is not recognized by the automaton.

This example illustrates how an FSA can be used to accept or recognize a string. The set of all strings that leave us in a final state is called the language accepted or defined by the FA. This means *ac* is not a word in the language defined by the automaton of Figure 3.1.

Now, consider the input acb. Again, we start with state q_0 and go to state q_1. The next input symbol is *c*, so we go to state q_3. The next input symbol is *b*, which leads to state q_4. No more input is left and we have reached to final state, i.e., this time we have a successful termination. Hence, the string *acb* is a word of the language defined by the automaton.

The language defined by this automaton can be described by the regular expression /abb|acb/.

The example considered here is quite simple. Typically, the list of transition rules can be quite long. Listing all transition rules may be inconvenient, so often we represent an automaton as a *state-transition table*. The rows in this table represent states and the columns correspond to input. The entries in the table represent the transition corresponding to a given state-input pair. A ϕ entry indicates missing transition. This table contains all the information needed by FSA. The state transition table for the automaton considered in Example 3.2 is shown in Table 3.5.

Table 3.5 The state-transition table for the DFA shown in Figure 3.1

	Input		
State	*a*	*b*	*c*
start: q_0	q_1	ϕ	ϕ
q_1	ϕ	q_2	q_3
q_2	ϕ	q_4	ϕ
q_3	ϕ	q_4	ϕ
final: q_4	ϕ	ϕ	ϕ

Now, consider a language consisting of all strings containing only *a*s and *b*s and ending with *baa*. We can specify this language by the regular expression /(a|b)*baa$/. The NFA implementing this regular expression is shown in Figure 3.3.

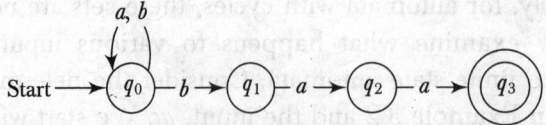

Figure 3.3 NFA for the regular expression /(a|b)*baa$/

Table 3.6 The state-transition table for the NFA shown in Figure 3.3.

State		Input	
		a	*b*
start:	q_0	$\{q_0\}$	$\{q_0, q_1\}$
	q_1	$\{q_2\}$	ϕ
	q_2	$\{q_3\}$	ϕ
final:	q_3	ϕ	ϕ

Two automata that define the same language are said to be *equivalent.* An NFA can be converted to an equivalent DFA and vice versa. The equivalent DFA for the NFA shown in Figure 3.3 is shown in Figure 3.4.

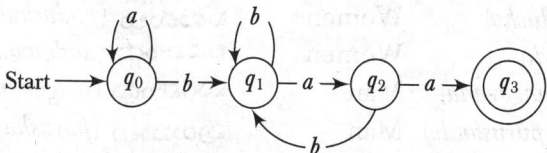

Figure 3.4 Equivalent DFA for NFA in Figure 3.3

3.4 MORPHOLOGICAL PARSING

Morphology is a sub-discipline of linguistics. It studies word structure and the formation of words from smaller units (morphemes). The goal of morphological parsing is to discover the morphemes that build a given word. Morphemes are the smallest meaning-bearing units in a language. For example, the word 'bread' consists of a single morpheme and 'eggs' consist of two: the morpheme egg and the morpheme -s. A morphological parser should be able to tell us that the word 'eggs' is the plural form of the noun stem 'egg'.

There are two broad classes of morphemes: stems and affixes. The stem is the main morpheme, i.e., the morpheme that contains the central meaning. Affixes modify the meaning given by the stem. Affixes are divided into prefix, suffix, infix, and circumfix. Prefixes are morphemes which appear before a stem, and suffixes are morphemes applied to the end of the stem. Circumfixes are morphemes that may be applied to either end of the stem while infixes are morphemes that appear inside a stem. Prefixes and suffixes are quite common in Urdu, Hindi, and English. For example, the Urdu word, بے وقت (*bewaqt*), meaning untimely, is

composed of the stem, *waqt*, and the prefix, *be-*, گھوڑوں (*ghodhon*) is composed of the stem, گھوڑا (*ghodha*) and the suffix, وں (*on*). Also, the word, ضرورت مند (*zarooratmand*), is composed of the stem, ضرورت

(*zaroorat*), and the suffix, مند (*mand*). Similarly, the English word, unhappy, is composed of the stem, happy, and the prefix, un-. The English word, birds, is composed of the stem, bird, and the suffix, -s. Likewise, the Hindi word, शीतलता is composed of a stem शीतल and the suffix – ता.

Telgu word గుర్రములు (*gurramulu*–plural form of *gurramu*, meaning horse) is composed of the stem గుర్రము (*gurramu*) and suffix -లు (*lu*). Some commonly used suffixes in plural forms of Telugu nouns are: లు (*lu*), ల్లు (*llu*), లు (*dlu*). Here is a list of singular and plural forms of a few Telugu words:

Singular	Meaning	Plural
పిల్లి (*pilli*)	Cat	పిల్లులు (*pillulu*)
పడుచు (*paduchu*)	Women	పడుచులు (*paduchulu*)
ఆడది (*aadadi*)	Women	ఆడవాండ్లు (*aadawandlu** or *aadawalu*)
మగవాడు (*magavadu*)	Man	మగవాండ్లు (*magavandlu* or *magavallu*)
పురుషుడు (*purushudu*)	Man	పురుషులు (*purushulu*)
చెవి (*chevi*)	Ear	చెవులు (*chevulu*)
ఇల్లు (*illu*)	House	ఇల్లులు (*illulu* or *illlu*)

*Another plural form of *aadadi* is *adawaru*, which is used to show respect.

There are three main ways of word formation: inflection, derivation, and compounding. In inflection, a root word is combined with a grammatical morpheme to yield a word of the same class as the original stem. Derivation combines a word stem with a grammatical morpheme to yield a word belonging to a different class, e.g., formation of the noun 'computation' from the verb 'compute'. The formation of a noun from a verb or adjective is called nominalization. Compounding is the process of merging two or more words to form a new word. For example, personal computer, desktop, overlook. Morphological analysis and generation deal with inflection, derivation and compounding process in word formation.

New words are continually forming a natural language. Many of these are morphologically related to known words. Understanding morphology is therefore important to understand the syntactic and semantic properties of new words. Morphological analysis and generation are essential to many NLP applications ranging from spelling corrections to machine translations. In parsing, e.g., it helps to know the agreement features of words. In information retrieval, morphological analysis helps identify the presence of a query word in a document in spite of different morphological variants.

Parsing, in general, means taking an input and producing some sort of structures for it. In NLP, this structure might be morphological, syntactic, semantic, or pragmatic. Morphological parsing takes as input the inflected *surface form* of each word in a text. As output, it produces the parsed form consisting of a canonical form (or *lemma*) of the word and a set of tags showing its syntactical category and morphological characteristics, e.g., possible part of speech and/or inflectional properties (gender, number, person, tense, etc.). Morphological generation is the inverse of this process. Both analysis and generation rely on two sources of information: a dictionary of the valid lemmas of the language and a set of inflection paradigms.

A morphological parser uses following information sources:

1. **Lexicon**

   A lexicon lists stems and affixes together with basic information about them.

2. **Morphotactics**

   There exists certain ordering among the morphemes that constitute a word. They cannot be arranged arbitrarily. For example, rest-less-ness is a valid word in English but not rest-ness-less. Morphotactics deals with the ordering of morphemes. It describes the way morphemes are arranged or touch each other.

3. **Orthographic rules**

   These are spelling rules that specify the changes that occur when two given morphemes combine. For example the $y \rightarrow ier$ spelling rule changes '*easy*' to '*easier*' and not to '*easyer*'.

Morphological analysis can be avoided if an exhaustive lexicon is available that lists features for all the word-forms of all the roots. Given a word, we simply consult the lexicon to get its feature values. For example, suppose an exhaustive lexicon for Hindi contains the following entries for the Hindi root-word *ghodhaa*.

Table 3.7 A sample lexicon entry

Word form	Category	Root	Gender	Number	Person
Ghodhaa	noun	GhoDaa	masculine	singular	3rd
Ghodhii	-do-	-do-	feminine	-do-	-do-
Ghodhon	-do-	-do-	masculine	plural	-do-
Ghodhe	-do-	-do-	-do-	-do-	-do-

Given a word, say *ghodhon*, we can look up the lexicon to get its feature values.

However, this approach has several limitations. First, it puts a heavy demand on memory. We have to list every form of the word, which results in a large number of, often redundant, entries in the lexicon.

Second, an exhaustive lexicon fails to show the relationship between different roots having similar word-forms. That means the approach fails to capture linguistic generalization, which is essential to develop a system capable of understanding unknown words.

Third, for morphologically complex languages, like Turkish, the number of possible word-forms may be theoretically infinite. It is not practical to list all possible word-forms in these languages.

These limitations explain why morphological parsing is necessary. The complexity of the morphological analysis varies widely among the world's languages, and is quite high even in relatively simple cases, such as English.

The simplest morphological systems are stemmers that collapse morphological variations of a given word (word-forms) to one lemma or stem. They do not require a lexicon. Stemmers have been especially used in information retrieval. Two widely used stemming algorithms have been developed by Lovins (1968) and Porter (1980). Stemmers do not use a lexicon; instead, they make use of rewrite rules of the form:

$$ier \rightarrow y \text{ (e.g., earlier} \rightarrow \text{early)}$$

$$ing \rightarrow \varepsilon \text{ (e.g., playing} \rightarrow \text{play)}$$

Stemming algorithms work in two steps:

(i) Suffix removal: This step removes predefined endings from words.

(ii) Recoding: This step adds predefined endings to the output of the first step.

These two steps can be performed sequentially as in Lovins's stemmer or simultaneously as in Porter's stemmer. For example, Porter's stemmer makes use of the following transformation rule:

$$ational \rightarrow ate$$

to transform word such as 'rotational' into 'rotate'.

It is difficult to use stemming with morphologically rich languages. Even in English, stemmers are not perfect. Krovitz (1993) pointed out errors of omissions and commissions in the Porter algorithm, such as transformation of the word 'organization' into 'organ' and 'noise' into 'noisy'. Another problem with Porter's algorithm is that it reduces only suffixes; prefixes and compounds are not reduced.

A more efficient *two-level morphological model*, first proposed by Koskenniemi (1983), can be used for highly inflected languages. In this model, a word is represented as a correspondence between its lexical level form and its surface level form. The surface level represents the actual spelling of the word while the lexical level represents the concatenation of its constituent morphemes. Morphological parsing is viewed as a mapping from the surface level into morpheme and feature sequences on the lexical level.

For example, the surface form 'playing' is represented in the lexical form as play + V + PP as shown in Figure 3.5. The lexical form consists of the stem 'play' followed by the morphological information +V +PP, which tells us that 'playing' is the present participle form of the verb.

Surface level | p | l | a | y | i | n | g |

Lexical level | p | l | a | y | +V | PP |

Figure 3.5 Surface and lexical forms of a word

Similarly, the surface form 'books' is represented in the lexical form as 'book + N + PL', where the first component is the stem and the second component (N + PL) is the morphological information, which tells us that the surface level form is a plural noun.

This model is usually implemented with a kind of finite-state automata, called *finite-state transducer* (FST). A transducer maps a set of symbols to another. A finite state transducer does this through a finite state automaton. An FST can be thought of as a two-state automaton, which recognizes or generates a pair of strings. FST passes over the input string by consuming the input symbols on the tape it traverses and consists it to the output string in the form of symbols. Formally, an FST has been defined by Hopcroft and Ullman (1979) as follows:

A finite-state transducer is a 6-tuple $(\Sigma_1, \Sigma_2, Q, \delta, S, F)$, where θ is set of states, S is the initial state, and $F \subseteq Q$ is a set of final states, Σ_1 is *input* alphabet, Σ_2 is *output* alphabet, and δ is a function mapping $Q \times (\Sigma_1 \cup \{\varepsilon\}) \times (\Sigma_2 \cup \{\varepsilon\})$ to a subset of the power set of Q.

Transducers can be seen as automata with transitions labelled with symbols from $\Sigma_1 \times \Sigma_2$, where Σ_1 and Σ_2 are the alphabets of input and output respectively. Thus, an FST is similar to an NFA except in that transitions are made on strings rather than on symbols and, in addition, they have outputs.

Figure 3.6 shows a simple transducer that accepts two input strings, hot and cat, and maps them onto cot and bat respectively. It is a common practice to represent a pair like *a:a* by a single letter.

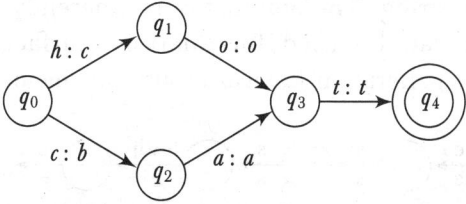

Figure 3.6 Finite-state transducer

Just as FSAs encode regular languages, FSTs encode regular relations. Regular relation is the relation between regular languages. The regular language encoded on the upper side of an FST is called upper language, and the one on the lower side is termed lower language. If T is a transducer, and s is a string, then we use $T(s)$ to represent the set of strings encoded by T such that the pair (s, t) is in the relation.

The FSTs are closed under union concatenation, composition, and Kleene closure. However, in general, they are not closed under intersection and complementation.

With this introduction, we can now implement the two-level morphology using FST. To get from the surface form of a word to its morphological analysis, we proceed in two steps as illustrated in Figure 3.7.

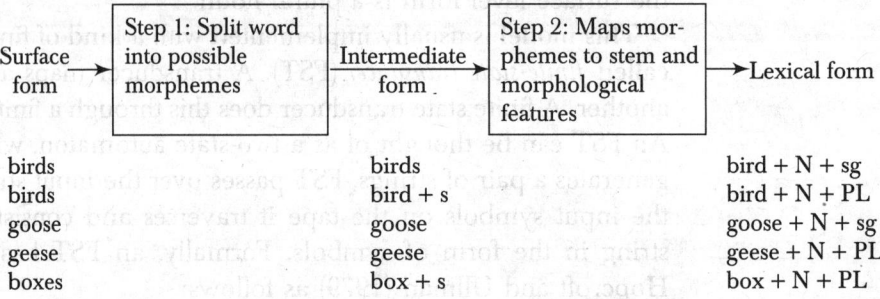

Figure 3.7 Two-step morphological parser

First, we split the words up into its possible components. For example, we make *bird + s* out of *birds*, where + indicates morpheme boundaries. In this step, we also consider spelling rules. Thus, there are two possible ways of splitting up *boxes*, namely *boxe + s* and *box + s*. The first one assumes that *boxe* is the stem and *s* the suffix, while the second assumes that *box* the stem is and that *e* has been introduced due to the spelling rule. The output of this step is a concatenation of morphemes, i.e., stems and affixes. There can be more than one representation for a given word. A transducer that does the mapping (translation) required by this step for the surface form 'lesser' might look like Figure 3.8. This FST represents the information that the comparative form of the adjective *less* is *lesser*, ε here is the empty string. The automaton is inherently bi-directional: the same transducer can be used for analysis (surface input, 'upward' application) or for generation (lexical input, 'downward' application).

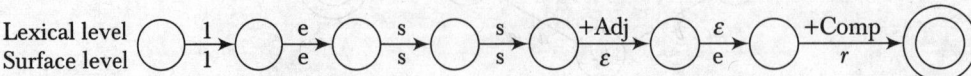

Figure 3.8 A simple FST

In the second step, we use a lexicon to look up categories of the stems and meaning of the affixes. So, *bird* + *s* will be mapped to *bird*+ *N*+ *PL*, and *box* + *s* to *box*+ *N* + *PL*. We also find out now that *boxe* is not a legal stem. This tells us that splitting *boxes* into *boxe* + *s* is an incorrect way of splitting *boxes*, and should therefore be discarded. This may not be the case always. We have words like *spouses* or *parses* where splitting the word into *spouse* + *s* or *parse* + *s* is correct. Orthographic rules are used to handle these spelling variations. For instance, one of the spelling rules says- add e after -s, -z, -x, -ch, -sh before the s (e.g., dish→dishes, box→boxes).

Each of these steps can be implemented with the help of a transducer. Thus, we need to build two transducers: one that maps the surface form to the intermediate form and another that maps the intermediate form to the lexical form.

We now develop an FST-based morphological parser for singular and plural nouns in English. The plural form of regular nouns usually end with -s or -es. However, a word ending in 's' need not necessarily be the plural form of a word. There are a number of singular words ending in s, e.g., miss and ass. One of the required translations is the deletion of the 'e' when introducing a morpheme boundary. This deletion is usually required for words ending in xes, ses, zes (e.g., suffixes and boxes). The transducer in Figure 3.9 does this. Figure 3.10 shows the possible sequences of states that the transducer undergoes, given the surface forms *birds* and *boxes* as input.

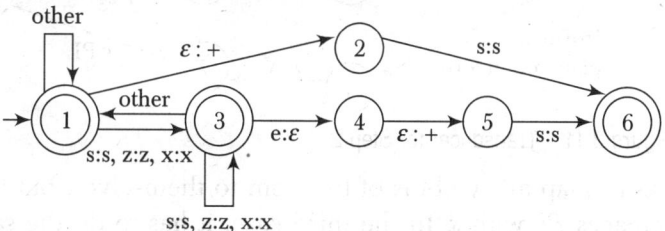

Figure 3.9 A simplified FST, mapping English nouns to the intermediate form

The next step is to develop a transducer that does the mapping from the intermediate level to the lexical level. The input to transducer has one of the following forms:

- Regular noun stem, e.g., *bird, cat*
- Regular noun stem + s, e.g., *bird + s*
- Singular irregular noun stem, e.g., *goose*
- Plural irregular noun stem, e.g., *geese*

In the first case, the transducer has to map all symbols of the stem to themselves and then output *N* and *sg* (Figure 3.7). In the second case, it

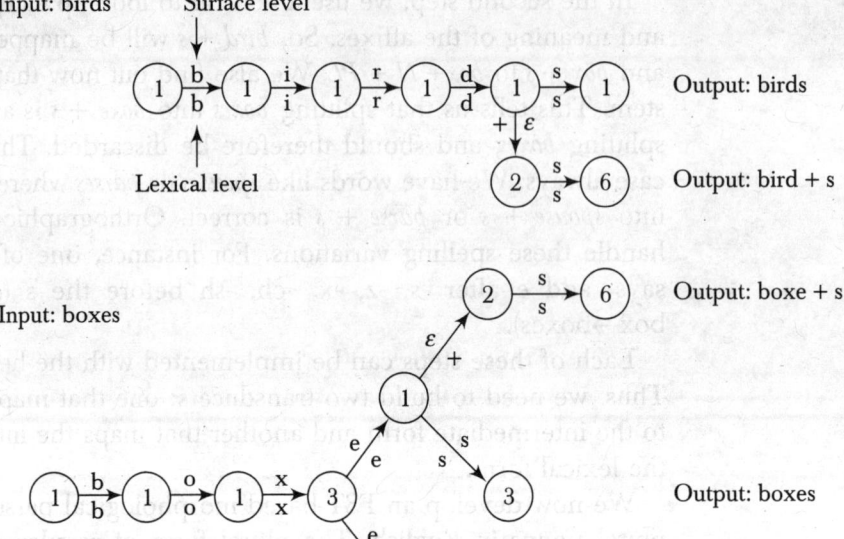

Figure 3.10   Possible sequences of states

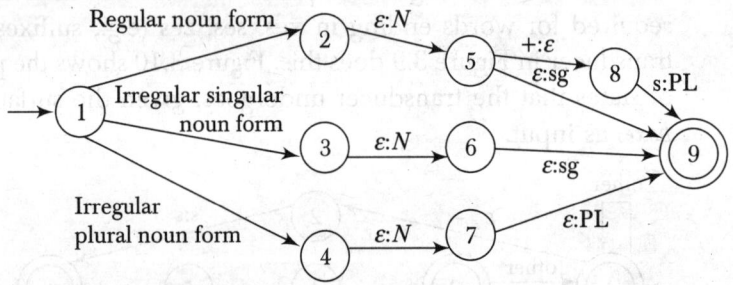

Figure 3.11   Transducer for Step 2

has to map all symbols of the stem to themselves, but then output *N* and replaces *PL* with *s*. In the third case, it has to do the same as in the first case. Finally, in the fourth case, the transducer has to map the irregular plural noun stem to the corresponding singular stem (e.g., *geese* to *goose*) and then it should add *N* and *PL*. The general structure of this transducer looks like Figure 3.11.

The mapping from State 1 to State 2, 3, or 4 is carried out with the help of a transducer encoding a lexicon. The transducer implementing the lexicon maps the individual regular and irregular noun stems to their correct noun stem, replacing labels like regular noun form, etc. This lexicon maps the surface form *geese*, which is an irregular noun, to its correct stem *goose* in the following way:

g:g e:o e:o s:s e:e

Mapping for the regular surface form of *bird* is b:b i:i r:r d:d. Representing pairs like *a:a* with a single letter, these two representations are reduced to g e:o e:o s e and b i r d respectively.

Composing this transducer with the previous one, we get a single two-level transducer with one input tape and one output tape. This maps plural nouns into the stem plus the morphological marker + pl and singular nouns into the stem plus the morpheme + sg. Thus a surface word form *birds* will be mapped to bird + N + pl as follows.

b:b i:i r:r d:d + ε:N + s:pl

Each letter maps to itself, while ε maps to morphological feature +N, and s maps to morphological feature pl. Figure 3.12 shows the resulting composed transducer.

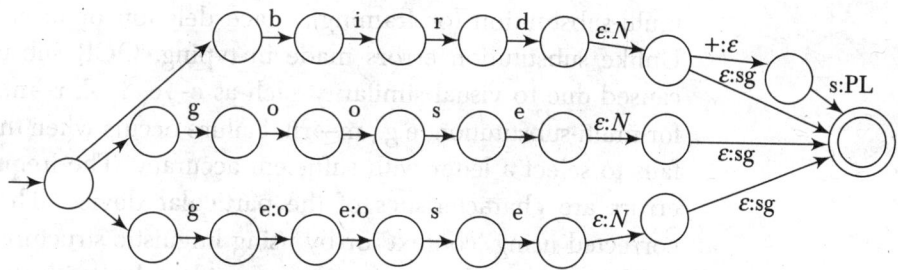

Figure 3.12 A transducer mapping nouns to their stem and morphological features

The power of the transducer lies in the fact that the same transducer can be used for analysis and generation. That is, we can run it in the downward direction (input: surface form and output: lexical form) or in the upward direction.

3.5 SPELLING ERROR DETECTION AND CORRECTION

In computer-based information systems, errors of typing and spelling constitute a very common source of variation between strings. These errors have been widely investigated. All investigations agree that single character omission, insertion, substitution, and reversal are the most common typing mistakes. In an early investigation, Damearu (1964) reported that over 80% of the typing errors were single-error misspellings:
(1) substitution of a single letter,
(2) omission of a single letter,
(3) insertion of a single letter, and
(4) transposition of two adjacent letters.

Shafer and Hardwick (1968) found that the most common type of single character error was substitution, followed by omission of a letter, and then insertion of a letter. Single character omission occurs when a single character is missed (deleted), e.g., when 'concept' is accidentally typed as 'concpt'. Insertion error refers to the presence of an extra character in a word, e.g. when 'error' is misspell as 'errorn'. Substitution error occurs when a wrong letter is typed in place of the right one, as in 'errpr', where 'p' appears in place of 'o'. Reversal refers to a situation in which the sequence of characters is reversed, e.g., 'aer' instead of 'are'. This is also termed transposition.

Optical character recognition (OCR) and other automatic reading devices introduce errors of substitution, deletion, and insertion but not of reversal. OCR errors are usually grouped into five classes: substitution, multi-substitution (or framing), space deletion or insertion, and failures. Unlike substitution errors made in typing, OCR substitution errors are caused due to visual similarity such as c→e, 1→l, r→n. The same is true for multi-substitution, e.g., m→rn. Failure occurs when the OCR algorithm fails to select a letter with sufficient accuracy. The frequency and type of errors are characteristics of the particular device. These errors can be corrected using 'context' or by using linguistic structures.

Many approaches to speech recognition deal with strings of phonemes (or symbols representing sounds), and attempt to match a spoken utterance with a dictionary of known utterances.

Unlike typing errors, spelling errors are mainly phonetic, where the misspell word is pronounced in the same way as the correct word. Phonetic errors are harder to set right because they distort the word by more than a single insertion, deletion, or substitution. Phonetic variations are common in transliteration. For example,

Spelling errors belong to one of two distinct categories: non-word errors and real word errors. When an error results in a word that does not appear in a given lexicon or is not a valid orthographic word form, it is termed a non-word error. Most of the early research on spelling errors focused on the detection of such non-words. The two main techniques used were *n*-gram analysis and dictionary lookup. Non-word error detection is now considered a solved problem.

A real-word error results in actual words of the language. It occurs due to typographical mistakes or spelling errors, e.g., substituting the spelling of a homophone or near-homophone, such as piece for peace or meat for meet. Real-word errors may cause local syntactic errors, global syntactic errors, semantic errors, or errors at discourse or pragmatic levels. It becomes impossible to decide that a word is wrong without some contextual information.

Spelling correction consists of *detecting* and *correcting* errors. Error detection is the process of finding misspelled words and error correction is the process of suggesting correct words to a misspelled one. These sub-problems are addressed in two ways:

1. Isolated-error detection and correction
2. Context-dependent error detection and correction

In isolated-word error detection and correction, each word is checked separately, independent of its context. Detecting whether or not a word is correct seems simple—why not to look up the word in a lexicon? Unfortunately, it is not as simple as it appears. There are a number of problems associated with this simple strategy.

- The strategy requires the existence of a lexicon containing all correct words. Such a lexicon would take a long time to compile and occupy a lot of space.
- Some languages are highly productive. It is impossible to list all the correct words of such languages.
- This strategy fails when spelling error produces a word that belongs to the lexicon, e.g., when 'theses' is written in place of 'these'. Such an error is called a *real-word error.*
- The larger the lexicon, the more likely it is that an error goes undetected, because the chance of a word being found is greater in a large lexicon.

Context dependent error detection and correction methods, utilize the context of a word to detect and correct errors. This requires grammatical analysis and is thus more complex and language dependent. Even in context dependent methods, the list of candidate words must first be obtained using an isolated-word method before making a selection depending on the context.

The spelling correction algorithm has been broadly categorized by Kukich (1992) as follows.

Minimum edit distance The minimum edit distance between two strings is the minimum number of operations (insertions, deletions, or substitutions) required to transform one string into another. Spelling correction algorithms based on minimum edit distance are the most studied algorithms.

Similarity key techniques The basic idea in a similarity key technique is to change a given string into a key such that similar strings will change into the same key. The SOUNDEX system (Odell and Russell 1918) is an example of a system that uses this technique in phonetic spelling correction applications.

n-gram based techniques The *n*-grams can be used for both non-word and real-word error detection because in the English alphabet, certain bi-grams and tri-grams of letters never occur or rarely do so; for example the tri-gram *qst* and the bi-gram *qd*. This information can be used to handle non-word error. Strings that contain these unusual *n*-grams can be identified as possible spelling errors. *n*-gram techniques usually require a large corpus or dictionary as training data, so that an *n*-gram table of possible combinations of letters can be compiled. In case of real-word error detection, we calculate the likelihood of one character following another and use this information to find possible correct word candidates.

Neural nets These have the ability to do associative recall based on incomplete and noisy data. They can be trained to adapt to specific spelling error patterns. The drawback of neural nets is that they are computationally expensive.

Rule-based techniques In a rule-based technique, a set of rules (heuristics) derived from knowledge of a common spelling error pattern is used to transform misspelled words into valid words. For example, if it is known that many errors occur from the letters *ue* being typed as *eu*, then we may write a rule that represents this.

3.5.1   Minimum Edit Distance

The minimum edit distance is the number of insertions, deletions, and substitutions required to change one string into another (Wagner and Fischer 1974). When we talk about distance between two strings, we are talking of the minimum edit distance. For example, the minimum edit distance between 'tutor' and 'tumor' is 2: We substitute 'm' for 't' and insert 'u' before 'r'. No smaller edit sequence can be found for this conversion. Therefore, the minimum edit distance is 2. Edit distance between two strings can be represented as a binary function, ed, which maps two strings to their edit distance. ed is symmetric. For any two strings, s and t, $ed(s, t)$ is always equal to $ed(t, s)$.

Edit distance can be viewed as a string alignment problem. By aligning two strings, we can measure the degree to which they match. There may be more than one possible alignment between two strings. The best possible alignment corresponds to the minimum edit distance between the strings. The alignment shown here, between *tutor* and *tumour*, has a distance of 2.

t   u   **t**   o   –   r

t   u   **m**   o   u   r

A dash in the upper string indicates insertion. A substitution occurs when the two alignment symbols do not match (shown in bold). We can associate a weight or cost with each operation. The *Levensthein* distance between two sequences is obtained by assigning a unit cost to each operation. Another possible alignment for this sequences is:

t   u   t   –   o   –   r

t   u   –   **m**   o   u   r

which has a cost of 3. We already have a better alignment than this one.

The problem of finding minimum edit distance seems quite simple but in fact is not so. A choice that seems good initially might lead to problems later. Dynamic programming algorithms can be quite useful for finding minimum edit distance between two sequences. Dynamic programming refers to a class of algorithms that apply a table-driven approach to solve problems by combining solutions to sub-problems. The dynamic programming algorithm for minimum edit distance is implemented by creating an edit distance matrix. This matrix has one row for each symbol in the source string and one column for each matrix in the target string. The (i, j)th cell in this matrix represents the distance between the first i character of the source and the first j character of the target string. Each cell can be computed as a simple function of its surrounding cells. Thus, by starting at the beginning of the matrix, it is possible to fill each entry iteratively. The value in each cell is computed in terms of three possible paths.

$$dist[i, j] = \begin{cases} dist[i-1, j] + \textit{insert\_cost}, \\ dist[i-1, j-1] + \textit{subst\_}\cos t[source_i, t\arg et_j] \\ dist[i, j-1] + \textit{delete\_cost} \end{cases}$$

The substitution will be 0 if the ith character in the source matches with jth character in the target. The minimum edit distance algorithm is shown in Figure 3.13. How the algorithm computes the minimum edit distance between *tutor* and *tumour* is shown in Figure 3.14.

```
Input: Two strings, X and Y
Output: The minimum edit distance between X and Y
m ← length(X)
n ← length(Y)
for i = 0 to m do
    dist[i,0] ← i
for j = 0 to n do
    dist[0,j] ← j
for i = 0 to m do
    for j = 0 to n do
        dist[i,j] = min{ dist[i–1,j] + insert_cost,
                         dist[i–1,j–1] + subst_cost(X_i, Y_j),
                         dist[i,j–1] + delet_cost }
```

Figure 3.13   Minimum edit distance algorithm

	#	t	u	m	o	u	r
#	0	1	2	3	4	5	6
t	1	0	1	2	3	4	5
u	2	1	0	1	2	3	4
t	3	2	1	1	2	3	4
o	4	3	2	2	1	2	3
r	5	4	3	3	2	2	2

Figure 3.14   Computing minimum edit distance

Minimum edit distance algorithms are also useful for determining accuracy in speech recognition systems. Kemal Oflazer (1996) proposed an efficient algorithm based on spelling correction with finite-state automata.

3.6   WORDS AND WORD CLASSES

Words are classified into categories called part-of-speech. These are sometimes called word classes or lexical categories. These lexical categories are usually defined by their syntactic and morphological behaviours. The most common lexical categories are nouns and verbs. Other lexical categories include adjectives, adverbs, prepositions, and conjunctions. Table 3.8 shows some of the word classes in English. Lexical categories and their properties vary from language to language. Word classes are further categorized as open and closed word classes. Open word classes constantly acquire new members while closed word classes do not (or only infrequently do so). Nouns, verbs (except auxiliary verbs), adjectives, adverbs, and interjections are open word classes. Prepositions, auxiliary verbs, delimiters, conjunction, and particles are closed word classes.

Table 3.8 Part-of-speech example

NN	noun	student, chair, proof, mechanism
VB	verb	study, increase, produce
ADJ	adj	large, high, tall, few,
JJ	adverb	carefully, slowly, uniformly
IN	preposition	in, on, to, of
PRP	pronoun	I, me, they
DET	determiner	the, a, an, this, those

3.7 PART-OF-SPEECH TAGGING

Part-of-speech tagging is the process of assigning a part-of-speech (such as a noun, verb, pronoun, preposition, adverb, and adjective), to each word in a sentence. The input to a tagging algorithm is the sequence of words of a natural language sentence and specified tag sets (a finite list of part-of-speech tags). The output is a single best part-of-speech tag for each word. Many words may belong to more than one lexical category. For example, the English word 'book' can be a noun as in '*I am reading a good book*' or a verb as in '*The police booked the snatcher*'. The same is true for other languages. For example, the Hindi word '*sona*' may mean 'gold' (noun) or 'sleep' (verb). However, only one of the possible meanings is used at a time. In tagging, we try to determine the correct lexical category of a word in its context. No tagger is efficient enough to identify the correct lexical category of each word in a sentence in every case. The tag assigned by a tagger is the most likely for a particular use of word in a sentence.

The collection of tags used by a particular tagger is called a *tag set*. Most part-of-speech tag sets make use of the same basic categories, i.e., noun, verb, adjective, and prepositions. However, tag sets differ in how they define categories and how finely they divide words into categories. For example, both *eat* and *eats* might be tagged as a verb in one tag set, but assigned distinct tags in another tag set. In addition, most tag sets capture morpho-syntactic information such as singular/plural, number, gender, tense, etc. Consider the following sentences:

Zuha *eats* an apple daily.

Aman *ate* an apple yesterday.

They have *eaten* all the apples in the basket.

I like to *eat* guavas.

The word *eat* has a distinct grammatical form in each of these four sentences. *Eat* is the base form, *ate* its past tense, and the form *eats* requires a third person singular subject. Similarly, *eaten* is the past participle form and cannot occur in another grammatical context. It is required after have or has. Thus, the following sentences are ungrammatical:

I like to *eats* guava.

They *eaten* all the apples.

The number of tags used by different taggers varies substantially. Some use 20, while others use over 400 tags. The Penn Treebank tag set contains 45 tags while C7 uses 164. For a language like English, which is not morphologically rich, the C7 tagset is too big. The tagging process would yield too many mistagged words and the result would have to be manually corrected. Despite this, bigger tag sets have been used, e.g., TOSCA-ICE for the International Corpus of English with 270 tags (Garside 1997), or TESS with 200 tags. The larger the tag set, the greater the information captured about a linguistic context. However, the task of tagging becomes complicated and requires manual correction. A bigger tag set can be used for morphologically rich languages without introducing too many tagging errors. A tag set that uses just one tag to denote all the verbs will assign identical tags to all the forms of a verb. Although this coarse-grained distinction may be appropriate for some tasks, a fine-grained tag set captures more information. This is useful for tasks like syntactic pattern detection. The Penn Treebank tag set captures finer distinctions by assigning distinct tags to distinct grammatical forms of a verb, as summarized in Table 3.9. Tags assigned to the four different forms of the word *eat* according to this tag set is shown in Table 3.10.

Table 3.9   Tags from Penn Treebank tag set

VB	Verb, base form
	Subsumes imperatives, infinitives, and subjunctives
VBD	Verb, past tense
	Includes the conditional form of the verb *to be*
VBG	Verb, gerund, or present participle
VBN	Verb, past participle
VBP	Verb, non-3rd person singular present
VBZ	Verb, 3rd person singular present

Table 3.10 Possible tags for the word to *eat*

eat	VB
ate	VBD
eaten	VBN
eats	VBP

Here is an example of a tagged sentence:

Speech/NN sounds/NNS were/VBD sampled/VBN by/IN a/DT microphone/NN.

The tag set used is Penn Treebank.

Another tagging possible for this sentence is as follows:

Speech/NN sounds/VBZ were/VBD sampled/VBN by/IN a/DT microphone/NN.

It is easy to see that the second tagged sequence is not correct. It leads to semantic incoherence. We resolve the ambiguity using the context of the word. The context is also utilized by automatic taggers.

Part-of-speech tagging is an early stage of text processing in many NLP applications including speech synthesis, machine translation, information retrieval, and information extraction. In information retrieval, part-of-speech tagging can be used for indexing (for identifying useful tokens like nouns and phrases) and for disambiguating word senses. Tagging is not as complex as parsing. In tagging, a complete parse tree is not built; part-of-speech is assigned to words using contextual information.

Part-of-speech tagging methods fall under the three general categories.

- Rule-based (linguistic)
- Stochastic (data-driven)
- Hybrid

Rule-based taggers use hand-coded rules to assign tags to words. These rules use a lexicon to obtain a list of candidate tags and then use rules to discard incorrect tags.

Stochastic taggers have data-driven approaches in which frequency-based information is automatically derived from corpus and used to tag words. Stochastic taggers disambiguate words based on the probability that a word occurs with a particular tag. The simplest scheme is to assign the most frequent tag to each word. An early example of stochastic tagger was CLAWS (constituent likelihood automatic word-tagging system). CLAWS is the Stochastic equivalent of TAGGIT. Hidden Markov model (HMM) is tshe standard Stochastic tagger.

Hybrid taggers combine features of both these approaches. Like rule-based systems, they use rules to specify tags. Like stochastic systems, they use machine-learning to induce rules from a tagged training corpus automatically. The transformation-based tagger or Brill tagger is an example of the hybrid approach.

3.7.1 Rule-based Tagger

Most rule-based taggers have a two-stage architecture. The first stage is simply a dictionary look-up procedure, which returns a set of potential tags (parts-of-speech) and appropriate syntactic features for each word. The second stage uses a set of hand-coded rules to discard contextually illegitimate tags to get a single part-of-speech for each word. For example, consider the noun-verb ambiguity in the following sentence:

The show must go on.

The potential tags for the word *show* in this sentence is {VB, NN}. We resolve this ambiguity by using the following rule.

IF *preceding word is determiner* THEN *eliminate VB tag.*

This rule simply disallows verbs after a determiner. Using this rule the word *show* in the given sentence can only be noun.

In addition to contextual information, many taggers use morphological information to help in the disambiguation process. An example of a rule that make use of morphological information is:

IF *word ends in –ing and preceding word is a verb* THEN *label it a verb (VB).*

Capitalization information can be utilized in the tagging of unknown nouns. Rule-based taggers usually require supervised training. Instead, rules can be induced automatically. To induce rules untagged text is run through a tagger. The output is then manually corrected. The corrected text is then submitted to the tagger, which learns correction rules by comparing the two sets of data. This process may be repeated several times.

The earlier systems for automatic tagging were all rule-based. An example is TAGGIT (Greene and Rubin 1971), which was used for the initial tagging of the Brown corpus (Francis and Kucera 1982). This was also rule-based system. It used 3,300 disambiguation rules and was able to tag 77% of the words in the Brown corpus with their correct part-of-speech. The rest was done manually over several years. Yet another rule-based tagger is ENGTWOL (Voutilainen 1995).

Speed is an advantage of the rule-based tagger, and unlike stochastic taggers, they are deterministic. One of the arguments against them is the skill and effort required in writing disambiguation rules. However, stochastic taggers also require manual work if good performance is to be achieved. In the rule-based system, time is spent in writing a rule-set. For stochastic taggers, time is spent developing restrictions on transitions and emissions to improve tagger performance. Another disadvantage of the rule-based tagger is that it is usable for only one language. Using it for another one requires a rewrite of most of the program; unlike a probabilistic tagger which would be usable with only slight changes and new training.

3.7.2 Stochastic Tagger

The standard stochastic tagger algorithm is the HMM tagger. A Markov model applies the simplifying assumption that the probability of a chain of symbols can be approximated in terms of its parts or n-grams. The simplest n-gram model is the unigram model, which assigns the most likely tag (part-of-speech) to each token.

The unigram model needs to be trained using a tagged training corpus before it can be used to tag data. The most likely statistics are gathered over the corpus and used for tagging. The context used by the unigram tagger is the text of the word itself. For example, it will assign the tag JJ for each occurrence of *fast*, since *fast* is used as an adjective more frequently than it is used as a noun, verb, or adverb. This results in incorrect tagging in each of the following sentences:

She had a *fast*. (3.1)

Muslims *fast* during Ramadan. (3.2)

Those who were injured in the accident need to be helped *fast*. (3.3)

In the first sentence, *fast* is used as a noun. In the second, it is a verb, and in the third, an adverb.

We would expect more accurate predictions if we took more context into account when making a tagging decision. A bi-gram tagger uses the current word and the tag of the previous word in the tagging process. As the tag sequence "DT NN" is more likely than the tag sequence "DT JJ", a bi-gram model will assign a correct tag to the word *fast* in sentence (3.1). Similarly, it is more likely that an adverb (rather than a noun or an adjective) follows a verb. Hence, in sentence (3.3), the tag assigned to *fast* will be RB.

In general, an n-gram model considers the current word and the tag of the previous $n-1$ words in assigning a tag to a word. The context considered by a tri-gram model is shown in Figure 3.15. The area shaded in grey represents the context.

Tokens	w_{n-2}	w_{n-1}	w_n	w_{n+1}
tags	t_{n-2}	t_{n-1}	t_n	t_{n+1}

Figure 3.15 Context used by a tri-gram tagger

So far, we have considered how a tag is assigned to a word given the previous tag(s). However, the objective of a tagger is to assign a tag sequence to a given sentence. We now discuss how the HMM tagger assigns the most likely tag sequence to a given sentence. We call this the HMM because it uses two layers of states: a visible layer corresponding to the input words, and a hidden layer learnt by the system corresponding to the tags. While tagging the input data, we only observe the words—the tags (states) are hidden. States of the model are visible in training, not during the tagging task.

As discussed earlier, the HMM makes use of lexical and bi-gram probabilities estimated over a tagged training corpus in order to compute the most likely tag sequence for each sentence. One way to store the statistical information is to build a probability matrix. The probability matrix contains both the probability that an individual word belongs to a word class as well as the n-gram analysis, e.g., for a bi-gram model, the probability that a word of class X follows a word of class Y. This matrix is then used to drive the HMM tagger while tagging an unknown text.

We now return to the original problem. Given a sequence of words (sentence), the objective is to find the most probable tag sequence for the sentence.

Let W be the sequence of words.

$$W = w_1, w_2, \dots, w_n$$

The task is to find the tag sequence

$$T = t_1, t_2, \dots, t_n$$

which maximizes $P(T|W)$, i.e.,

$$T' = \text{argmax}_T \ P(T|W)$$

Applying Bayes Rule, $P(T|W)$ can be estimated using the expression:

$$P(T|W) = P(W|T) * P(T)/P(W)$$

As the probability of the word sequence, $P(W)$, remains the same for each tag sequence, we can drop it. The expression for the most likely tag sequence becomes:

$$T' = \text{argmax}_T \, P(W|T) * P(T)$$

Using the Markov assumption, the probability of a tag sequence can be estimated as the product of the probability of its constituent n-grams, i.e.,

$$P(T) = P(t_1) * P(t_2|t_1) * P(t_3|t_1t_2) \dots * P(t_n|t_1 \dots t_{n-1})$$

$P(W/T)$ is the probability of seeing a word sequence, given a tag sequence. For example, it is asking the probability of seeing 'The egg is rotten' given 'DT NNP VB JJ'. We make the following two assumptions:

- The words are independent of each other.
- The probability of a word is dependent only on its tag.

Using these assumptions, we obtain

$$P(W/T) = P(w_1/t_1) * P(w_2/t_2) \dots P(w_i/t_i) * \dots P(W_n/t_n)$$

i.e., $\qquad P(W/T) \approx \prod_{i=1}^{n} P(w_i|t_i)$

So, $\qquad P(W/T) * P(T) = \prod_{i=1}^{n} P(w_i|t_i)$

$$\times P(t_1) \times P(t_2|t_1) * P(t_3|t_1t_2) \times \dots \times P(t_n|t_1 \dots t_{n-1})$$

Approximating the tag history using only the two previous tags, the transition probability, $P(T)$, becomes

$$P(T) = P(t_1) \times P(t_2|t_1) \times P(t_3|t_1t_2) \times \dots \times P(t_n|t_{n-2} \, t_{n-1})$$

Hence, $P(T/W)$ can be estimated as

$$P(W/T) * P(T) = \prod_{i=1}^{n} P(w_i|t_i)$$

$$\times P(t_1) \times P(t_2|t_1) * P(t_3|t_2t_1) \times \dots \times P(t_n|t_{n-2} \, t_{n-1})$$

$$= \prod_{i=1}^{n} P(w_i|t_i) \times P(t_1) \times P(t_2/t_1) \times \prod_{i=3}^{n} P(t_i|t_{i-2}t_{i-1})$$

We estimate these probabilities from relative frequencies via Maximum Likelihood Estimation.

$$P(t_i|t_{i-2} \, t_{i-1}) = \frac{c(t_{i-2}, t_{i-1}, t_i)}{c(t_{i-2}, t_{i-1})}$$

$$P(w_i|t_i) = \frac{c(w_i, t_i)}{c(t_i)}$$

where $c(t_{i-2}, t_{i-1}, t_i)$ is the number of occurrences of t_i followed by $t_{i-2} \, t_{i-1}$.

Stochastic models have the advantage of being accurate and language independent. Most stochastic taggers have an accuracy of 96–97%. The accuracy seems to be quite high but it should be noted that this is measured as a percentage of words. An accuracy of 96% means that for a sentence containing 20 words, the error rate per sentence will be $1-0.96^{20} = 56\%$. This corresponds to approximately one word per sentence.

One of the drawbacks of stochastic taggers is that they require a manually tagged corpus for training. Kupiec (1992), Cutting *et al.* (1992), and others have demonstrated that the HMM tagger can be trained from unannotated text. This makes it possible to use the model for languages in which a manually tagged corpus is not available. However, a tagger trained on a hand-coded corpus performs better than one trained on an unannotated text. In order to achieve good performance a tagged corpus is required.

We now consider an example demonstrating how the probability of a particular part-of-speech sequence for a given sentence can be computed.

Example 3.3   Consider the sentence

The bird can fly.

and the tag sequence

DT   NNP   MD   VB

Using bi-gram approximation, the probability

$$
P\begin{pmatrix} \text{DT} & \text{NNP} & \text{MD} & \text{VB} \\ | & | & | & | \\ \text{The} & \text{bird} & \text{can} & \text{fly} \end{pmatrix}
$$

can be computed as

$$
= P(\text{DT}) \times P(\text{NNP}|\text{DT}) * P(\text{MD}|\text{NNP}) \times P(\text{VB}|\text{MD})
$$
$$
\times P(\text{the}/\text{DT}) \times P(\text{bird}|\text{NNP}) \times P(\text{can}|\text{MD}) \times P(\text{fly}|\text{VB})
$$

3.7.3   Hybrid Taggers

Hybrid approaches to tagging combine the features of both the rule-based and stochastic approaches. They use rules to assign tags to words. Like the stochastic taggers, this is a machine learning technique and rules are automatically induced from the data. Transformation-based learning (TBL) of tags, also known as Brill tagging, is an example of hybrid approach. TBL is a machine learning method introduced by E. Brill (in 1995). Transformation-based error-driven learning has been applied to a number of natural language problems, including part-of-speech tagging, speech generation, and syntactic parsing (Brill 1993, 1994, Huang et al. 1994).

Figure 3.16 illustrates the TBL process. Like most HMM taggers, TBL is also a supervised learning technique. The steps involved in the TBL tagging algorithm are shown in Table 3.11. The input to Brill's TBL tagging algorithm is a tagged corpus and a lexicon (with most frequent information as indicated in the training corpus). The initial state annotator uses the lexicon to assign the most likely tag to each word as the start state. An ordered set of transformation rules are applied sequentially. The rule that results in the most improved tagging is selected. A manually tagged corpus is used as reference for truth. The process is iterated until some stopping criterion is reached, such as when no significant information is achieved over the previous iteration. At each iteration, the transformation that results in the highest score is selected. The output of the algorithm is a ranked list of learned transformation that transform the initial tagging close to the correct tagging. New text can then be annotated by first assigning the most frequent tag and then applying the ranked list of learned transformations in order.

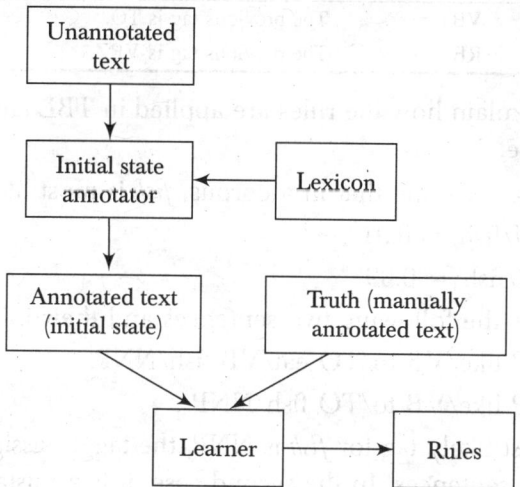

Figure 3.16   TBL learner

Table 3.11   TBL tagging algorithm

INPUT:	Tagged corpus and lexicon (with most frequent information)
Step 1:	Label every word with most likely tag (from dictionary)
Step 2:	Check every possible transformation and select one which most improves tagging
Step 3:	Re-tag corpus applying the rules
Repeat 2–3	Until some stopping criterion is reached
RESULT:	Ranked sequence of transformation rules

Each transformation is a pair of a re-write rule of the form $t_1 \rightarrow t_2$ and a contextual condition. In order to limit the set of transformations, a small set of templates is constructed. Any allowable transformation is an instantiation of these templates. Some of the transformation templates and transformations learned by TBL tagging are listed in Table 3.12.

Table 3.12   Examples of transformation templates and rules learned by the tagger

Change tag a to tag b when:
1. The preceding (following) word is tagged z.
2. The preceding (following) word is w.
3. The word two before (after) is w.
4. One of the preceding two words is w.
5. One of the two preceding (following) words is tagged z.
6. The current word is w and the preceding (following) word is x.
7. One of the previous three words is tagged z.

#	Change tags from	to	Contextual condition	Example
1.	NN	VB	The previous tag is TO.	To/TO fish/NN
2.	JJ	RB	The previous tag is VBZ.	runs/VBZ fast/JJ

We now explain how the rules are applied in TBL tagger with the help of an example.

Example 3.4   Assume that in a corpus, *fish* is most likely to be a noun.

$$P(\text{NN/fish}) = 0.91$$

$$P(\text{VB/fish}) = 0.09$$

Now consider the following two sentences and their initial tags.

I/PRP like/VB to/TO eat/VB fish/NNP.

I/PRP like/VB to/TO fish/NNP.

As the most likely tag for *fish* is NNP, the tagger assigns this tag to the word in both sentences. In the second case, it is a mistake.

After initial tagging when the transformation rules are applied, the tagger learns a rule that applies exactly to this mis-tagging of *fish*:

Change NNP to VB if the previous tag is TO.

As the contextual condition is satisfied, this rule will change fish/NN to fish/VB:

like/VB to/TO fish/NN $\rightarrow$ like/VB to/TO fish/VB

The algorithm can be made more efficient by indexing the words in a training corpus using potential transformation. Recent works have involved the use of finite state transducers to compile pattern-action rules, combining them to yield a single transducer representing the simultaneous application of all rules. Roche and Schabes (1997) have applied this approach to Brill's tagger. The resulting tagger is larger than Brill's original tagger and significantly faster.

Most of the work in part-of-speech tagging is done for English and some European languages. In other languages, part-of-speech tagging, and NLP research in general, is constrained by the lack of annotated corpuses. This is true for Indian language as well. A few part-of-speech tagging systems reported in recent years use morphological analysers along with a tagged corpus, e.g. a Bengali tagger based on HMM developd by Sandipan et al. (2004) and a Hindi tagger developed by Smriti et al. (2006). Smriti et al. used a decision tree based learning algorithm. A number of other part-of-speech taggers for Hindi, Bengali, and Telugu were developed as a result of the NLPAI-2006 machine learning contest on part-of-speech and chunking for Indian languages.

Tagging Urdu is more difficult. A number of factors contribute to this complexity. Among these is the right to left directionality of the written script and the presence of grammatical forms borrowed from Arabic and Persian. Little had been done to develop an extensive tag set for Urdu before Hardie (2003). His work was a part of the EMILLE—Enabling Minority Language Engineering project (see http://www.emille.lancs.ac.uk/about.php)—which focuses on the development of corpus and tools for South Asian languages.

3.7.4 Unknown Words

Unknown words are words that do not appear in dictionary or a training corpus. They create a problem during tagging. There are several potential solutions to this problem. One is to assign the most frequent tag (which occurs with most word types in the training corpus) to the unknown word. Another solution is to assume that the unknown words can be of any part-of-speech and initialize them by assigning them open class tags. Then proceed to disambiguate them using the probabilities of those tags. We can also use morphological information, such as affixes, to guess the possible tag of an unknown word. In this approach, the unknown word is assigned a tag based on the probability of the words belonging to a specific part-of-speech in the training corpus having the same suffix or prefix. A similar approach is used in Brill's tagger.

SUMMARY

This chapter has dealt with word level analysis. The topics covered include methods for characterizing word sequences, identifying morphological variants, detecting and correcting misspelled words, and identifying the correct part-of-speech for a word. The main points are as follows.

- Regular expressions can be used for specifying words. They can be encoded as a finite automaton.
- The goal of morphological parsing is to find out morphemes using which a given word is built. Morphemes are the smallest meaning-bearing units in a language.
- Morphological analysis and generation are essential to many NLP applications ranging from spelling error corrections to machine translations.
- The simplest morphological systems are stemmers. They do not use a lexicon. Instead, they use re-write rules. However, stemmers are not perfect.
- A two-level morphological model is more efficient. Both of its steps can be implemented using a finite state transducer.
- Word errors belong to one of two distinct categories, namely, *non-word errors* and *real word errors*. The latter category requires the context of the word to detect and correct errors.
- Words are classified into categories called part-of-speech or word classes. Word classes can be open or closed.
- Part-of-speech tagging is the process of assigning a part-of-speech like noun, verb, pronoun, preposition, adverb, adjective, etc., to each word in a sentence.
- Part-of-speech tagging methods fall under the following three categories:
  1. Rule-based (linguistic)
  2. Stochastic (data-driven)
  3. Hybrid

  Rule-based taggers use hand-coded rules to assign tags to words. *Stochastic taggers* require a pre-tagged corpus for training. *Hybrid taggers* combine features of both these approaches. Like rule-based systems, they use rules to specify tags. Like stochastic methods, they use machine learning to automatically induce rules from a tagged training corpus.
- Unknown words can be assigned the most frequent tags. We can also use morphological information to guess the correct tag.

REFERENCES

Brill, E., 1993, 'Transformation-based error-driven parsing,' *Proceedings of the Third International Workshop on Parsing Technologies,* Tilburg, The Netherlands.

Brill, E., 1994, 'Some advances in rule-based part of speech tagging,' *Proceedings of the Twelfth National Conference on Artificial Intelligence (AAAI-94),* Seattle.

Brill, E., 1995, 'Error-driven learning and natural language processing: a case study in part-of-speech tagging,' *Computational Linguistics.*

Cutting, D., J. Kupiec, J. Pederson, and P. Sibun, 1992, 'A practical part-of-speech tagger,' *Proceedings of the Third Conference on Applied Natural Language Processing, ACL.*

Damerau, F. J., 1964, 'A technique for computer detection and correction of spelling errors,' *Communications of the ACM,* 7(3) pp. 171–76.

Francis, W. Nelson and Henry Kucera, 1982, *Frequency Analysis of English Usage: Lexicon and Grammar,* Houghton Mifflin, Boston.

Garside, R., G. Leech, and G. Sampson, 1987, *The Computational Analysis of English: A Corpus-Based Approach,* Longman, London.

Green, B. and G. Rubin, 1971, 'Automated grammatical tagging of English,' Department of Linguistics, Brown University.

Hardie, A., 2003, 'Developing a tag-set for automated part-of-speech tagging in Urdu,' *Proceedings of the Corpus Linguistics 2003 Conference,* D. Archer, P. Rayson, A. Wilson, and T. McEnery, (Eds.), UCREL Technical Papers, 16, Department of Linguistics, Lancaster University.

Hopcroft, J.E. and J.D. Ullman, 1979, *Introduction to Automata Theory, Languages, and Computation,* Addison-Wesley, Reading, Massachusetts.

Huang et al., 1994, 'Generation of pronunciation from orthographies using transformation-based error-driven learning,' *Proceedings of Int. Conference on Speech and Language Processing (ICSLP),* Yokohama, Japan.

Kleene, S.C., 1956, 'Representation of events in nerve nets and finite automata,' *Automata Studies,* C. Shannon and J. McCarthy (Eds.), Princeton University Press, Princeton, NJ, pp. 3–41.

Koskenniemi, K., 1983, 'Two-level morphology: A general computational model of word-form recognition and production, Technical Report *Publication No. 11,* Department of General Linguistics,' University of Helsinki.

Krovetz, R., 1993, 'Viewing morphology as an inference process,' In *SIGIR-93,* pp. 191–202.

Kukich, K., 1992, 'Techniques for automatically correcting words in text,' *ACM Computing Surveys*, 24, pp. 377–439.

Kupiec, J., 1992, 'Robust part-of-speech tagging using a Hidden Markov Model,' *Computer Speech and Language*, vol. 6, pp. 225–42.

Lovins, J. B., 1968, 'Development of a stemming algorithm,' *Mechanical Translation and Computational Linguistics*, 11(1–2), pp. 22–31.

Odell, M. K. and R. C. Russell, US Patents 1261167/1435663 (1918/1922).

Oflazer, Kemal, 1996, 'Error-tolerant finite state recognition with applications to morphological analysis and spelling correction,' *Computational Linguistics*, 22(1).

Porter, M. F., 1980, 'An algorithm for suffix stripping program,' 14(3), pp. 130–37.

Roche, E. and Y. Schabes, 1995, 'Deterministic part of speech tagging with finite state transducers,' *Computational Linguistics*.

Sandipan, D., Kumar Nagraj, and Uma Sawant, 2004, 'A hybrid model for part-of-speech tagging and its application to Bengali,' *Proceedings of International Conference on Computational Intelligence*.

Shaffer, L. and J. Hardwick, 1968, 'Typing performance as a function of text,' *Quarterly Journal of Experimental Psychology*, 20, pp. 360–69.

Singh, Smriti, Kuhoo Gupta, Manish Shrivastava, and Pushpak Bhattacharyya, 2006, 'Morphological richness offsets resource demand–experiences in constructing a POS tagger for Hindi,' *Proceedings of the COLING/ACL 2006 Main Conference Poster Sessions*, Association for Computational Linguistics, Sydney, pp. 779–86.

Voutilainen, A., 1996, 'Morphological disambiguation,' *Constraint Grammar: A language-independent system for parsing uncertainty text*, F. Karlsson, A. Voutilainen, J. Heikkila, and A. Anttila (Eds.), Berlin, pp. 165–284.

Wagner, R.A. and M.J. Fischer, 1974, 'The string-to-string correction problem,' *Journal of the Association for Computing Machinery*, 21, 168–73.

EXERCISES

1. Define a finite automaton that accepts the following language: (aa)*(bb)*.

2. A typical URL is of the form:

http	://	www.abc.com	/nlppaper/public	/xxx.html
1	2	3	4	5

In this table, 1 is a protocol, 2 is name of a server, 3 is the directory, and 4 is the name of a document. Suppose you have to write a program that takes a URL and returns the protocol used, the DNS name of the server, the directory and the document name. Develop a regular expression that will help you in writing this program.

3. Distinguish between non-word and real-word error.
4. Compute the minimum edit distance between *paecflu* and *peaceful*.
5. Comment on the validity of the following statements:
   (a) Rule-based taggers are non-deterministic.
   (b) Stochastic taggers are language independent.
   (c) Brill's tagger is a rule-based tagger.
6. How can unknown words be handled in the tagging process?

LAB EXERCISES

1. Write a program to find minimum edit distance between two input strings.
2. Use any tagger available in your lab to tag a text file. Now write a program to find the most likely tag in the tagged text.
3. Write a program to find the probability of a tag given previous two tags, i.e., $P(t3/t2\ t1)$.

CHAPTER 4

SYNTACTIC ANALYSIS

CHAPTER OVERVIEW

This chapter introduces a number of important notions about syntax. Syntactic parsing deals with the syntactic structure of a sentence. In many languages, words are brought together to form larger groups termed constituents or phrases, which can be modelled using context-free grammar. Context-free grammar is a set of rules or productions that tell which elements can occur in a phrase and in what order. This chapter discusses phrase structural rules for various phrases and introduces feature structures to efficiently capture the properties of grammatical categories. The parsing strategies covered here include top-down and bottom-up parsing, dynamic programming parsing algorithms (namely, Earley and CYK), and probabilistic CYK parsing. Finally, a framework based on Paninian grammar is discussed for Hindi.

4.1 INTRODUCTION

In Chapter 3, we talked about word level analysis. We now move on to higher level constituents like phrases and sentences. The word 'syntax' refers to the grammatical arrangement of words in a sentence and their relationship with each other. The objective of syntactic analysis is to find the syntactic structure of the sentence. This structure is usually depicted as a tree, as shown in Figure 4.1. Nodes in the tree represent the phrases and leaves correspond to the words. The root of the tree is the whole sentence. Identifying the syntactic structure is useful in determining the meaning of the sentence. The identification is done using a process known as parsing. Syntactic parsing can be considered as the process of assigning 'phrase markers' to a sentence (Charniak 1997). One must, therefore, have a clear understanding of what phrases are. We will describe various phrases that could appear in a language so that we can specify grammar rules for them.

Researchers have proposed a number of parsing methods for natural language sentences. It is not possible to discuss all of them in a chapter, so we focus on a few widely-known ones. Most text books use only English for their discussions. We differ by including Indian languages as well. In addition to an introduction to grammar formalism, this chapter also provides an introduction to Hindi grammar.

Two important ideas in natural language are those of constituency and word order. Constituency is about how words are grouped together and how we know that they are really grouping together. Word order is about how, within a constituent, words are ordered with respect to one another, and also how constituents are ordered with respect to one another.

A widely used mathematical system for modelling constituent structure in natural language is context-free grammar (CFG) also known as phrase structure grammar. We begin our discussion with an overview of CFG in Section 4.2. In Section 4.3, we discuss various types of phrases and phrase structure rules. We then introduce feature structures in Section 4.4, to capture certain properties of grammatical categories which cannot be handled efficiently using CFG. We discuss probabilistic grammar and CYK parser based on this in Section 4.5. The CFG is basically a positional grammar and is not suitable for Indian languages, which are free word order languages. We provide a brief overview of Paninian grammar (PG), which is suitable for modelling free word order languages, in Section 4.6. We also discuss a parsing framework proposed by Bharti and Sangal (1990) for Indian languages based on this grammar. Finally, we provide a brief summary of the chapter.

4.2 CONTEXT-FREE GRAMMAR

Context-free grammar (CFG) was first defined for natural language by Chomsky (1957) and used for the Algol programming language by Backus (1959) and Naur (1960). A CFG (also called phrase-structure grammar) consists of four components:

1. A set of non-terminal symbols, N
2. A set of terminal symbols, T
3. A designated start symbol, S, that is one of the symbols from N.
4. A set of productions, P, of the form:

   $$A \rightarrow \alpha$$

where $A \in N$ and α is a string consisting of terminal and non-terminal symbols. The rule $A \rightarrow \alpha$ says that constituent A can be rewritten as α. This is also called the phrase structure rule. It specifies which elements (or constituents) can occur in a phrase and in what order. For example, the rule S → NP VP states that S consists of NP followed by VP, i.e., a sentence consists of a noun phrase followed by a verb phrase.

A language is usually defined through the concept of derivation. The basic operation is that of rewriting a symbol appearing on the left hand side of production by its right hand side. A CFG can be used to generate a sentence or to assign a structure to a given sentence. When used as a generator, the arrows in the production rule may be read as 'rewrite the symbol on the left with symbols on the right'. Consider the toy grammar shown in Figure 4.1. The symbol S can be rewritten as NP VP using Rule 1, then using rules R2 and R4, NP and VP are rewritten as N and V NP respectively. NP is then rewritten as Det N (R3). Finally, using rules R6 and R7, we get the sentence:

Hena reads a book.   (4.1)

We say that the sentence (4.1) can be derived from S. The representation of this derivation is shown in Figure 4.1. Sometimes, a more compact bracketed notation is used to represent a parse tree. The parse tree in Figure 4.1 can be represented using this notation as follows:

$$[_S [_{NP} [_N \text{Hena}]] [_{VP} [_V \text{reads}] [_{NP} [_{Det} \text{a}] [_N \text{book}]]]]$$

The set of all the strings containing terminal symbols which can be derived from the start symbol of the grammar, defines the language generated by that grammar. The parse tree shown in Figure 4.1 essentially represents a mapping of a string to its parse tree. This mapping process is called parsing. We will come back to this issue in Section 4.4.

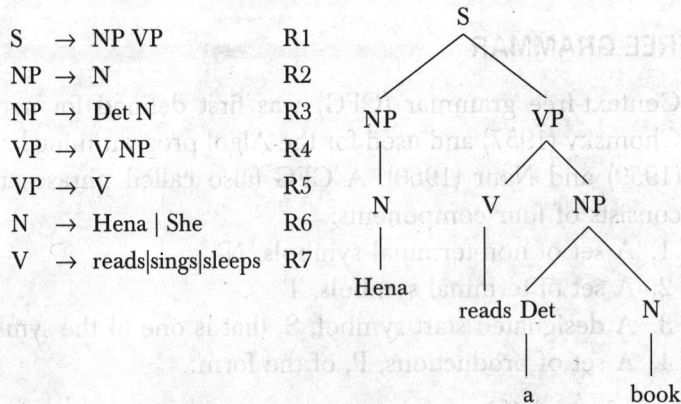

S	→	NP VP	R1
NP	→	N	R2
NP	→	Det N	R3
VP	→	V NP	R4
VP	→	V	R5
N	→	Hena	She	R6
V	→	reads|sings|sleeps	R7

Figure 4.1   Toy CFG and sample parse tree

4.3 CONSTITUENCY

Words in a sentence are not tied together as a sequence of part-of-speech. Language puts constraints on word order. For example, certain words go together with each other more than with others, and seem to behave as a unit. The fundamental idea of syntax is that words group together to form constituents (often termed phrases), each of which acts as a single unit. They combine with other constituents to form larger constituents, and eventually, a sentence. *The bird, The rain, The Wimbledon court, The beautiful garden* are all noun phrases that can occur in the same syntactic context. For example, they can all function as the subject or the object of a verb. These constituents combine with others to form a sentence constituent. For example, the noun phrase, *The bird,* can combine with the verb phrase, *flies,* to form the sentence, *The bird flies.* Different types of phrases have different internal structures. In this section, we discuss some of the major phrase types and try to build phrase structure rules to identify them.

4.3.1 Phrase Level Constructions

As discussed earlier, a fundamental notion in natural language is that certain groups of words behave as constituents. These constituents are identified by their ability to occur in similar contexts. One of the simplest ways to decide whether a group of words is a phrase, is to see if it can be substituted with some other group of words without changing the meaning. If such a substitution is possible then the set of words forms a phrase. This is called the substitution test. Consider sentence (4.1). We can substitute a number of other phrases:

Hena reads a book.
Hena reads a storybook.
Those girls read a book.
She reads a comic book.

We can easily identify the constituents that can be replaced for each other in these sentences. These are *Hena, she,* and *Those girls* and *a book, a storybook,* and *a comic book.* These are the words that form a phrase. In linguistics, such constituents represent a paradigmatic relationship. Elements that can substitute each other in certain syntactic positions are said to be members of one paradigm.

Phrase types are named after their head, which is the lexical category that determines the properties of the phrase. Thus, if the head is a noun, the phrase is called a noun phrase, if the head is a verb, the phrase is called a verb phrase; and so on for other lexical categories such as adjective and preposition. Figure 4.2 shows a sentence with a noun phrase, verb phrase, and preposition phrase.

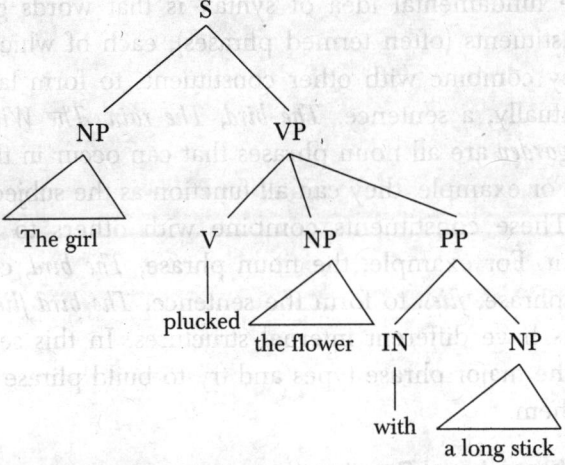

Figure 4.2 A sentence with NP, VP, and PP

Noun Phrase

A noun phrase is a phrase whose head is a noun or a pronoun, optionally accompanied by a set of modifiers. It can function as subject, object, or complement. The modifiers of a noun phrase can be determiners or adjective phrases. The obligatory constituent of a noun phrase is the noun head—all other constituents are optional. These structures can be represented using the phrase structure rule. As discussed earlier, phrase structure rules are of the form A → BC, which states that constituent A can be rewritten as two constituents B and C. These rules specify which elements can occur in a phrase and in what order. Using this notation, we can represent the phrase structure rules for a noun phrase as follows.

NP → Pronoun
NP → Det Noun
NP → Noun
NP → Adj Noun
NP → Det Adj Noun

We can combine all these rules in a single phrase structure rule as follows:

NP → (Det) (Adj) Noun|Pronoun

The constituents in parentheses are optional. This rule states that a noun phrase consists of a noun, possibly preceded by a determiner and an adjective (in that order). This rule does not cover all possible NPs. A noun phrase may include post-modifiers and more than one adjective. For example, it may include a prepositional phrase (PP). More than one adjective is handled by allowing an adjective phrase (AP) for the adjective in the rule. After incorporating PP and AP in the phrase structure rule, we get the following.

$$NP \rightarrow (Det) \; (AP) \; Noun \; (PP)$$

The following are a few examples of noun phrases:

They	(4.2a)
The foggy morning	(4.2b)
Chilled water	(4.2c)
A beautiful lake in Kashmir	(4.2d)
Cold banana shake	(4.2e)

Let us see how the phrases (4.2a–e) can be generated using phrase structure rules. The phrase (4.2a) consists only of a pronoun; (4.2b) consists of a determiner, an adjective (foggy) that stands for an entire adjective phrase, and a noun; (4.2c) comprises an adjective phrase and a noun; (4.2d) consists of a determiner (the), an adjective phrase (beautiful), a noun (lake), and a prepositional phrase (in Kashmir); and (4.2e) consists of an adjective followed by a sequence of nouns. A noun sequence is termed as nominal. None of the phrase structure rules discussed so far are able to handle nominals. So, we modify our rules to cover this situation.

$$NP \rightarrow (Det) \; (AP) \; Nom \; (PP)$$
$$Nom \rightarrow Noun \mid Noun \; Nom$$

A noun phrase can act as a subject, an object, or a predicate. The following sentences demonstrate each of these uses.

The foggy damped weather disturbed the match.	(4.3a)
I would like a *nice cold banana shake*.	(4.3b)
Kula botanical garden is a *beautiful location*.	(4.3c)

In (4.3a), the noun phrase acts as a subject. In (4.3b), it acts as an object, and in (4.3c), it is a predicate.

Verb Phrase

Analogous to the noun phrase is the verb phrase, which is headed by a verb. There is a fairly wide range of phrases that can modify a verb. This makes verb phrases a bit more complex. The verb phrase organizes various elements of the sentence that depend syntactically on the verb.

The following are some examples of verb phrases:

Khushbu *slept*.	(4.4a)
The boy *kicked the ball*.	(4.4b)

Khushbu *slept in the garden*.　　　　　　　　　　　(4.4c)

The boy *gave the girl a book*.　　　　　　　　　　　(4.4d)

The boy *gave the girl a book with blue cover*.　　　　(4.4e)

As you can see from these examples a verb phrase can have a verb [VP → Verb in (4.4a)]; a verb followed by an NP [VP → Verb NP in (4.4b)]; a verb followed by a PP [VP → Verb PP in (4.4c)]; a verb followed by two NPs [VP → Verb NP NP in (4.4d)]; or a verb followed by two NPs and a PP [VP → Verb NP NP PP in (4.4e)]. In general, the number of NPs in a VP is limited to two, whereas it is possible to add more than two PPs.

$$VP \rightarrow Verb\ (NP)\ (NP)\ (PP)^*$$

Things are further complicated by the fact that objects may also be entire clauses as in the sentence, *I know that Taj is one of the seven wonders*. Hence, we must also allow for an alternative phrase statement rule, in which NP is replaced by S.

$$VP \rightarrow Verb\ S$$

Prepositional Phrase

Prepositional phrases are headed by a preposition. They consist of a preposition, possibly followed by some other constituent, usually a noun phrase.

We played volleyball *on the beach*.

We can have a preposition phrase that consists of just a preposition.

John went *outside*.

The phrase structure rule that captures the above eventualities is as follows.

$$PP \rightarrow Prep\ (NP)$$

Adjective Phrase

The head of an adjective phrase (AP) is an adjective. APs consist of an adjective, which may be preceded by an adverb and followed by a PP. Here are few examples.

Ashish is *clever*.

The train is *very late*.

My sister is *fond of animals*.

The phrase structure rule for adjective phrase is

$$AP \rightarrow (Adv)\ Adj\ (PP)$$

Adverb Phrase

An adverb phrase consists of an adverb, possibly preceded by a degree adverb. Here is an example.

Time passes *very quickly.*

AdvP → (Intens) Adv

4.3.2 Sentence Level Constructions

Having discussed phrase structures, we now focus our attention on sentences. A sentence can have varying structure. The four commonly known structures are declarative structure, imperative structure, yes-no question structure, and wh-question structure.

Sentences with a declarative structure have a subject followed by a predicate. The subject of a declarative sentence is a noun phrase and the predicate is a verb phrase, e.g., *I like horse riding.* The phrase structure rule for declarative sentences is

S → NP VP

Sentences with an imperative structure usually begin with a verb phrase and lack subject. The subject of these types of sentence is implicit and is understood to be 'you'. These types of sentences are used for commands and suggestions, and hence are called imperative. The grammar rule for this kind of sentence structure is

S → VP

Examples of this kind of sentences are as follows:

> *Look at the door.*
> *Give me the book.*
> *Stop talking.*
> *Show me the latest design.*

Sentences with the yes-no question structure ask questions which can be answered using yes or no. These sentences begin with an auxiliary verb, followed by a subject NP, followed by a VP. Here are some examples:

> *Do you have a red pen?*
> *Is there a vacant quarter?*
> *Is the game over?*
> *Can you show me your album?*

We expand our grammar by adding another rule for the expansion of S, as follows:

S → Aux NP VP

Sentences with wh-question structure are more complex. These sentences begin with a wh-words—who, which, where, what, why, and how. A wh-question may have a wh-phrase as a subject or may include another subject. Consider the following wh-question:

Which team won the match?

This sentence is similar to a declarative sentence except that it contains a wh-word. A simple rule to handle this type of sentence structure is

S → Wh-NP VP

Another type of wh-question structure is one that involves more than one NP. In this type of questions, the auxiliary verb comes before the subject NP, just as in yes-no question structures.

Which cameras can you show me in your shop?

The rule for this type of wh-questions is

S → Wh-NP Aux NP VP

A simplified view of the grammar rules discussed so far is summarized in Table 4.1.

Table 4.1 Summary of grammar rules

S	→	NP VP
S	→	VP
S	→	Aux NP VP
S	→	Wh-NP VP
S	→	Wh-NP Aux NP VP
NP	→	(Det) (AP) Nom (PP)
VP	→	Verb (NP) (NP) (PP)*
VP	→	Verb S
AP	→	(Adv) Adj (PP)
PP	→	Prep (NP)
Nom	→	

Coordination

The grammar rules in Table 4.1 are not exhaustive. There are other sentence-level structures that cannot be modelled by the rules discussed here. Coordination is one such structure. It refers to conjoining phrases with conjunctions like 'and', 'or', and 'but'. For example, a coordinate noun phrase can consist of two other noun phrases separated by a conjunction 'and', as in

I ate [NP [NP an apple] and [NP a banana]].

Similarly, verb phrases and prepositional phrases can be conjoined as follows:

It is [VP [VP dazzling] and [VP raining]].

Not only that, even a sentence can be conjoined.

[S [S I am reading the book] and [S I am also watching the movie]]

We need to devise rules to handle these constructions. Conjunction rules for NP, VP, and S can be built as follows:

NP → NP and NP

VP → VP and VP

S → S and S

Agreement

Most verbs use two different forms in present tense—one for third person, singular subjects, and the other for all other kinds of subjects. The third person singular (3sg) form ends with a -s whereas the non-3sg does not. Whenever there is a verb that has some noun acting as a subject, this agreement has to be confirmed. Here are a few examples that demonstrate how the subject NP affects the form of the verb.

Does [NP Priya] sing? (4.5a)

Do [NP they] eat? (4.5b)

In the first sentence, the subject NP is singular. Hence, the -es form of 'do', i.e. 'does' is used. The second sentence has a plural NP subject. Hence, the form 'do' is being used. Sentences in which subject and verb do not agree are ungrammatical. The following sentences are ungrammatical:

[Does] they eat? (4.5 c)

[Do] she sings? (4.5 d)

Sentences (4.5c) and (4.5d) point out that a grammatical phenomenon can lead to ungrammatical sentences—a problem known as over-generation. Rules that handle the yes-no questions of Example 4.5 are as follows:

S → Aux NP VP

To take care of the subject-verb agreement, we replace this rule with a pair of rules as follows:

S → 3sgAux 3sgNP VP

S → Non3sgAux Non3sgNP VP

These rules ensure appropriate subject-verb agreement. We could add rules for the lexicon like these:

3sg Aux → does| has| can

Non3sg Aux → do | have | can

Similarly, rules for 3sgNP and Non3sgNP need to be added. So we replace each of the phrase structure rules for noun phrase by a pair of rules as follows:

$$
\begin{array}{ll}
\textit{3sgNP} & \rightarrow \textit{(Det) (AP) SgNom (PP)} \\
\textit{Non3sgNP} & \rightarrow \textit{(Det) (AP) PlNom (PP)} \\
\textit{SgNom} & \rightarrow \textit{SgNoun | SgNoun SgNom} \\
\textit{PlNom} & \rightarrow \textit{PlNoun | PlNoun PlNom} \\
\textit{SgNoun} & \rightarrow \textit{Priya | lake | banana | sister | ...} \\
\textit{PlNoun} & \rightarrow \textit{Children | ...}
\end{array}
$$

We also have to add rules for the first and second person pronouns. This method of dealing with number agreement doubles the size of the grammar. Each rule that makes use of a noun or verb phrase results in the introduction of a pair of rules—one to handle singular form and another to handle plural form. We also need to introduce new versions of NP and noun rules for various cases, e.g., nominative (I, she, they, he) and accusative (me, her, him, them) cases of pronoun. Languages like Hindi and Urdu, which have not only noun-verb agreements but also gender agreements, further aggravate the problem by adding another multiplier. Clearly, CFG cannot handle this problem efficiently.

We solve the problem of over-generation by introducing new grammatical categories corresponding to each such constraint. This results in an explosion in the number of grammar rules and loss of generality. An alternative solution is to associate each non-terminal of the grammar with feature structures. Feature structures are able to capture grammatical properties without increasing the size of the grammar. We can think of grammatical categories (and the grammar rules that make use of them) as objects having properties associated with them. The information in these properties can be thought of as constraints imposed by the grammatical categories. Models based on this idea are called constraint-based formalisms. We now introduce feature structures and discuss how they are used to represent the constraints imposed by grammatical categories without the loss of generality.

Feature Structures

Feature structures are sets of feature-value pairs. They can be used to efficiently capture the properties of grammatical categories. Features are simply symbols representing properties that we wish to capture. For example, the number property of a noun phrase can be represented by NUMBER feature. The value that a NUMBER feature can take is SG (for singular) and PL (for plural). Values can be either atomic symbols or feature structures. Feature structures are represented by a matrix-like diagram called attribute value matrix (AVM).

$$\begin{bmatrix} FEATURE_1 & VALUE_1 \\ FEATURE_2 & VALUE_2 \\ ... & ... \\ FEATURE_n & VALUE_n \end{bmatrix}$$

Figure 4.3 An attribute value matrix (AVM)

An AVM consisting of a single NUMBER feature with the value SG is represented as follows:

$$\begin{bmatrix} NUMBER & SG \end{bmatrix}$$

The value of a feature can be left unspecified and represented by an empty pair of square brackets, as in the following example:

$$\begin{bmatrix} NUMBER & [] \end{bmatrix}$$

The feature structure can be used to encode the grammatical category of a constituent and the features associated with it. For example, the following structure represents the third person singular noun phrase.

$$\begin{bmatrix} CAT & NP \\ NUMBER & SG \\ PERSON & 3 \end{bmatrix}$$

Similarly, a third person plural noun phrase can be represented as follows:

$$\begin{bmatrix} CAT & NP \\ NUMBER & PL \\ PERSON & 3 \end{bmatrix}$$

The values of CAT and PERSON features remain the same in both structures. This explains how feature structures aid in generalization while making the necessary distinction possible. As mentioned earlier, feature values are not limited to atomic symbols. A feature can have another feature structure as its value. Consider the case of combining the NUMBER and PERSON features into a single AGREEMENT feature. This makes sense because grammatical subjects must agree with their predicates in NUMBER as well as PERSON properties. Using this new feature, we represent the grammatical category third person plural noun phrase by the following structure:

$$\begin{bmatrix} CAT & NP \\ AGREEMRENT & \begin{bmatrix} NUMBER & PL \\ PERSON & 3 \end{bmatrix} \end{bmatrix}$$

In order for feature structures to be useful, we must be able to perform operations on them. The two most important operations we need to perform are merging the information content of the two structures that are similar and rejecting structures that are incompatible. The computational technique that is used to perform these operations is called unification. Unification is implemented as a binary operator ($\sqcup$) that takes two feature structures as arguments and returns a merged feature structure if they are compatible, otherwise reports a failure. Here is a simple application of the unification operator for performing an equality check.

$$[NUMBER \quad PL] \sqcup [NUMBER \quad PL] = [NUMBER \quad PL]$$

The unification succeeds as the two structures have the same value for the NUMBER feature. A feature with an unspecified value in one structure, can be successfully matched with any value in a corresponding feature in another structure. In such cases, the unification operation produces a structure with the value provided by the structure having non-null value. For example,

$$[NUMBER \quad PL] \sqcup [NUMBER \quad []] = [NUMBER \quad PL]$$

In this example, the two structures are considered compatible and merged into a structure with PL as its value for the NUMBER feature. The value PL of the first structure matches the value [] of the second structure and becomes the value of the NUMBER feature of the output structure.

However, the following application of unification results in failure as the NUMBER features of the first and second structures have incompatible values.

$$[NUMBER \quad PL] \sqcup [NUMBER \quad SG] \text{ Fails}$$

The CFG rules can have feature structures attached to them to realize constraints on the constituents of the sentence.

4.4   PARSING

A CFG defines the syntax of a language but does not specify how structures are assigned. The task that uses the rewrite rules of a grammar to either generate a particular sequence of words or reconstruct its derivation (or phrase structure tree) is termed parsing. A phrase structure tree constructed from a sentence is called a parse. The syntactic parser is thus responsible for recognizing a sentence and assigning a syntactic structure to it. It is possible for many different phrase structure trees to derive the same sequence of words. This means a sentence can have multiple parses. This phenomenon is called syntactic ambiguity.

Garden pathing is another phenomenon related to syntactic parsing. It refers to the process of constructing a parse by exploring the parse tree along different paths, one after the other till, eventually, the right one is found. The popular example given for garden pathing is the sentence

The horse ran past the barn fell. (4.6)

In the first attempt, most of us come up with a parse corresponding to the sentence *The horse ran past the barn,* leaving no possibility for the word *fell* to be incrementally added in the sentence. In order to complete the parse of the sentence, we have to backtrack.

Finding the right parse can be viewed as a search process. The search finds all trees whose root is the start symbol S and whose leaves cover exactly the word in the input. The search space in this conception corresponds to all possible parse trees defined by the grammar. The following constraints guide the search process.

1. *Input*: The first constraint comes from the words in the input sentence. A valid parse is one that covers all the words in a sentence. Hence, these words must constitute the leaves of the final parse tree.

2. *Grammar*: The second kind of constraint comes from the grammar. The root of the final parse tree must be the start symbol of the grammar.

These two constraints give rise to the two most widely used search strategies by parsers, namely, top-down or goal-directed search and bottom-up or data-directed search.

4.4.1   Top-down Parsing

As the name suggests, top-down parsing starts its search from the root node S and works downwards towards the leaves. The underlying assumption here is that the input can be derived from the designated start symbol, S, of the grammar. The next step is to find all sub-trees which can start with S. To generate the sub-trees of the second-level search, we expand the root node using all the grammar rules with S on their left hand side. Likewise, each non-terminal symbol in the resulting sub-trees is expanded next using the grammar rules having a matching non-terminal symbol on their left hand side. The right hand side of the grammar rules provide the nodes to be generated, which are then expanded recursively. As the expansion continues, the tree grows downward and eventually reaches a state where the bottom of the tree consist only of part-of-speech categories. At this point, all trees whose leaves do not match words in the input sentence are rejected, leaving only trees that represent successful parses. A successful parse corresponds to a tree which matches exactly with the words in the input sentence.

Table 4.2   Sample grammar

S → NP VP	VP → Verb NP
S → VP	VP → Verb
NP → Det Nominal	PP → Preposition NP
NP → Noun	Det → this | that | a | the
NP → Det Noun PP	Verb → sleeps | sings | open | saw | paint
Nominal → Noun	Preposition → from | with | on | to
Nominal → Noun Nominal	Pronoun → she | he | they

Consider the grammar shown in Table 4.2 and the sentence

Paint the door.   (4.7)

A top-down search begins with the start symbol of the grammar. Thus, the first level (ply) search tree consists of a single node labelled S. The grammar in Table 4.2 has two rules with S on their left hand side. These

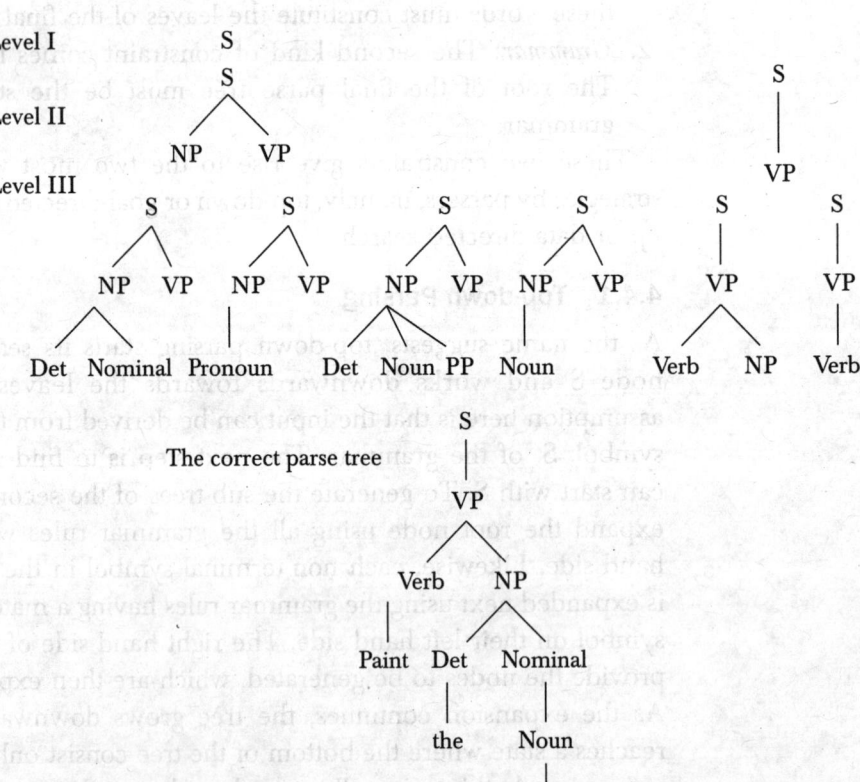

Figure 4.4   A top-down search space

rules are used to expand the tree, which gives us two partial trees at the second level search, as shown in Figure 4.4. The third level is generated by expanding the non-terminal at the bottom of the search tree in the previous ply. Due to space constraints, only the expansion corresponding to the left-most non-terminals has been shown in the figure. The subsequent steps in the parse are left, as an exercise, to the readers. The correct parse tree shown in Figure 4.4 is obtained by expanding the fifth parse tree of the third level.

4.4.2 Bottom-up Parsing

A bottom-up parser starts with the words in the input sentence and attempts to construct a parse tree in an upward direction towards the root. At each step, the parser looks for rules in the grammar where the right hand side matches some of the portions in the parse tree constructed so far, and reduces it using the left hand side of the production. The parse is considered successful if the parser reduces the tree to the start symbol of the grammar. Figure 4.5 shows some steps carried out by the bottom-up parser for sentence (4.7).

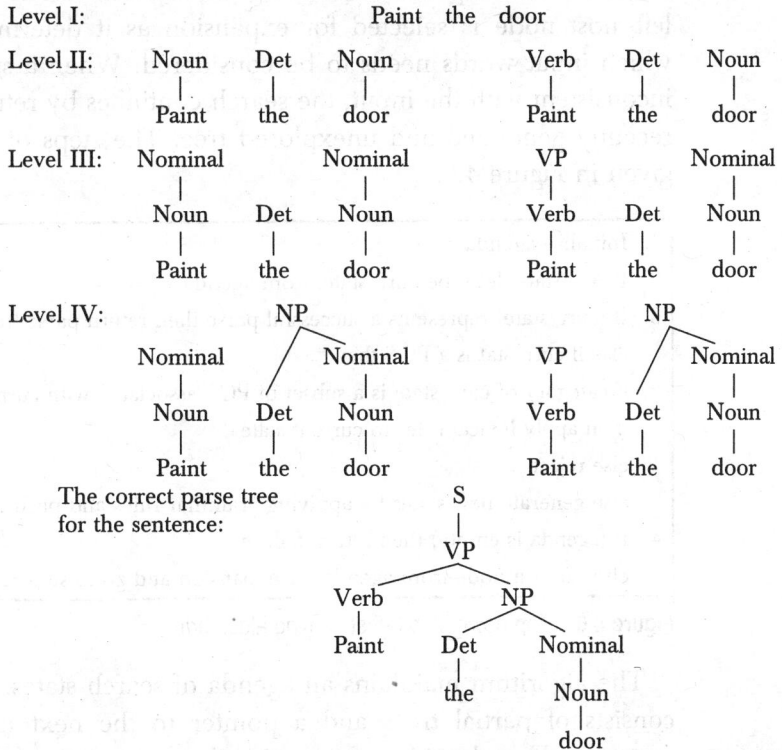

Figure 4.5   A bottom-up search space for sentence (4.7)

Each of these parsing strategies has its advantages and disadvantages. As the top-down search starts generating trees with the start symbol of the grammar, it never wastes time exploring a tree leading to a different root. However, it wastes considerable time exploring S trees that eventually result in words that are inconsistent with the input. This is because a top-down parser generates trees before seeing the input. On the other hand, a bottom-up parser never explores a tree that does not match the input. However, it wastes time generating trees that have no chance of leading to an S-rooted tree. The left branch of the search space in Figure 4.5 that explores a sub-tree assuming *paint* as a noun, is an example of wasted effort. We now present a basic search strategy that uses the top-down method to generate trees and augments it with bottom-up constraints to filter bad parses.

4.4.3   A Basic Top-down Parser

The approach presented here is essentially a depth first, left to right search. The depth first approach expands the search space incrementally by one state at a time. At each step, the left-most unexpanded leaf nodes of the tree are expanded first using the relevant rule of the grammar. The left-most node is selected for expansion as it determines the order in which input words needs to be considered. When a state arrives that is inconsistent with the input, the search continues by returning to the most recently generated and unexplored tree. The steps of the algorithm are given in Figure 4.6.

```
1. Initialize agenda
2. Pick a state, let it be curr_state, from agenda
3. If (curr_state) represents a successful parse then return parse tree
     else if curr_stat is a POS then
     if category of curr_state is a subset of POS associated with curr_word
     then apply lexical rules to current state
     else reject
     else generate new states by applying grammar rules and push them into agenda
4. If (agenda is empty) then return failure
     else select a node from agenda for expansion and go to step 3.
```

Figure 4.6   Top-down, depth-first parsing algorithm

The algorithm maintains an agenda of search states. Each search state consists of partial trees and a pointer to the next input word in the sentence. The algorithm starts with the state at the front of the agenda and generates a set of new states by applying grammar rule to the left-most unexpanded node of the tree associated with it. The newly generated states are put on the front of the agenda in the order defined by the textual order of the grammar rules used to create them. The process continues until either a successful parse tree is discovered or the agenda is empty, indicating a failure.

Figure 4.7 shows the trace of the algorithm on the sentence, *Open the door.* The algorithm starts with the node S and input word *Open.* It first expands S using the grammar rule S → NP VP. It then expands the left-most unexpanded non-terminal NP using the rule NP → Det Nominal. But the word *Open* cannot be derived from Det. Hence, the parser eliminates the rule and tries the second alternative, i.e., NP → noun, which again leads to a failure. The next search space on the agenda corresponds to the S → VP rule. The expansion of VP using the rule VP → Verb NP, successfully matches the first input words. The algorithm proceeds in a depth-first, left-to-right manner, to match the rest of the input words.

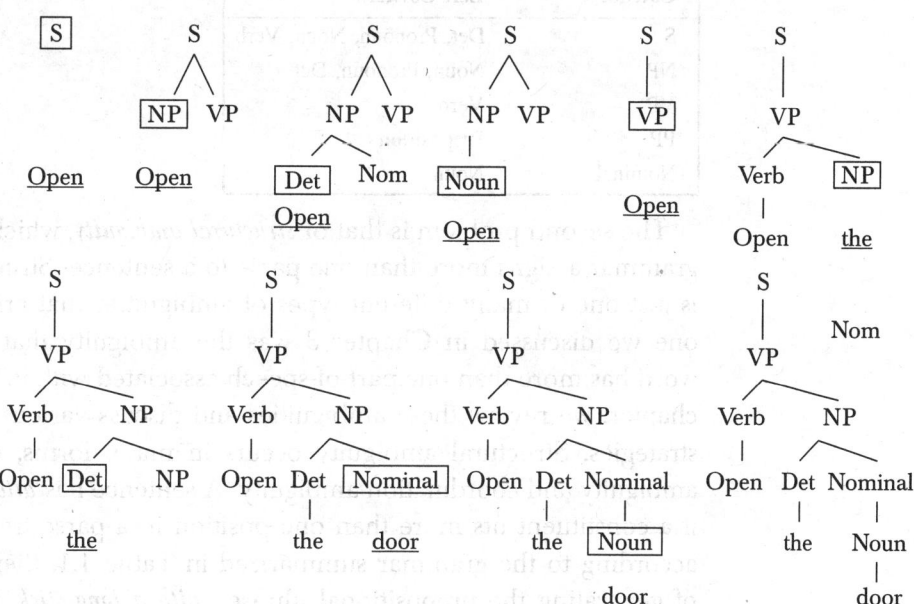

Figure 4.7 A derivation using top-down, depth-first algorithm

In any successful parse, the current input word must match the first word in the derivation of the node that is being expanded. This information can be utilized in eliminating spurious parses. A grammar rule that cannot lead to the input word as the first word along the left side of a derivation, should not be considered for expansion. The first word along the left side of the derivation is called the left corner of the tree. Using the left corner notion, we see that in our example, only the rule S → VP is applicable, as the word *Open* cannot be a left corner of the NP. In order to utilize this filter, we create a table containing a list of all the valid left corner categories for each non-terminal of the grammar. While selecting a rule for expansion, the table is consulted to see if the non-terminal associated with the rule has a part-of-speech associated with the current input. If not, then the rule is not considered. The left corner table for grammar is shown in Table 4.3.

The top-down, depth-first, left-to-right search algorithm suffers from certain disadvantages. The first is that of left recursion, which causes the search to get stuck in an infinite loop. This problem arises if the grammar is left recursive that, is, it contains a non-terminal A, which derives, in one or more steps, a string beginning with the same non-terminal, i.e., $A^* \Rightarrow A\beta$ for some β.

Table 4.3 Left corner for each grammar category

Category	Left Corners
S	Det, Pronoun, Noun, Verb
NP	Noun, Pronoun, Det
VP	Verb
PP	Preposition
Nominal	Noun

The second problem is that of *structural ambiguity,* which occurs when a grammar assigns more than one parse to a sentence. Structural ambiguity is just one of many different types of ambiguities that arise in NLP. The one we discussed in Chapter 3 was the ambiguity that occurs when a word has more than one part-of-speech associated with it. In the following chapter, we review these ambiguities and discuss various disambiguation strategies. Structural ambiguity occurs in many forms, e.g., attachment ambiguity and coordination ambiguity. A sentence has *attachment ambiguity* if a constituent fits more than one position in a parse tree. For example, according to the grammar summarized in Table 4.1, there are two ways of generating the prepositional phrase, *with a long stick,* in the sentence, *The girl plucked the flower with a long stick.* It can be generated from the verb phrase, as in the parse tree shown in Figure 4.2, or from the noun phrase, as in the parse tree shown in Figure 4.8. The first parse leads to the interpretation that the stick is used to pluck the flower, whereas the second parse gives the interpretation that the flower being plucked has a long stick.

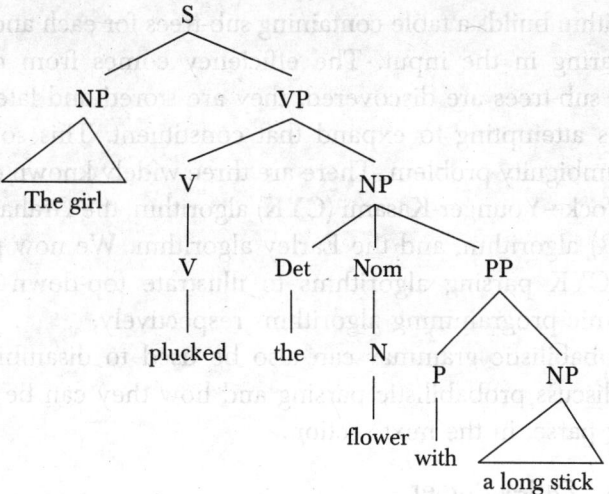

Figure 4.8 PP-attachment ambiguity

Coordination ambiguity occurs when it is not clear which phrases are being combined with a conjunction like *and*. For example, the phrase *beautiful hair and eyes* may have the structure [*beautiful hair*] and [*eyes*] or [*beautiful hair*] and [*beautiful eyes*]. Identifying the correct parse from a number of possible parses is known as disambiguation. A parser may utilize statistical and semantic knowledge to disambiguate the parse tree, or it may return all possible parses and leave the disambiguation for subsequent processing. The basic top-down parser we discussed, returns the first successful parse without exploring other possibilities. It needs to be modified to return all possible parses.

A sentence may have *local ambiguity* resulting in inefficient parsing. Local ambiguity occurs when certain parts of a sentence are ambiguous. For example, the sentence *Paint the door* is unambiguous, but during parsing it is not known whether the first word *Paint* is a verb or a noun. Hence, the parser makes a few incorrect expansions before discovering that Paint is a verb. Thus, it must use backtracking, or parallelism, to consider both parses.

Yet another problem associated with our basic top-down strategy is that of repeated parsing. The parser often builds valid trees for portions of the input that it discards during backtracking. These have to be rebuilt during subsequent steps in the parse. If we could avoid this extra effort, we would have more efficient parsers.

Dynamic programming algorithms can solve these problems. These algorithms construct a table containing solutions to sub-problems, which, if solved, will solve the whole problem. In parsing, a dynamic programming algorithm builds a table containing sub-trees for each and every constituent appearing in the input. The efficiency comes from the fact that once these sub-trees are discovered, they are stored and later consulted by all parses attempting to expand that constituent. This solves the reparsing and ambiguity problem. There are three widely known dynamic parsers—the Cocke-Younger-Kasami (CYK) algorithm, the Graham-Harrison-Ruzzo (GHR) algorithm, and the Earley algorithm. We now present the Earley and CYK parsing algorithms to illustrate top-down and a bottom-up dynamic programming algorithms respectively.

Probabilistic grammar can also be used to disambiguate parse trees. We discuss probabilistic parsing and how they can be used to identify a likely parse, in the next section.

4.4.4 Earley Parser

The Earley parser implements an efficient parallel top-down search using dynamic programming. It builds a table of sub-trees for each of the constituents in the input. This way, the algorithm eliminates the repetitive parse of a constituent which arises from backtracking, and successfully reduces the exponential-time problem to polynomial time. The Earley parser can handle recursive rules such as $A \rightarrow AC$ without getting into an infinite loop.

The most important component of this algorithm is the Earley chart that has $n+1$ entries, where n is the number of words in the input. The chart contains a set of states for each word position in the sentence. The algorithm makes a left to right scan of input to fill the elements in this chart. It builds a set of states, one for each position in the input string (starting from 0), that describe the condition of the recognition process at that point in the scan. The states in each entry provide the following information.

1. A sub-tree corresponding to a grammar rule.
2. Information about the progress made in completing the sub-tree.
3. Position of the sub-tree with respect to input.

A state is represented as a dotted rule and a pair of numbers representing starting position and the position of dot. This representation takes the form

$$A \rightarrow X_1 \ldots \bullet C \ldots X_m, \, [i, j]$$

where the dot ($\bullet$) represents the position in the rule's right hand side, and the two numbers (i and j) represent where the state begins and where the dot lies. A dot at the right end of the rule represents a successful parse of the associated non-terminal.

```
Earley Parsing
Input: Sentence and the Grammar
Output: Chart
    chart[0] ← S' → S, [0,0]
    n ← length (sentence)  // number of words in the sentence
    for i = 0 to n do
        for each state in chart[i] do
            if (incomplete (state) and next category is not a part of speech) then
            predictor (state)
            else if (incomplete (state) and next category is a part of speech)
                    scanner (state)
                else
                    completer (state)
                end-if
            end-if
        end for
    end for
    return
Procedure predictor (A → X₁ ... •B...Xₘ, [i, j] )
for each rule (B → α) in G do
insert  the state B → • α, [j, j] to chart [j]
End
Procedure scanner (A → X₁ ... •B...Xₘ, [i, j] )
If B is one of the part of speech associated with word[j] then
Insert the state B → word [j] •, [j, j + 1] to chart [j + 1]
End
Procedure Completer (A → X₁ ... •, [j, k] )
for each B → X1...•A ...,[i, j] in chart[j] do
insert the state B → X1...A• ...[i, k] to chart[k]
End
```

Figure 4.9 The Earley Parsing Algorithm

The algorithm uses three operations to process states in the chart. These are:

- Predictor
- Scanner
- Completer

The algorithm sequentially constructs the sets for each of the $n+1$ chart entries. Chart [0] is initialized with a dummy state S' → •S, [0,0]. At each step one of the three operations are applicable depending on the state.

Application of these operators result in addition of new states to either the current or the next set of states. The presence of a state S → α, [0,N] indicates a successful parse. The algorithm is shown in Figure 4.9.

We now explain the function of the three operators.

Predictor

As the name suggests, the predictor generates new states representing potential expansion of the non-terminal in the left-most derivation. A predictor is applied to every state that has a non-terminal to the right of the dot, when the category of that non-terminal is different from the part-of-speech. The application of this operator results in the creation of as many new states as there are grammar rules for the non-terminal. These new states are placed into the same chart entry as the generating state. Their start and end positions are at the point where the generating state ends. If

$$A \rightarrow X_1 \ldots \bullet B \ldots X_m, [i, j]$$

Then for every rule of the form B → α, the operation adds to chart [j], the state

$$B \rightarrow \bullet \, \alpha, [j, j]$$

For example, when the generating state is S → • NP VP, [0,0], the predictor adds the following states to chart [0]:

        NP → • Det Nominal, [0,0]
        NP → • Noun, [0,0]
        NP → • Pronoun, [0,0]
        NP → • Det Noun PP, [0,0]

Scanner

A scanner is used when a state has a part-of-speech category to the right of the dot. The scanner examines the input to see if the part-of-speech appearing to the right of the dot matches one of the part-of-speech associated with the current input. If yes, then it creates a new state using the rule that allows generation of the input word with this part-of-speech. It advances the pointer over the predicted input category and adds it to the next chart entry. If the state is A → …• a …, [i, j] and 'a' is one of the part-of-speech associated with w_j, then it adds a → … w_j •, [i, j] to chart [j+1].

Returning to our example, when the state NP → • Det Nominal, [0,0] is processed, the parser finds a part-of-speech category next to the dot. It checks if the category of the current word (curr_word) matches with the expectation in the current state. If yes, then it adds the new state Det → curr_word •, [0,1] to the next chart entry.

Completer

The completer is used when the dot reaches the right end of the rule. The presence of such a state signifies successful completion of the parse of some grammatical category. The completer identifies all previously generated states that expect this grammatical category at this position in the input and creates new states by advancing the dots over the expected category. All these newly generated states are inserted in the current chart entry. More formally, if A → ...•, $[j,k]$, then the completer adds B → ...A• ...$[i,k]$ to chart $[k]$ for all states B → ...•A ...,$[i, j]$ in chart $[j]$. An item is added to a set only if it is not already in the set.

Example 4.1 Let us trace the algorithm using sentence (4.7). The sequence of states created by the parser is shown in Figure 4.10.

Chart [0]	S0	S' → • S	[0,0]
Dummy		start	state
	S1	S → • NP VP	
	S2	S → • VP	
	S3	NP → • Det Nominal	
	S4	NP → • Noun	
	S5	NP → • Pronoun	
	S6	NP → Det Noun PP	
	S7	VP → • Verb NP	
	S8	VP → • Verb	
Chart [1]	S9	Noun → paint •	[0,1]
	S10	Verb → paint •	[0,1]
	S11	NP → Noun •	[0,1]
	S12	VP → Verb• NP	[0,1]
	S13	VP → Verb •	[0,1]
	S14	S → NP • VP	[0,1]
	S15	NP → • Det Nominal	[1,1]
	S16	NP → • Noun	[1,1]
	S17	S → VP •	[0,1]
	S18	VP → • Verb NP	[1,1]
	S19	VP → • Verb	[1,1]
Chart [2]	S20	Det → the •	[1,2]
	S21	NP → Det • Nominal	[1,2]
	S22	Nominal → • Noun	[2,2]
	S23	Nominal → • Noun Nominal	[2,2]
Chart [3]	S24	Noun → door •	[2,3]
	S25	Nominal → Noun •	[2,3]
	S26	NP → Det Nominal•	[1,3]
	S27	S → NP • VP	[0,3]
	S28	VP → Verb NP •	[0,3]
	S29	VP → • Verb NP	[3,3]
	S30	VP → • Verb	[3,3]
	S31	S → VP •	[0,3]

Figure 4.10 Sequence of states created in parsing sentence (4.7) using Earley algorithm

The presence of the state S → VP •, [0,3] in the chart indicates successful completion of the parse of the sentence, but it does not return the exact parse. Thus, the algorithm in this form can be used only to recognize a sentence. However, it is possible to utilize the entries appearing in the table to construct the parse tree of a valid sentence. In order to do so the algorithm needs to be modified. This modification is carried out by the completer which creates new states by advancing earlier incomplete states whenever it finds a state having a dot at the end of the rule. We add a pointer to the previous states of the new state. A list of previous states is thus maintained. Extracting a parse tree from the chart involves following these pointers, beginning with the state marking the successful completion of the parse. Though the Earley algorithm fills the chart entry in polynomial time, extracting all parse trees still requires an exponential amount of time.

The Earley parser can be augmented with unification structures to eliminate ill-formed structures as they are introduced. This requires certain modification in the algorithm. The first change involves the addition of a feature structure derived from their unification constraints. The second change is the addition of a new field to the chart. This new field is a directed acyclic graph representing the feature structure corresponding to the state. The details of the modified Earley algorithm can be found in Jurafsky and Martin (2000).

4.4.5 The CYK Parser

Like the Earley algorithm, the CYK (Cocke-Younger-Kasami) is a dynamic programming parsing algorithm. However, it follows a bottom-up approach in parsing. It builds a parse tree incrementally. Each entry in the table is based on previous entries. The process is iterated until the entire sentence has been parsed. The CYK parsing algorithm assumes the grammar to be in Chomsky normal form (CNF). A CFG is in CNF if all the rules are of only two forms:

A → BC

A → w, where w is a word.

The algorithm first builds parse trees of length one by considering all rules which could produce words in the sentence being parsed. Then, it constructs the most probable parse for all the constituents of length two. The parse of shorter constituents constructed in earlier iterations can be used in constructing the parse of longer constituents.

Like the Earley algorithm, the basic CYK algorithm is also a chart-based algorithm. A non-terminal is stored in the $[i, j]$th entry of the chart if, and only if, $A \Rightarrow w_i \cdot w_{i+1} \cdots w_{i+j-1}$. The chart is triangular. A sentence is recognized if the start symbol S is in the entry $[1, n]$ of the chart. Beginning with the start symbol of the grammar, we are able to derive the entire sequence of words appearing in the sentence. The algorithm builds smaller constituents before attempting to construct larger ones. First, the terminal derivation rules of the grammar are used to generate the $[i,1]$th entries. These entries represent non-terminals that derive the individual words appearing in the sentence, $w_{i1} = w_i$, for $1 \leq i \leq n$, where n is the length (number of words) of the sentence. $A \Rightarrow w_{i1}$ if $A \rightarrow w_i$ is a rule in the grammar. It then continues with sub-string of length two, three, and so on. For every non-terminal A in the grammar, the algorithm determines if $A^* \Rightarrow w_{ij}$. As the grammar is in CNF, A could derive if there existed a rule of the form $A \rightarrow B \ C$ such that B derives the first k words of the w_{ij} (i.e. $B \rightarrow w_{ik}$) and C derives the remaining $j-k$ words ($C \rightarrow w_{kj}$) as shown in Figure 4.11. More formally,

$$A^* \Rightarrow w_{ij} \text{ if}$$

1. $A \rightarrow B \ C$ is a rule in grammar
2. $B^* \Rightarrow w_{ik}$ and
3. $C^* \Rightarrow w_{kj}$

For a sub-string w_{ij} of length j starting at i, the algorithm considers all possible ways of breaking it into two parts w_{ik} and w_{kj}. Finally, since $s = w_{1n}$, we have to verify that $S^* \Rightarrow w_{1n}$, i.e., the start symbol of the grammar derives w_{1n}.

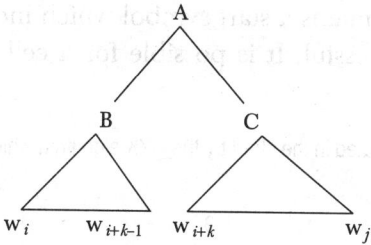

Figure 4.11 Breaking a string

The steps involved in the algorithm are shown in Figure 4.12. In order to use the information contained in the table to construct parse trees, we need to maintain back-pointers to the table entries we combine. This allows us to construct all possible parse trees by following back pointers.

```
Let w = w₁ w₂ w₃ wᵢ …wⱼ…wₙ
    and wᵢⱼ = wᵢ …wᵢ₊ⱼ₋₁
// Initialization step
for i := 1 to n do
    for all rules A→ wᵢ do
        chart [i,1] = {A}
// Recursive step
for j = 2 to n do
    for i = 1 to n−j+1 do
    begin
        chart [i, j] = φ
        for k = 1 to j −1 do
        chart [i, j] := chart[i, j] ∪{A | A → BC is a production and
                         B ∈ chart[i, k] and  C ∈ chart [i+k, j−k]}
    end
if S ∈ chart[1, n] then accept else reject
```

Figure 4.12   The CYK algorithm

To give a better understanding of the whole idea, we work out an example. Consider the following simplified grammar in CNF:

S → NP VP	Verb → wrote
VP → Verb NP	Noun → girl
NP → Det Noun	Noun → essay
Det → an | the	

The sentence to be parsed is: *The girl wrote an essay.*

Table 4.4 contains entries after a complete scan of the algorithm. The entry in the [1,n]th cell contains a start symbol which indicates that $S^* \Rightarrow w_{1n}$, i.e., the parse is successful. It is possible for a cell to have multiple entries.

Table 4.4   Sequence of states created in the chart by the CYK algorithm while parsing the sentence, *Sana wrote an essay*

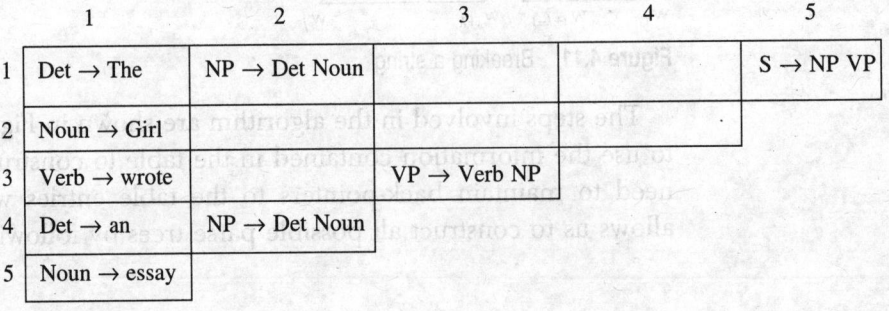

	1	2	3	4	5
1	Det → The	NP → Det Noun			S → NP VP
2	Noun → Girl				
3	Verb → wrote		VP → Verb NP		
4	Det → an	NP → Det Noun			
5	Noun → essay				

4.5 PROBABILISTIC PARSING

Statistical parser, like statistical tagging (Chapter 3), requires a corpus of hand-parsed text. There are such corpora available, the most notably being the Penn tree-bank (Marcus et al. 1993). The Penn tree-bank is a large corpus of articles from the *Wall Street Journal* that have been tagged with Penn tree-bank tags and then parsed according to a simple set of phrase structure rules conforming to Chomsky's government and binding syntax. The parsed sentences are represented in the form of properly bracketed trees. A statistical parser works by assigning probabilities to possible parses of a sentence and returning the most likely parse as the final one. More formally, given a grammar G, sentence s, and a set of possible parse trees of s which we denote by $\tau(s)$, a probabilistic parser finds the most likely parse φ' of s as follows:

$$\varphi' = \text{argmax}_{\varphi \in \tau(s)} \, P(\varphi \mid s)$$
$$= \text{argmax}_{\varphi \in \tau(s)} \, P(\varphi, s)$$
$$= \text{argmax}_{\varphi \in \tau(s)} \, P(\varphi)$$

In order to construct a statistical parser, we have to first find all possible parses of a sentence, then assign probabilities to them, and finally return the most probable parse. We discuss an implementation of statistical parsing based upon probabilistic context-free grammars (PCFGs). But before proceeding ahead to this discussion, let us have a look at why we need probabilistic parsing at all. What advantages do these parsers offer?

The first benefit that a probabilistic parser offers is removal of ambiguity from parsing. We have seen earlier, that sentences can have multiple parse trees. The parsing algorithm discussed so far in this chapter has no means to decide which parse is the correct or most appropriate one. Probabilistic parsers assign probabilities to parses. These probabilities are then used to decide the most likely parse tree structure of an input sentence.

Another benefit this parser offers is related to efficiency. The search space of possible tree structures is usually very large. With no information on which sub-trees are more likely to be a part of the final parse tree, the search can be quite time consuming. Using probabilities to guide the process, the search becomes more efficient.

A probabilistic context-free grammar (PCFG) is a CFG in which every rule is assigned a probability (Charniak 1993). It extends the CFG by augmenting each rule $A \rightarrow \alpha$ in set of productions P, with a conditional probability p:

$$A \rightarrow \alpha \, [p]$$

where p gives the probability of expanding a constituent using the rule. $A \rightarrow \alpha$.

Let us now define PCFG. A PCFG is defined by the pair (G, f), where G is a CFG and f is a positive function defined over the set of rules such that, the sum of the probabilities associated with the rules expanding a particular non-terminal is 1 (Infante-Lopez and Maarten de Rijke 2006).

$$\sum_{\alpha} f(A \rightarrow \alpha) = 1$$

An example of PCFG is shown in Figure 4.13. We can verify that for each non-terminal, the sum of probabilities is 1.

$f(S \rightarrow NP\ VP) + f(S \rightarrow VP) = 1$

$f(NP \rightarrow Det\ Noun) + f(NP \rightarrow Noun) + f(NP \rightarrow Pronoun) + f(NP \rightarrow Det\ Noun\ PP) = 1$

$f(VP \rightarrow Verb\ NP) + f(NP \rightarrow Verb) + f(VP \rightarrow VP\ PP) = 1.0$

$f(Det \rightarrow this) + f(Det \rightarrow that) + f(Det \rightarrow a) + f(Det \rightarrow the) = 1.0$

$f(Noun \rightarrow paint) + f(Noun \rightarrow door) + f(Noun \rightarrow bird) + f(Noun \rightarrow hole) = 1.0$

S → NP VP	0.8	Noun → door	0.25
S → VP	0.2	Noun → bird	0.25
NP → Det Noun	0.4	Noun → hole	0.25
NP → Noun	0.2	Verb → sleeps	0.2
NP → Pronoun	0.2	Verb → sings	0.2
NP → Det Noun PP	0.2	Verb → open	0.2
VP → Verb NP	0.5	Verb → saw	0.2
VP → Verb	0.3	Verb → paint	0.2
VP → VP PP	0.2	Preposition → from	0.3
PP → Preposition NP	1.0	Preposition → with	0.25
Det → this	0.2	Preposition → on	0.2
Det → that	0.2	Preposition → to	0.25
Det → a	0.25	Pronoun → she	0.35
Det → the	0.35	Pronoun → he	0.35
Noun → paint	0.25	Pronoun → they	0.25

Figure 4.13   A probabilistic context-free grammar (PCFG)

Similarly, the condition is satisfied for the other non-terminals.

4.5.1   Estimating Rule Probabilities

The next question is how are probabilities assigned to rules? One way to estimate probabilities for a PCFG is to manually construct a corpus of a parse tree for a set of sentences, and then estimate the probabilities of each rule being used by counting them over the corpus. The MLE estimate for a rule $A \rightarrow \alpha$ is given by the expression

$$P_{MLE}(A \to \alpha) = \frac{\text{Count}(A \to \alpha)}{\sum_{\alpha} \text{Count}(A \to \alpha)}$$

If our training corpus consists of two parse trees (as shown in Figure 4.14), we will get the estimates as shown in Table 4.5 for the rules.

Table 4.5 The MLE for the grammar rules used in trees of Figure 4.14

Rule	Count $(A \to \alpha)$	Count A	MLE estimates
S → VP	2	2	1
NP → Det Noun PP	1	4	0.25
NP → Det Noun	3	4	0.75
VP → Verb NP	2	3	0.66
VP → VP PP	1	3	0.33
Det → the	2	2	1
Noun → hole	2	4	0.5
Noun → door	2	4	0.5
Prep → with	1	1	1
Verb → Paint	1	1	1

We now turn to another important question—what do we do with these probabilities? We assign a probability to each parse tree φ of a sentence s. The probability of a complete parse is calculated by multiplying the probabilities for each of the rules used in generating the parse tree:

$$P(\varphi, s) = \prod_{n \in \phi} p\{r(n)\}$$

where n is a node in the parse tree φ and r is the rule used to expand n.

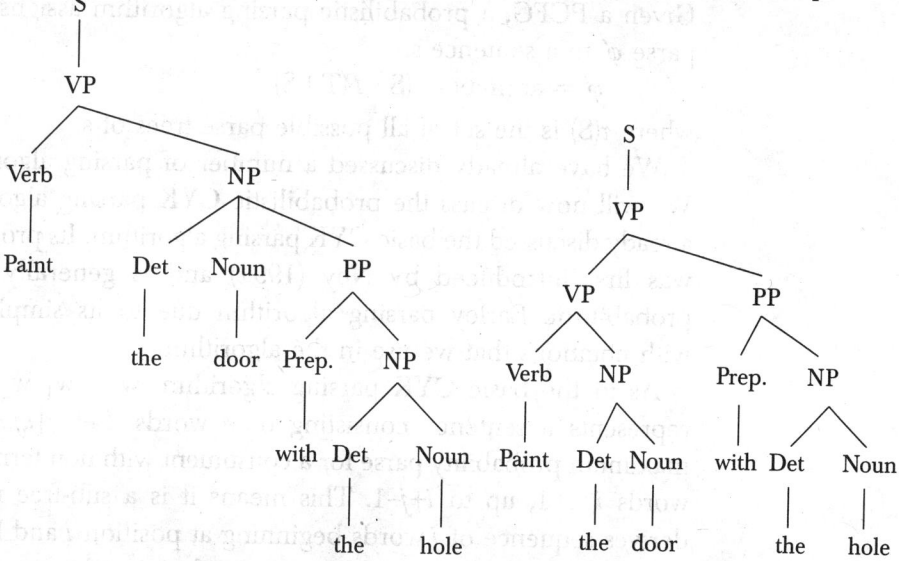

Figure 4.14 Two possible parse trees

For a sentence, the parse trees generated by a PCFG are the same as those generated by a corresponding CFG. However, the PCFG assigns a probability to each parse. The probability of the two parse trees of the sentence *Paint the door with the hole* (shown in Figure 4.14) can be computed as follows:

$$P(t1) = 0.2 * 0.5 * 0.2 * 0.2 * 0.35 * 0.25 * 1.0 * 0.25 * 0.4 * 0.35 * 0.25$$
$$= 0.0000030625$$
$$P(t2) = 0.2 * 0.2 * 0.5 * 0.2 * 0.4 * 0.35 * 0.25 * 1 * 0.25 * 0.4 * 0.35 * 0.25$$
$$= 0.000001225$$

In this example, the first tree has a higher probability leading to correct interpretation.

We can calculate probability to a sentence *s* by summing up probabilities of all possible parses associated with it.

$$P(s) = \sum_{t \in \tau(s)} P(t, s)$$

$$= \sum_{t \in \tau(s)} P(t)$$

Thus, the sentence will have the probability

$$P(t1) + P(t2) = 0.0000030625 + 0.000001225$$
$$= 0.0000042875$$

4.5.2   Parsing PCFGs

Given a PCFG, a probabilistic parsing algorithm assigns the most likely parse φ' to a sentence s.

$$\varphi' = \operatorname{argmax}_{T \in \tau}(S) \ P(T \mid S)$$

where $\tau(S)$ is the set of all possible parse trees of s.

We have already discussed a number of parsing algorithms for CFG. We will now discuss the probabilistic CYK parsing algorithm. We have already discussed the basic CYK parsing algorithm. Its probabilistic version was first introduced by Ney (1991) and is generally preferred over probabilistic Earley parsing algorithm due to its simplicity. We begin with notations that we use in the algorithm.

As in the basic CYK parsing algorithm, $w = w_1 \ w_2 \ w_3 \ w_i \ ... w_j ... w_n$ represents a sentence consisting of *n* words. Let $\varphi[i,j,A]$ represent the maximum probability parse for a constituent with non-terminal A spanning words i, $i+1$, up to $i+j-1$. This means it is a sub-tree rooted at A that derives sequence of *j* words beginning at position *i* and has a probability greater than all other possible sub-trees deriving the same word sequence.

An array named BP is used to store back pointers. These pointers allow us to recover the best parse. The output for a successful parse is the maximum probability parse $\varphi[1,n,S]$, which corresponds to a parse tree rooted at S and spanning the sequence of words $w_1\ w_2\ w_3...w_n$, i.e. the whole sentence.

Initialization:

    for $i := 1$ to n do

        for all rules $A \rightarrow w_i$ do

            $\varphi[i,1,A] = P(A \rightarrow w_i)$

Recursive Step:

    for $j = 2$ to n do

        for $i = 1$ to $n-j+1$ do

        begin

            $\varphi[i,1,A] = \phi$

            for $k = 1$ to $j-1$ do

                $\varphi'[i,j,A] = \max_k \varphi'[i,k,B] \times \varphi'[k,j,C] \times P(A \rightarrow BC),$

                such that $A \rightarrow BC$ is a production rule in grammar

                $BP[i,j,A] = \{\ k,\ A,\ B\}$

        end

Figure 4.15 Probabilistic CYK algorithm

The algorithm is given in Figure 4.15. Like the basic CYK algorithm, the first step is to generate items of length 1. However, in this case, we have to initialize the maximum probable parse trees deriving a string of length 1, with the probabilities of the terminal derivation rules used to derive them.

$$\varphi[i,1,A] = P(A \rightarrow w_i)$$

Similarly, recursive step involves breaking a string into all possible ways and identifying the maximum probable parse.

$$\varphi'[i,j,A] = \max_k \varphi'[i,k,B] \times \varphi'[k,j,C] \times P(A \rightarrow BC)$$

The rest of the steps follow those of basic CYK parsing algorithm.

4.5.3 Problems with PCFG

The PCFG is not without disadvantages. Its first problem lies in the independence assumption. We calculate the probability of a parse tree assuming that the rules are independent of each other. However, this is not true. How a node expands depends on its location in the parse tree. For example, Francis et al. (1999) showed that pronouns occur more frequently as subjects rather than objects. These dependencies are not captured by a PCFG, as the probability of, say, expanding an NP as a pronoun versus a lexical NP, is independent of whether the NP appears as a subject or an object.

Another problem associated with a PCFG is its lack of sensitivity to lexical information. Lexical information plays a major role in determining correct parse in case of PP attachment ambiguities and coordination ambiguities (Collins 1999). Two structurally different parses that use the same rules will have the same probability under a PCFG, making it difficult to identify the correct or most probable parse. The words appearing in a parse may make certain parses unnatural. This however, requires a model which captures lexical dependency statistics for different words. Such a model is presented next.

Lexicalization

In PCFG, the chance of a non-terminal expanding using a particular rule is independent of the actual words involved. However, this independence assumption does not seem reasonable. Words do affect the choice of the rule. Investigations made on tree bank data suggest that the probabilities of various common sub-categorization frames differ depending on the verb that heads the verb phrase (Manning and Schutze 1999). This suggests the need for lexicalization, i.e., involvement of actual words in the sentences, to decide the structure of the parse tree. Lexicalization is also helpful in choosing phrasal attachment positions. This model of lexicalization is based on the idea that these are strong lexical dependencies between heads and their dependents, for example, between a head noun and its modifiers, or between a verb and a noun phrase object, where the noun phrase object in turn can be approximated by its head noun.

One way to achieve lexicalization is to mark each phrasal node in a parse tree by its head word. Figure 4.16 is an example of such a tree. One way to implement this model is to use a lexicalized grammar. A lexicalized grammar is a grammar in which every finite structure is associated with one or more lexical heads. A context free grammar is not lexicalized as no lexical item is associated with its rule, such as, S → NP VP. Nor is the PCFG lexicalized. In order to convert a PCFG into lexicalized grammar, each of its rules must identify a head daughter, which is one of the constituents appearing on its right hand side, for example head daughter of S → NP VP rule is VP. The head word of a node in the parse tree is set to the head word of its head daughter. For example, the head of a verb phrase is the main verb. Hence, the head of the node VP in the Figure 4.16 is jumped. Similarly, the head of a constituent expanded using the rule S → NP VP is the head of VP.

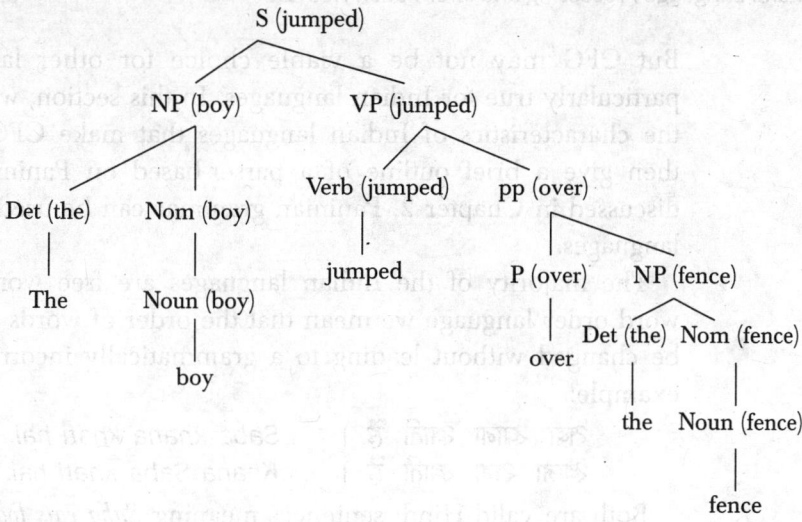

Figure 4.16 A lexicalized tree

The probability in a lexicalized PCFG is calculated for each rule or head of a phrase and head of the sub-phrase or head of the phrase (i.e., head/mother head) combination. For example, we need to collect the following rule/head probability:

VP (jumped) → verb (jumped) PP (over)
VP (jumped) → verb (jumped) PP (into)
VP (jumped) → verb (jumped) PP (to)

An example of the head or mother head combination is the following type of probability:

"What is the probability that an NP whose mother's head is *over* has the head *fence*?"

With a lexicalized PCFG, the probability of the parse tree is computed as the product of the probability of the head of the constituent c given the probability of the mother m(c) and the probability of the rule used to expand constituent c given the head of constituent (Charniak 1997):

$$P(\varphi, s) = \prod_{c \in \phi} p[(h(c)|m(c)] \cdot p[(r(c)|h(c)]$$

where h(c) = head of the constituent c
       r(c) = rule used to expand constituent c
and   m(c) = mother of the constituent c

4.6 INDIAN LANGUAGES

Not all natural languages have the same characteristics. So far we have talked only of the English language, using CFG as the grammar formalism.

But CFG may not be a viable choice for other languages. This is particularly true for Indian languages. In this section, we discuss some of the characteristics of Indian languages that make CFG unsuitable. We then give a brief outline of a parser-based on Paninian grammar. As discussed in Chapter 2, Paninian grammar can be used to model Indian languages.

The majority of the Indian languages are free word order. By free word order language we mean that the order of words in a sentence can be changed without leading to a grammatically incorrect sentence. For example:

सबा खाना खाती है ।   *Saba khana khati hai.*
खाना सबा खाती है ।   *Khana Saba khati hai.*

Both are valid Hindi sentences meaning *Saba eats food.*

The same is true for Urdu and most other Indian languages. The CFG we used for parsing English is basically positional. It can be used to model language in which a position of the constituents carries useful information, but it fails to model free word order languages.

Extensive and productive use of complex predicates (CPs) is another property that most Indian languages have in common. A complex predicate combines a light verb with a verb, noun, or adjective, to produce a new verb. For example,

(a) सबा आयी ।
(*Saba Ayi.*)
Saba came.

(b) सबा आ गयी ।
(*Saba a gayi.*)
Saba come went.
Saba arrived.

(c) सबा आ पडी ।
Saba a pari.
Saba come fell.
Saba came (suddenly).

The complex predicates change the functional structure of the sentence; however they do not change the sub-categorization frame of the verb.

The use of post-position case markers (Vibhakti) and verb complexes consisting of sequences of verbs, e.g., खा रही है (*kha rahi hai*) are other properties common to Indian languages. The auxiliary verbs in this sequence provide information about tense, aspect, and modality. Paninian grammar provides a framework to model Indian languages. It focuses on the extraction of Karak relations from a sentence.

Bharti and Sangal (1990) described an approach for parsing of Indian languages based on Paninian grammar formalism. Their parser works in two stages. The first stage is responsible for identifying word groups and the second for assigning a parse structure to the input sentence. The input to the first stage comes from the morphological analysis phases (as discussed in Chapter 3). The output of the first stage is word groups, which are sequences of words that act as a unit. For example, a verb group consists of a main verb and a sequence of auxiliaries. For the sentence

लड़कियाँ मैदान में हाकी खेल रही हैं । (4.8)
Ladkiyan maidaan mein hockey khel rahi hein.

The word *ladkiyan* forms one unit, the words *maidaan* and *mein* are grouped together to form a noun group, and the word sequence *khel rahi hein* forms a verb group.

The choice of noun and verb groups over noun and verb phrases adds computational simplicity to the approach. The concept of verb phrase is not natural to Indian languages and computing a noun phrase is difficult.

In the second stage, the parser takes the word groups formed during first stage and identifies (i) Karaka relations among them, and (ii) senses of words. A data structure, called Karaka chart, stores additional information like Karaka–Vibhakti mapping, Karaka necessity (mandatory or optional), and transformation rules for Karaka relations, needed in this step. Transformation rules tell us how to create a Karaka chart for a verb group using the default Karaka chart. The form of the default Karaka chart is shown in Table 4.6.

Table 4.6 Default Karaka chart

Karaka (case relations)	Vibhakti (case markers or post-positions)	Necessity
Karta	ϕ	Mandatory
Karma	*Ko* or ϕ	Mandatory
Adhikaran	*Mein* or *par*	Optional
Sampradan	*Ko* or *ke liye*	Optional

Once the Karaka chart for the verb groups are ready, noun groups are tested against them. A noun group satisfying the Vibhakti restriction for a verb group becomes a candidate for its Karaka. The Karaka relation between a verb group and a noun group can be depicted using a constraint graph. Nodes in the graph represent word groups and an arc represents a Karaka restriction between word groups. We have earlier identified the word groups in sentence (4.8). Its constraint graph is given in Figure 4.17.

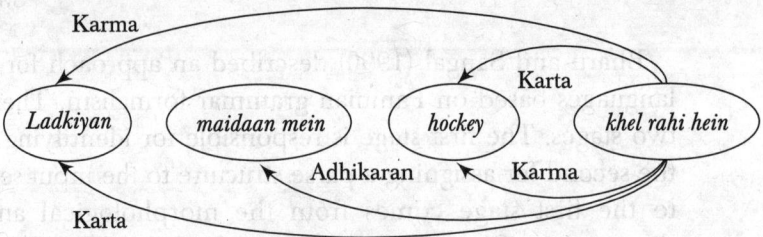

Figure 4.17   Constraint graph for sentence (4.8)

Each sub-graph of the constraint graph that satisfies the following constraints yields a parse of the sentence.

1. It contains all the nodes of the graph.
2. It contains exactly one outgoing edge from a verb group for each of its mandatory Karakas. These edges are labelled by the corresponding Karaka.
3. For each of the optional Karaka in Karaka chart, the sub-graph can have at most one outgoing edge labelled by the Karaka from the verb group.
4. For each noun group, the sub-graph should have exactly one incoming edge.

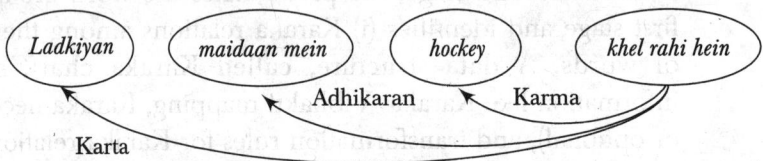

Figure 4.18   A parse of the sentence (4.8)

A sub-graph of the constraint graph (Figure 4.17) satisfying these constraints is shown in Figure 4.18. This represents one of the possible parse of sentence (4.8). More than one sub-graph may satisfy these constraints in which case the sentence is ambiguous and has multiple parse associated with it. If no sub-graph satisfies these constraints, then the grammar fails to assign any parse structure to the sentence. Disambiguation among various senses of verbs and nouns is carried out with the help of the lakshan chart, which contains features that are useful for discriminating among various senses. Readers are referred to Bharti and Sangal (1990) and Bharti et al. (1995) for detailed treatment of the algorithm.

SUMMARY

This chapter has introduced a number of important notions about syntax.

- Syntactic parsing deals with the syntactic structure of a sentence.
- In many languages, words are grouped to form larger groups termed constituent or phrases, which can be modelled by context-free grammar (CFG).
- A CFG consists of a set of rules or productions stating which elements can occur in a phrase and in which order.
- The set of all the strings containing terminal symbols that can be derived from the start symbol of the grammar, defines the language generated by that grammar.
- Phrase types are named after their head, which is the important lexical category that determines the properties of a phrase. If the head is a noun the phrase is called noun phrase, if head is a verb, the phrase is called a verb phrase, and so forth.
- Grammatical categories impose certain constraints: subject-verb agreement is one of them. The subjects in a sentence must agree with the main verb in number and person. Grammar rules should be able to ensure this agreement.
- CFG can't handle subject-verb agreement efficiently. Feature structures can be used to efficiently capture the properties of grammatical categories.
- Top-down and bottom-up are two commonly used parsing approaches.
- The Earley parser and the CYK parser are dynamic programming algorithms for parsing. The Earley algorithm takes a top-down approach whereas the CYK takes a bottom-up approach.
- Probabilistic parsing assigns the most likely parse to a sentence and hence, can be used to handle parsing ambiguities. However, it is insensitive to lexical information.
- Indian languages are free word order languages and cannot be modelled by CFG.

REFERENCES

Backus, J.W., 1959, 'The syntax and semantics of the proposed international algebraic language of the Zurich ACM-GAMM conference,' *Proceedings of the International Conference on Information Processing*, UNESCO, pp. 125–32.

Bharti, Akshar and Rajeev Sangal, 1990, 'A Karaka-based approach to parsing of Indian languages,' *Proceedings of the 13th Conference on Computational Linguistics,* Association for Computational Linguistics, 3.

Bharti, Akshar, Vineet Chaitanya, and Rajeev Sangal, 1995, *Natural Language Processing: A Paninian Perspective,* Prentice-Hall of India.

Charniak, Eugene, 1993, *Statistical Language Learning,* MIT Press, Cambridge.

_____1997, 'Statistical techniques for natural language parsing,' *AI Magazine.*

Chomsky, N., 1957, *Syntactic Structures,* Mouton, The Hague.

Collins, M.J., 'Head-driven statistical parsing for natural language processing,' *Ph.D. Thesis,* University of Pennsylvania, Philadelphia.

Infante-Lopez, Gabriel and Maarten de Rijke, 2006, 'A note on the expressive power of probabilistic context free grammars,' *Journal of Logic, Language and Information,* Kluwer Academic Publisher, 15(3).

Jurafsky, Daniel and James H. Martin, 2000, *Speech and Language Processing: An Introduction to Natural Language Processing, Computational Linguistics and Speech Recognition,* Prentice Hall, NJ.

Manning, C. and H. Shutze, 1999, *Foundations of Statistical Natural Language Processing,* MIT Press, Cambridge.

Marcus, Mitchell P., Beatrice Santorini, and Mary Ann Marcinkiewicz, 1993, 'Building a large annotated corpus of English: the Penn treebank,' *Computational Linguistics,* 19, pp. 313–30.

Naur, Peter, J.W. Backus , F.L. Bauer, J. Green , C. Katz, J. McCarthy, A. J. Perlis, H. Rutishauser, K. Samelson, B. Vauquois, J. H. Wegstein, A. van Wijngaarden, and M. Woodger, 1960, 'Report on the algorithmic language ALGOL 60,' *Communications of the ACM,* 3(5), pp. 299–314.

Ney, H., 1991, 'Dynamic programming parsing for context-free grammars in continuous speech recognition,' *IEEE Transactions on Signal Processing,* 39(2), pp. 336–40.

EXERCISES

1. Give two possible parse trees for the sentence, *Stolen painting found by tree.*
2. Identify the noun and verb phrases in the sentence, *My soul answers in music.*
3. Give the correct parse of sentence (4.6).

4. Discuss the disadvantages of the basic top-down parser with the help of an appropriate example.
5. Tabulate the sequence of states created by CYK algorithm while parsing, *The sun rises in the east.* Augment the grammar in section 4.4.5 with appropriate rules of lexicon.
6. Discuss the disadvantages of probabilistic context free grammar.
7. What does lexicalized grammar mean? How can lexicalization be achieved? Explain with the help of suitable examples.
8. List the characteristics of a garden path sentence. Give an example of a garden path sentence and show its correct parse.
9. What is the need of lexicalization?
10. Use the following grammar:

S → NP VP	S → VP	NP→ Det Noun
NP → Noun	NP → NP PP	VP → VP NP
VP → Verb	VP → VP PP	PP → Preposition NP

Give two possible parse of the sentence: '*Pluck the flower with the stick.*' Introduce lexicon rules for words appearing in the sentence. Using these parse trees obtain maximum likelihood estimates for the grammar rules used in the tree. Calculate probability of any one parse tree using these estimates.

LAB EXERCISES

1. Write a program to extract all the noun phrases from a text file. Use the phrase structure rule given in this chapter.
2. Write a program to check whether a given grammar is context free grammar or not.
3. Write a program to convert a given CFG grammar in CNF.
4. Write a program to implement a basic top-down parser.
5. Implement Earley parsing algorithm.

CHAPTER 5

SEMANTIC ANALYSIS

CHAPTER OVERVIEW

This chapter deals with the meaning of written text. The general idea of semantic interpretation is to take natural language sentences or utterances and map them onto some representation of meaning. The chapter begins with an introduction to semantic analysis. Next, general characteristics of meaning representations and meaning structures are described. A general approach to semantic analysis based on semantic compositionality is then discussed. Lexical semantics is the next topic of discussion in this chapter. Finally, word sense disambiguation is discussed. This includes a discussion of selectional restriction-based and context-based disambiguation approaches, and word sense disambiguation application and its evaluation.

5.1 INTRODUCTION

The goal of computational linguistics has long been to produce a semantic analysis. Semantics is associated with the meaning of language. The general idea of semantic interpretation is to take natural language sentences or utterances and map them onto some representation of meaning. Semantic analysis is concerned with creating representations for the meaning of linguistic inputs. This chapter deals with the meaning of written text. We can divide semantics into two parts as follows:

- The study of the meaning of individual words (*lexical semantics*)
- The study of how individual words combine to give meaning to a sentence (or larger units)

Once we have the meaning of the words, we need to combine them into the meaning of the whole sentence. The principle of semantic compositionality (sometimes called Freg's principle) states that the meaning of the whole is comprised of the meanings of its parts (Keenan and Faltz 1985), i.e., the meaning of a sentence can be composed from the meaning of its constituent words. Natural languages do not always obey this principle of compositionality. Often, the meaning of the whole is only partially dependent on the meaning of its constituents (as in collocates) and sometimes entirely different from the meanings of individuals (e.g., in idioms).

Hence, this decomposition of semantics may not be realistic. Word meaning is just one component of semantics. The relationship that exists between words, the domain, the word order, the syntactic structure, the underlying context, and the real world knowledge, all contribute to the meaning of a sentence. Still, compositional and lexical (word) semantics remains the dominant approach to semantics in computational linguistics. There are theories supporting the idea that linguistic knowledge is knowledge about words (Joshi 1985). This means that lexicon contains all the knowledge about language. These theories completely dispense with grammar as an independent entity, which has been accepted for so many years. The idea that syntax and lexicon can not be described adequately without reference to each other is gaining increasing support from the results of corpus analysis, which attempts to investigate role of context in syntax and semantics.

Lexical semantics has been the starting point for all the early theories of semantics. The most common paradigm involves decomposing lexical meanings in terms of semantic primitives or atomic units of meaning. However, this theory turns out to be inadequate in handling compositional semantics. This has led to the development of model theoretic semantics, which, along with the structural semantics, is the dominant approach to semantics within linguistics. Model theoretic semantics is inspired by the semantics that logicians use for formal logical languages. Logicians focus on 'logical words' (and, or, not, if, all, some, only, etc.) and do not pay attention to non-logical words (most nouns, verbs and adjectives). The primary advantage of this theory is its ability to explain compositional semantics—how the meaning of a sentence is determined from the meanings of its parts. One important aspect of meaning is that it relates sentences to the outside world. However, we do not know what the world is. A model theoretic approach to semantics attempts to create a model of the world and determines the truth of a sentence using this model. In this theory, the truth of a sentence does not mean that the sentence is actually true; it simply means the sentence is true in the world being modelled. Model theoretic semantics is effective in studying pragmatics as well as semantics.

We organize the discussion in this chapter around two main topics: the meaning representation of the whole sentence and the meaning of words (i.e., lexical semantics). Section 5.2 deals with sentence level meaning representation. In particular, we discuss the general characteristics of meaning representation languages (Section 5.2.1), the meaning structure of the language (Section 5.2.2), and computational approaches to semantic analysis (syntax-driven semantic analysis and semantic grammars in Section 5.2.3). Next, we discuss the internal structure of words, their relationships, and their meanings in Section 5.3. The meaning of the words themselves cannot be derived by separating them from their context; especially the correct sense of a word cannot be identified unless we examine the words in context. This task of word sense disambiguation is of considerable theoretical and practical interest. We give a special mention to the various types of ambiguities and approaches to word sense disambiguation in this chapter (Sections 5.4 and 5.5).

5.2   MEANING REPRESENTATION

Many language tasks require non-linguistic knowledge of the world besides phonological, morphological, and syntactic knowledge. The semantic representation bridges the gap from linguistic inputs to the non-linguistic knowledge involving the meaning of linguistic inputs. The language used to describe the syntax and semantics of these representations are called meaning representation languages. The process of creating such representation and assigning it to linguistic inputs is called semantic analysis. Commonly used meaning representation languages are first order predicate calculus (FOPC) and semantic networks and conceptual dependency. We hope you are already familiar with these knowledge representation formalisms; a review of them is provided in the Appendix. Conceptual graphs were introduced to capture semantics of natural language, and have also been used for knowledge representation formalism. We discuss this formalism and its use in information retrieval in Chapter 10. We now focus on the general characteristics of meaning representation languages, the meaning structure of languages, and how meaning representations are created and assigned to sentences.

5.2.1   Characteristics of Meaning Representation Languages

There are certain characteristics that a meaning representation language must possess.

Verifiability

Meaning representations must be verifiable, i.e., we must be able to determine the truth of our representation. We determine the truth by comparing the meaning representation of an input with the repository of facts existing in the domain (i.e., representations in knowledge base). It is important, therefore, that our meaning representation language should facilitate such a comparison. Consider the following question:

Does Kingfisher serve Hyderabad?   (5.1)

The underlying proposition of this question is that Kingfisher serves Hyderabad. We can represent this as

Serves (Kingfisher, Hyderabad)

This representation will be matched against a knowledge base of facts. If the system finds a matching proposition, it returns an affirmative reply. It answers no if its knowledge of domain is complete and do not know if it has reason to believe that its knowledge base lacks information. This example illustrates the meaning of verifiability. To summarize, verifiability is a system's ability to compare the input representation to the situation that exists in the world as modelled in the knowledge base.

Unambiguous

The second requirement of a meaning representation language is that it should be unambiguous, i.e., it should support representations that have only one possible interpretation. Like other domains, the domain of semantics is also full of ambiguities–these are discussed in Sections 5.4 and 5.5. As our reasoning and subsequent action is based upon the semantic content of inputs, it is important that the final meaning representation of an input be unambiguous, regardless of ambiguity in the raw input.

A concept related to ambiguity is vagueness. Vagueness refers to lack of precision. Unlike ambiguity, it does not lead to multiple representations but makes it difficult to determine what to do with meaning representations. Consider the following sentence:

I want to go to a hill station.   (5.2)

This input provides enough information for a tour organizer to provide reasonable response, but it is quite vague as to where the user really wants to go. For some purposes, this vague representation of inputs may be sufficient, while a more specific representation may be needed for other purposes. A meaning representation language should support a certain level of vagueness.

Canonical Form

Consider the following alternative ways of expressing the content of sentence (5.1):

Does Kingfisher offer a flight to Hyderabad?   (5.3a)
Does Kingfisher have a flight to Hyderabad?   (5.3b)

These representations use different words and have different syntactic structures from sentence (5.1). This gives rise to different meaning representations. But we know that all three mean the same. Such a situation creates a problem during matching. If a system's knowledge base contains only a single representation of the fact, then all but one sentence will fail to match. One solution is to store all possible alternatives in the knowledge base, but this may lead to inconsistency in the knowledge base. What we want is to have the same meaning representations of inputs that mean the same. This necessitates the use of canonical forms. Canonical forms complicate the task of semantic analysis. To assign the same representation to sentences (5.1), (5.3a), and (5.3b), the system must know that 'having a flight to', 'offering a flight to', and 'serving to' Hyderabad all mean the same thing in this context. Hence, the different syntactic parses underlying these requests must lead to the same meaning representation. There exist systematic meaning relationships among word senses and among grammatical constituents. We use them to assign the same representation to various inputs. Words have different senses. The process of choosing the right sense in context is called word sense disambiguation and is discussed in Section 5.5.

Inference and Variables

Given a set of facts about the world in the knowledge base, and the input representation, we can derive new facts that logically follow the known facts. Inference refers to system's ability to draw valid conclusions based on the meaning representation of inputs and representation of facts in its knowledge base. A meaning representation language should support this type of derivation (inferencing). Another requirement for the meaning representation language is that it should allow the use of variables. Consider the following, more complex, scenario:

I would like to catch a flight to Hyderabad.   (5.4)

This sentence does not mention any particular flight; instead it refers to an unknown and unnamed flight that goes to Hyderabad. The simple matching we discussed so far does not work in this case. Answering this request requires a more complex matching that involves the use of variables. We represent it as

Goes (x, Hyderabad)

To satisfy this request, we have to search for a known object in the knowledge base such that, substituting x by it matches the whole proposition.

Expressiveness

Natural languages cover a wide variety of content (or subject matter). A meaning representation language must be able to represent the meaning of this wide range of content. Therefore, to be of any use a meaning representation language must be equipped with expressive power.

5.2.2 Meaning Structure of the Language

The previous section focused on the role of meaning representation languages without discussing the meaning structure of language. This section focuses on fundamental predicate argument structure.

Predicate Argument Structure

It seems that the semantic structure of all human languages have an underlying predicate argument structure which contributes significantly to its semantic structure. This structure states that specific relationships exist among the concepts expressed through the words and phrases of a sentence. It is mainly this structure that makes it possible to create a single composite meaning representation from the meanings of the various parts of an input. One of the important tasks of a grammar is to help organize this predicate argument structure. Therefore, a meaning representation language should support the representation of the predicate argument structures.

The predicate argument structure is mainly manifested by the constraints that various grammatical categories put on other constituents appearing with them in some linguistic unit. For example, a verb prescribes specific constraints on the number, syntactic category, and location of the phrases that are expected to occur with them in syntactic structures.

Example 5.1   Consider the following sentences:

Murthy likes music.
Murthy likes watching movies.
Murthy likes to perform in the theatre.

The syntactic argument frames for these examples are

NP likes NP
NP likes VP
NP likes inf-VP

These syntactic frames tell the number, position, and syntactic category of the arguments that are expected to occur with a verb. For example, the frame for the first sentence in Example 5.1 specifies the following facts:

1. There are two arguments to this predicate.
2. Both arguments are NPs.

The first argument occurs before the verb and plays the role of the subject.

The second argument occurs after verb and plays the role of the direct object.

These syntactic frames provide useful information about syntax and also carry useful semantic information that can be utilized in creating meaning representation. For example, in each of the three examples given earlier, the pre-verbal argument plays the role of the entity doing the act of liking, and the post-verbal argument plays the role of the thing being liked. These generalizations help in associating the surface arguments of a verb with the semantic roles the arguments play in its underlying meaning representation. The semantic role that a noun phrase plays in a sentence is termed *thematic-role* or *theta-role*. The term case-role denotes the same concept. Often, the verbs place some semantic restriction, called selectional restriction or selectional preference, on the type of arguments that can play these semantic roles. These restrictions help in arriving at a correct meaning representation. For instance, the verb eat requires that its pre-verbal argument, i.e., the eater, must be some animate entity, and the post-verbal argument, the thing being eaten, must be some edible item. Verbs are not the only objects that can have a predicate-argument structure. Prepositions and nouns may also be specified using a predicate argument structure. For example, the phrase *a cup on the table* represents a relationship between a *cup* and a *table* expressed through the preposition *on*. The meaning representation of this phrase is given as

On (cup, table)

Any useful meaning representation language must support the representation of predicate argument structure. More specifically, it must support the representation of predicate argument structure with a variable number of predicates, the semantic labelling of arguments to predicates, and the semantic restrictions placed on those arguments. Knowledge representation formalisms, such as first-order predicate calculus, semantic networks, conceptual graphs, etc., all support representation of predicate argument structure.

5.2.3 Syntax-driven Semantic Analysis

We now turn our attention to how meaning representations are created and assigned to linguistic units. The first approach we discuss is syntax-driven semantic analysis.

Syntax-driven semantic analysis is a computational approach to semantic analysis that utilizes static knowledge from the lexicon and the grammar.

This type of meaning representation corresponds to literal meaning, which is context independent and inference free. The basic notion that drives this approach is the principle of compositionality, which states that the meaning of the whole can be composed from the meaning of its parts. What this principle tells us is that we can create meaning representations for sentences from the meanings of their consistent words. Syntax-driven semantic analysis uses syntactic analysis to guide the process of arranging semantic representations.

```
Input                    Parse              Semantic        Output
       →   Parser    →    tree      →       analyser    →   Meaning representation
```

Figure 5.1   A simple approach to syntax-driven semantic analysis

Figure 5.1 shows the schematic diagram of this approach. The output produced by syntactic analysis is passed as input to a semantic analyser to produce a meaning representation. Usually, a semantic analyser produces multiple ambiguous meaning representations as output. This is due to the ambiguities arising from the syntax (as discussed in Chapter 4) and the lexicon. A semantic analyser creates a representation for all possible interpretations, leaving the task of resolving this ambiguity, which requires access to domain-specific and contextual knowledge, to a subsequent stage of processing. We can use part-of-speech tagger, PP-attachment mechanism, and word sense disambiguation mechanism to reduce the number of possible meaning representations produced by the semantic analyser.

We now illustrate how a semantic analyser actually builds the meaning representation of a sentence, with the help of an example. Consider the parse tree for the sentence (5.5) shown in Figure 5.2.

President nominates speaker. (5.5)

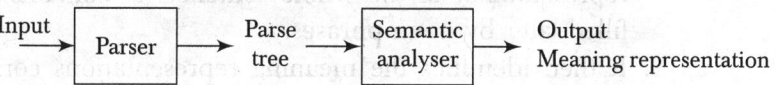

$\exists e$ is_a $(e,$ nomination$) \land$ nominator $(e,$ President$) \land$ nominee $(e,$ speaker$)$

Figure 5.2   Mapping syntactic constituents to meaning representation

Using the parse tree, a semantic analyser produces a semantic representation in following steps.

1. First, it retrieves a meaning representation from the sub-tree corresponding to the verb *nominates*. The program that interprets the parse tree must have the knowledge that it is the verb whose template will define the meaning representation of the whole sentence, and about position of its arguments and their roles. The meaning representation of the verb acts as a template for the meaning representation of the whole sentence. It contains variables that are filled later by noun phrases.

2. It then identifies the meaning representations corresponding to the two noun phrases.

3. And finally, it associates or binds the meaning representations of noun phrases to the variables appearing in the meaning representation of the verb, to give the meaning representation for the sentence as a whole (as shown in Figure 5.2).

As any reasonable grammar will have infinite number of such trees, mapping every possible tree to its semantic representation is not a viable approach. We need to identify general mappings, which is achieved by augmenting the lexicon and the grammar rule with semantic attachment, and devising a mapping between rules of the grammar and the rules of semantic representation. This is known as rule-to-rule hypothesis. The semantic attachments are instructions on mapping components of a rule to a semantic representation. An augmented rule takes the following form:

$$A \rightarrow \alpha_1 \alpha_2 \alpha_3 ... \alpha_n \{f(\alpha_{i \cdot \text{sem}}, ..., \alpha_{k \cdot \text{sem}})\}$$

The text appearing in {...} specifies that the meaning representation assigned to construction A (represented as $A_{.\text{sem}}$) is a function of the semantic attachments of A's constituents. To achieve a better understanding, we take an example. Consider sentence (5.5) as shown in Figure 5.2. Our first step is to associate the constants *President* and *speaker* with constituents (rules) that introduce them into sentence:

> Noun ← President {President}
> Noun ← speaker {speaker}

These two augmented rules state that the meaning associated with the sub-trees produced by the application of these rules consist of the constants *President* and *speaker*. However, sub-trees corresponding to these constants do not directly contribute to the final meaning representation. Instead, these meaning representations are passed on to their parents who contribute to the final meaning representation as indicated by the dotted arrows in the Figure 5.2. The augmented rule for noun phrase is

NP → Noun {Noun.$_{sem}$}

This rule states that the meaning representation of the noun phrase is the same as the meaning representation of its constituents. Now, we need to specify the semantics of the event underlying the sentence, i.e., the verb *nominates*. The *nomination* event involves a *nominator* and a *nominee*. The semantics for *nominate* can be represented by the logical formula:

$$\exists e, x, y \text{ is\_a } (e, \text{nomination}) \land \text{nominator } (e, x) \land \text{nominee } (e, y)$$

(5.6)

This logical formula is used as the semantic attachment of *nominate*, resulting in the following augmented rule.

verb → nominates {$\exists e, x, y$ is_a (nomination) $\land$ nominator (e, x) $\land$ nominee (e, y)}

The next constituent to be considered is VP, which dominates both *nominates* and the *speaker*. The grammar rule used is

VP → verb NP {verb$_{.sem}$ (NP$_{.sem}$)}

However, to create the semantic representation of VP, we cannot simply copy the meaning representations of its children. The goal for VP semantics of '*nominate*' is represented by the logical formula 5.6. To achieve this goal, we need to incorporate the meaning of NP into the meaning of the verb and assign it to the VP. This requires that the variable y be replaced with the logical term *speaker*, as the second argument of the nominee role of *nomination* event. Hence, VP$_{.sem}$ must tell us two things:

1. Which variable in the verb's semantic attachment is to be replaced by which argument.
2. How the replacement is to be performed.

But the semantic attachment of verb does not tell us how to replace each of its quantified variables. So, we need to revise the verb's semantic attachment.

We can use lambda calculus as the 'glue-language' to combine semantic representations systematically. Lambda calculus is an extension of FOPC. A lambda expression consists of a λ-operator, a list of variables (parameters), and an FOPC expression in those variables.

$\lambda x P(x)$

The parameter list in the lambda expression makes available the variable within the body of logical expressions, for binding to external arguments provided by the semantics of other constituents. This process of binding is known as *lambda reduction.* We illustrate this with the help of an example.

Example 5.2   Consider the application of $\lambda x P(x)$ to the constant *Taj.*

$$\lambda x P(x)\ (Taj)$$

The result of performing a λ-reduction on this expression is

$$P(Taj)$$

The reduction process replaces the variable in the body of the expression with *Taj* and removes λ. Thus, lambda calculus is able to satisfy the two requirements of VP semantics as identified earlier. The formal parameter list tells us variables within the body that are available for replacement and λ-reduction tells how to perform replacement. Changing the semantic attachment of the verb to a λ-expression, we get

$$\text{Verb} \rightarrow \text{nominates } \{\lambda y \exists\ e,\ x \text{ is\_a } (e, \text{nomination}) \wedge \text{nominator } (e, x)$$
$$\wedge \text{ nominee } (e, y)\}$$

Now y is externally available, and can be bound by the application of this expression to a logical term.

Let us go back to the VP semantics in our example sentence (5.5).

$$\text{VP} \rightarrow \text{verb NP } \{\text{verb}_{.sem}\ (\text{NP}_{.sem})\}$$

The value of $\text{NP}_{.sem}$ is *speaker.* Application of the $\text{verb}_{.sem}$ to the $\text{NP}_{.sem}$ results in the substitution of *speaker* for each occurrence of variable y in the expression. The meaning of a verb phrase *nominates speaker* is thus:

$$\exists\ e,\ x \text{ is a } (e, \text{nomination}) \wedge \text{nominator } (e, x) \wedge \text{nomine } (e, \text{speaker})\}$$

We now create the semantic attachment for the S rule.

$$\text{S} \rightarrow \text{NP VP} \qquad \{\text{VP}_{.sem}\ (\text{NP}_{.sem})\}$$

This rule says that the semantic representation of S ($S_{.sem}$) can be constructed by applying $\text{VP}_{.sem}$ to $\text{NP}_{.sem}$. As in the VP rule, a simple copying does not work. This rule requires the nominator role in the event representation to be bound to $\text{NP}_{.sem}$. However, the value of $\text{VP}_{.sem}$ does not contain any λ-expression. So, let us revise the semantic attachment of verb as follows:

$$\text{Verb} \rightarrow \text{nominates } \{\lambda y\ \lambda x \exists\ e \text{ is\_a } (e, \text{nomination}) \wedge \text{nominator } (e, x)$$
$$\wedge \text{ nominee } (e, y)\}$$

The attachment is now a nested λ-expression. In a nested λ-expression, $\lambda x\ \lambda y\ P(x,y)$ the first reduction binds the variable x and removes the outer λ. The resulting λ-expression can be applied to another term to arrive at a fully specified formula.

$\lambda x\ \lambda y\ P(x,\ y)(A)$ results in $\lambda y\ P(A,\ y)$, which in turn can be applied to another term, say B, as $\lambda y\ P(A,\ y)\ (B)$, resulting in $P(A,\ B)$.

With the revised attachment of the verb, the value of the $VP_{.sem}$ is $\lambda x \exists e$ is_a $(e, \text{nomination}) \wedge \text{nominator} (e, x) \wedge \text{nominee} (e, \text{speaker})\}$, which is applied to the $NP_{.sem}$. The value of the $NP_{.sem}$ is President. After λ-reduction we get,

$$\exists e \text{ is\_a} (e, \text{nomination}) \wedge \text{nominator} (e, \text{President}) \wedge$$
$$\text{nominee} (e, \text{speaker})\}$$

The parse tree for this example, with each node annotated with semantic value, is shown in Figure 5.3.

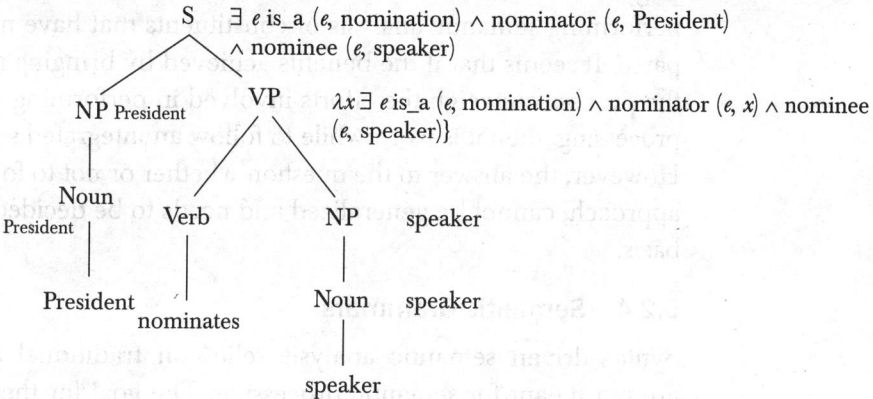

Figure 5.3 Parse tree with semantic values of each node for *sentence (5.5)*

As pointed pout earlier, idioms and collocates represent sentences in which the meaning does not depend on the meaning of the constituents, or depends on it only partially. This poses a challenge to the principle of compositionality. For example, consider the following sentence:

The old man finally kicked the bucket. (5.7)

The phrase *kick the bucket* has nothing to do with a *kick* or a *bucket*. To handle these situations, we need to introduce new grammar rules with semantic attachments that introduce logical terms and predicates, which are not related to any of the constituents of the rule. For example, for the idiom *kick the bucket*, we can introduce following grammar rule.

VP → kicked the bucket {died}

The lower case in the right hand side, represents words in the input and the semantic attachment is the meaning of the phrase. This illustrates that the meaning of the idiom is not based on the meaning of any of its parts. A similar problem is presented by collocates which represent sentences in which the meaning depends only partially on the constituents.

There are two approaches to semantic analysis: the pipeline approach and the integrated semantic approach. The schematic diagram we presented at the beginning of the chapter depicts a pipeline approach. In it, syntactic analysis is performed to give a parse tree. Then we walk through the parse tree applying semantic attachments in a bottom-up fashion.

In integrated semantic analysis, we modify the parse to include operations on semantic attachments. A semantic representation of the input fragments is created as they are being parsed. The advantage of this approach is that it can block a state as soon as an ill-formed semantic fragment is detected. This means, however, that effort is wasted in performing semantic analysis of constituents that have no role in the final parse. It seems that if the benefits achieved by bringing semantics early in the process outweigh the efforts involved in performing spurious semantic processing, then it is worthwhile to follow an integrated semantic approach. However, the answer to the question whether or not to follow an integrated approach, cannot be generalized and needs to be decided on an individual basis.

5.2.4   Semantic Grammars

Syntax-driven semantic analysis relies on traditional grammars, which are not meant for semantic processing. The goal for these grammars is to capture syntactic generalization and avoid over-generation; it is not to facilitate meaning representations. Often, the syntactic structures produced by these grammars do not fit semantic structures very well. Some of the obvious problems are as follows (adapted from Jurafsky and Martin 2000):

- Important semantic elements are often widely distributed across parse trees.
- Parse trees often contain constituents not important to making semantic distinctions.
- The general nature of many syntactic constituents results in semantic attachments that create meaningless representations.

An alternative approach is to bring semantic capabilities into the grammar itself. A grammar developed with the intention of handling semantics is called *semantic grammar*. Semantic grammars were first introduced in the domain of question-answering and intelligent tutoring. They are constructed specifically in terms of semantics information. Rules and constituents correspond directly to entities and activities in the domain. As an example, consider the following request:

I want to go from Delhi to Chennai on 24th December.   (5.8)

We can create a rule of the form:

InfoRequest: User wants to go to City from City TimeExpr.

Like the rules for idioms, these rules also mix terminals and non-terminals on their right hand side. In this rule, User, City, and TimeExpr are non-terminals from the domain. The presence of these non-terminals eliminates the need for lambda expression. The immediate constituents of the rule provide the necessary information to build semantic representation.

One of the main motivations behind the use of semantic grammar is dealing with anaphors and ellipsis, which is discussed in the following chapter. The advantage of semantic grammar is that we get exactly the semantic rules we need. However, it also has its drawbacks. The first is lack of generality. As the rules are domain specific, we need to develop a new grammar for each new domain. Second, as the semantic grammar lacks syntactic generalizations, the number of rules needed becomes quite large. For example, a traditional grammar rule for a noun phrase can cover both *Indian hotel* and *Chinese food*, but a semantic grammar must introduce two semantic grammar rules to handle these phrases. We also need to introduce separate rules to handle phrases like *expensive hotel*, *vegetarian hotel*, etc.

5.3 LEXICAL SEMANTICS

So far, we have focused on the creation of meaning representation of whole sentence. The approach we took supports the view that words contribute to the meaning of a sentence but do not themselves have meaning. It is perhaps this view that has led some to consider the lexicon as a simple list of words. However, this is narrow concept, too far from reality. Words have meaning; they have internal structure, and are involved in different relationships with other words. All this information can be captured in a systematic structure by a lexicon. In this section, we focus on this and other issues related to words. More precisely, we focus on lexical semantics, which is concerned with the linguistic study of systematic, meaning-related structure of words or lexemes (the minimal unit in a lexicon).

5.3.1 Relationships

One way to approach lexical semantics is to study the relationship among lexemes (an abstract representation of a 'word', the lexical entry in a dictionary). Semantics of a lexeme can be understood by analysing the relationships of lexemes with other lexemes. Lexical semantic information is useful for a wide variety of NLP applications. This section discusses a variety of relationship that holds among lexemes and their senses.

Homonymy

The first relationship that we discuss is homonymy which is perhaps the simplest relationship that exists among lexemes. Homonyms are words *that have the same form but have different, unrelated meanings.* A classic example of homonym is bank (river bank or financial institution). A related idea is that of *homophones* which refers to words that are pronounced in the same way but differ in meaning or spelling or both (e.g., be and bee, bear and bare).

Polysemy

Many words have more than one meaning or *sense.* Unlike homonyms, polysemes are words with related meanings. This linguistic phenomenon is called polysemy or lexical ambiguity. Words that have several senses are ambiguous and called polysemous. For example, the word 'chair' can refer to a piece of furniture, a person, the act of presiding over a discussion, etc. The word 'employ' is a polyseme as its two meanings—to hire (employ a person) and to accept (employ an idea)—are related.

In a particular use, only one of these meaning is correct. How we distinguish between these multiple senses is discussed in Section 5.5.

Hypernymy, Hyponymy, and Antonymy

A *hypernym* is a word with a more general sense. The word automobile is a hypernym for a car and a truck. A *hyponym* is a word with a more specific meaning. In the relationship between car and automobile, car is hyponym of automobile. *Antonymy* is a semantic relationship that holds between words that express opposite meanings. The word Dark is an antonym of light, and white is an antonym of black. Words may also be involved in a part-whole relationship called *meronymy*, e.g., 'wall', 'ceiling', 'floor' all are meronyms of 'room'.

Synonymy defines the relationship between different words that have a similar meaning. A simple way to decide whether two words are synonymous is to check for substitutability. Two words are synonyms in a context if they can be substituted for each other without changing the meaning of the sentence.

These relationships are useful in organizing words in lexical databases. One widely known lexical database is WordNet (see Chapter 12).

5.3.2 Internal Structure of Words

Having discussed the relationship between words, we now turn our attention to the structure of words. In our earlier discussions, we pointed out that there is a fundamental predicate argument structure that governs the meaning representation of a sentence. We have also seen that there is certain class of words that defines predicate and predicate structure, and another class of words that provides arguments to predicates. In this section, we discuss how the meanings of words are structured to support this idea.

Thematic Role

Thematic role is the semantic relationship between a predicate (e.g., a verb) and an argument (e.g., the noun phrases) of a sentence. For example, in sentence 5.6, the role of subject of the verb *nominate* is *nominator*. Similarly, an *eating* event will have an *eater*, a *reading* event will have a *reader*, a *singing* event will have a *singer*, and so on. *Eater, singer*, and *reader* have something in common. They are all performers of some event, are animate, and have a direct causal responsibly for their events. The thematic role provides a means to capture the semantic commonality between the actors of the events. In terms of thematic role, the subjects of the verb *eat, sing*, and *read* are agents. Agent is a thematic role which refers to the deliberate performers of an event or action. Similarly, the arguments that are affected by the action are assigned the thematic role Theme.

Thematic roles were first introduced by Gruber (1965) and Fillmore (1968), though the theory of semantic roles dates back to the work of Panini in Sanskrit grammar in 500–400 BC (Kearns 2000). The notion of thematic role was called Karaka in Paninian grammar. Panini established classes of NPs according to the broad interpretation of their grammatical form. Panini proposed six Karakas: Karma (action, object that is desired), Karana (means, instrument), Karta (agent, that which acts), Sampradana (transmission, what we aim at with the object), Apadana (removal, what is left when we move away), and Adhikarana (locative). Although thematic roles have widely appealed to linguists, they remain the subjects of disagreements, such as what roles to include and exactly how to define them. Some of the commonly used thematic roles along with their rough definitions are listed in Table 5.1. The thematic role has been used as shallow semantic language and as interlingua in machine translation.

Table 5.1   Commonly used thematic roles

Thematic role	Definition
Agent	Deliberate performer of the event or action, e.g., *I* opened the lock with a key.
Theme	The arguments that are most directly affected by an event, e.g., I opened the *lock* with a key.
Instrument	The instrument used to carry out the action, e.g., I opened the lock with a *key*.
Experiencer	Arguments that undergo a sensory, cognitive, or emotional experience, e.g., *Shweta* hates cricket.
Force	Unconscious performer of the event, e.g., The *storm* blown over many trees.
Location	Places where the event occurs, e.g., Children are playing in the *garden*.
Source	Origin of the object of a transfer event, e.g., Flight will take off in a short while from *Chennai*.
Goal	The destination of a transfer event.

Selectional Restrictions

Selectional restrictions are semantic constraints that are placed on arguments for a given word (sense). As an example, consider the verb *eat*, which requires food items as an object (theme) and an animate as a subject (agent). Selectional restrictions can be observed easily in cases where they are violated as in the following sentence:

A table eats grass. (5.9)

The semantics of selection restrictions can be captured in a meaning representation using thematic roles:

$$\exists e, x, y \; \text{eating}(e) \wedge \text{agent}(e, x) \wedge \text{theme}(e, y)$$

This representation means that there is an eating event, and x and y is associated with this via agent and theme relation respectively. To express that y must be edible we add new terms as follows:

$$\exists e, x, y \; \text{eating}(e) \wedge \text{agent}(e, x) \wedge \text{theme}(e, y) \wedge \text{is a } (y, \text{Edible Thing})$$

When a phrase like *ate an apple* is encountered, a semantic analyser can use the following representation:

$$\exists e, x, y \; \text{eating}(e) \wedge \text{agent}(e, x) \wedge \text{theme}(e, y) \wedge \text{is a } (y, \text{apple})$$

This is a reasonable representation, as category *apple* is consistent with its membership in the category *Edible Thing*. A phrase such as *ate a book* will be semantically ill-formed, as *book* is inconsistent with the membership in *Edible Thing*. However, this approach requires that the knowledge base contains representations of facts about the concepts that make up selectional restrictions.

Yet another approach to specify selectional restrictions is through WordNet hyponym relation. In this approach, WordNet synsets (synonym sets) are used to specify selectional restrictions on the arguments. A meaning representation is considered well-formed if one of the hyponyms of the word that fills a thematic role, matches the synset used to specify selectional restriction. Let us elaborate on this approach.

Example 5.3 Suppose the synset used to specify selectional restriction on the Theme role of the verb *eat* is {food}. This means that fillers of the theme role can only be words that belong to this synset and its hyponyms.

Now consider the phrase *ate an apple.*

In order to decide whether *apple* is valid filler for theme role, we need to examine the hypernym chain of *apple*, as shown in Figure 5.4. As *food* is in the chain of hypernyms, it is considered an appropriate filler.

Sense 1

apple—(fruit with red, yellow, or green skin, sweet to tart, and crisp whitish flesh)

⇒ edible fruit—(edible reproductive body of a seed plant especially one having sweet flesh)

⇒ produce, green goods, green groceries, garden truck—(fresh fruits and vegetables grown for the market)

⇒ food—(any solid substance that is used as a source of nourishment; "food and drink")

⇒ solid—(a substance that is solid at room temperature and pressure)

⇒ substance, matter—(that which has mass and occupies space; "an atom is the smallest indivisible unit of matter")

⇒ entity—(that which is perceived, known, or inferred to have its own distinct existence (living or nonliving))

⇒ fruit—(the ripened reproductive body of a seed plant)

Figure 5.4 Part of WordNet hypernym hierarchy of word *'apple'*

There can be valid sentences violating selectional restrictions. Perhaps you recall news stories about people demonstrating their skill in eating glass or bulbs, or pulling cars or planes with their teeth instead of cranes. Though, these are atypical examples involving violation of selectional restrictions, they are perfectly well-formed and interpretable. Adhering to selection restriction in these cases will eliminate all possible literal senses, leading the semantic processing to a dead end. One way to avoid this is to consider semantic constraints as preferences instead of restrictions.

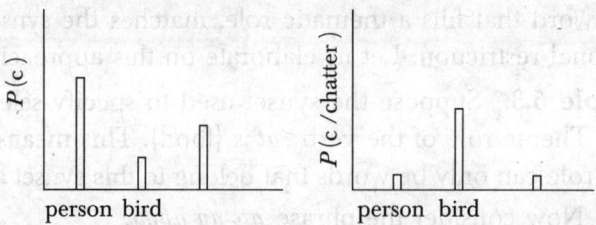

Figure 5.5 Prior and posterior distribution over argument classes

Selectional restrictions have been used in parsing to eliminate a parse that violates these restrictions. It can also be used to infer the meaning of words missing in a dictionary. Traditionally, selectional restriction has been associated with word sense disambiguation. We now discuss a model of selectional preferences proposed by Resnik (1996). The model is general enough to be applicable to any class of words that impose semantic constraints on its arguments. A few applicable instances are: verb-object, verb-subject, noun-noun, adjective-noun, and verb-prepositional phrase. The model is probabilistic in nature and attempts to capture co-occurrence behaviour of predicates and conceptual classes filling the roles of arguments. Figure 5.5 depicts the basic idea. The prior distribution $P(c)$ captures the probability of a class 'c' occurring as the argument in predicate argument relation R, regardless of the identity of the predicate. For example, given the verb-subject relationship, the prior probability for *person* tends to be significantly higher than the prior probability for *bird*. However, once the predicate is known, the probability can change. Take the case of the verb *chatter*. Given this verb, the probability of *bird* as a subject is expected to be significantly higher and the probability of *person* to be lower. In probabilistic terms, it is the difference between conditional (or posterior) and the prior distribution that determines selectional preferences. One way to quantify this difference is to use relative entropy. The model defines a selectional preference strength (SPS) of a predicate p as

$$S_R(p) = D(P(c/p) \parallel P(c))$$
$$= \sum_c P(c/p) \log \frac{P(c/p)}{P(c)}$$

where $P(c)$ is probability distribution of class c.

$P(c/p)$ is the probability distribution of class c in the direct object or subject position of p. $S_R(p)$ is a measure of the amount of information that predicate p provides about the conceptual class of its argument. The better $P(c)$ approximates $P(c/p)$, the less influence p has on its argument, and therefore the weaker is its selectional preference.

With this discussion, it follows naturally that the semantic well-formedness of an argument to a predicate can be determined using its relative contribution to SPS. Arguments that fit well are expected to show strong association, i.e., they tend to have higher posterior probability compared to prior probability, as is the case of *bird* in Figure 5.5. Formally, selectional association between a predicate p and a class c is defined as

$$A(\text{p, c}) = \frac{P(\text{c/p}) \log \dfrac{P(\text{c/p})}{P(\text{c})}}{S_R(\text{p})}$$

While assigning association strength to a word instead of word class, we can simply define it as $A(\text{p, c})$, or if it is a member of several classes, we can define its association strength as the highest association strength of any of its classes.

The probability that a direct object in class c occurs given the predicate p can be calculated as

$$P(\text{c/p}) = \frac{P(\text{p, c})}{P(\text{c})}$$

$P(\text{p})$ can be estimated by $C(\text{p})/\sum_{\text{p}'} C(\text{p}')$, its relative frequency with respect to other predicates. Resnik (1996) proposed the following estimate for $P(\text{p, c})$

$$P(\text{p, c}) \approx \frac{1}{N} \sum_{\text{w}\in\text{c}} \frac{\text{count}_R(\text{p, w})}{|\text{classes}(\text{w})|}$$

where $\text{count}_R(\text{p, w})$ is the number of times the word w occurs as argument of p with respect to relation R; class(w) is the number of classes to which the word w belongs; and N is the total number of predicate-object pairs in the corpus.

This model has been used to predict valid arguments as well as to assign the correct sense to a word in its context.

5.4 AMBIGUITY

Most of us find it difficult to understand why computers are not able to understand language the way people are. This is because we do not realize how vague and ambiguous natural languages are. Ambiguity, i.e., having more than one meaning, can be the result of syntax or semantics. Vagueness is not the same as ambiguity. In vagueness, words or phrases have only one meaning but they lack clarity, which makes it difficult to arrive at a precise meaning.

In many cases, a single word in a language corresponds to more than one thought, for example, the noun कलम (pen vs graft), चंदा (contribution, as in the sentence 'सभी सदस्यों को चंदा जमा करना है।' or moon as in (चन्द्रमा, चाँद); the adjective *light* (not heavy or not dark); the noun *bank* (financial institution or bank of a river); and the verb *run* (to move fast or to direct and manage). But this does not create a problem for us. We hardly give any thought to understanding what constitutes the correct meaning of a word or phrase, but we generally arrive at the correct one. The process is almost effortless. People are good at resolving ambiguity by considering the context of the written text or spoken utterances. Except for jokes and puns, where it is intended, ambiguity is not perceived as such. However, ambiguities existing at different levels are one of the major challenges in computational linguistics.

Let us try to understand why ambiguity is difficult. This is because it increases the range of possible interpretations of natural language. Suppose each word in an eight word sentence is ambiguous and has three possible interpretations. The total number of interpretations of the whole sentence is $3^8 = 6561$. Further, the syntactic and pragmatic ambiguities make the actual number of interpretations even larger. Resolving all these interpretations in a reasonable amount of time is difficult. There are words with a much larger number of senses than considered in this example. For example, the noun *pass* has 16 senses and the verb *fall* has 32 senses in WordNet. This gives a clear picture of the difficulty involved in the automatic interpretation of natural languages.

Ambiguity is a property of linguistic expressions. Ambiguity means *capable of being understood in more than one way* or *having more than one meaning*. It refers to a situation where an expression (word/phrase/sentence) can have more than one interpretation. Ambiguity can occur at four different levels:

- Lexical
- Syntactic
- Semantic
- Pragmatic

Lexical ambiguity is the ambiguity of a single word. A word can be ambiguous with respect to its internal structure or to its syntactic class. For example, in the sentence, *Look at the sky*, *look* is verb, whereas in *Lubna gave me a warm look*, it is noun. Similarly the word *silver* can be used as a noun, an adjective, or a verb.

She bagged two *silver* medals.
She made a *silver* speech.
His worries had *silvered* his hair.

However, this type of ambiguity is viewed as part-of-speech tagging in NLP and is considered to have been solved with reasonable accuracy. The type of lexical ambiguity with which we are concerned in this chapter is lexical semantic ambiguity, which occurs when a single word is associated with multiple senses. Two different occurrences of a single word belonging to the same syntactic category, may also have different meanings. As in the following sentences:

There is a hike in price of gold. (5.10a)
She has a heart of gold. (5.10b)

The occurrence of gold in both sentences corresponds to the syntactic category noun, but their meanings are different. The noun 'gold' has five senses listed in WordNet as shown in Figure 5.6. The use of gold in sentence (5.10a) corresponds to sense 3, and its use in sentence (5.10b) corresponds to sense 5.

1. (8) gold—(coins made of gold)
2. (8) amber, gold—(a deep yellow colour; "an amber light illuminated the room"; "he admired the gold of her hair")
3. (6) gold, Au, atomic number 79—(a soft yellow malleable ductile (trivalent and univalent) metallic element; occurs mainly as nuggets in rocks and alluvial deposits; does not react with most chemicals but is attacked by chlorine and aqua regia)
4. gold—(great wealth; "Whilst that for which all virtue now is sold, and almost every vice—almighty gold"—Ben Jonson)
5. gold—(something likened to the metal in brightness or preciousness or superiority etc.; "the child was as good as gold"; "she has a heart of gold")

Figure 5.6 WordNet senses of noun 'gold'

Ambiguity occurs across all languages. For example, consider the Urdu words قریب (*kareeb*) and گاڑھے (*gaadhe*) in the following sentences:

Indian pacers have improved a lot in **approximately** two years.

قریب دو سالوں میں بھارتی پیسروں نے خاصی ترقی کی ہے۔ (5.11a)

(*hai ki tarakki khasi ne paceron bhartiya mein saalon do kareeb*)
He lives near mosque.

وہ مسجد کے قریب رہتا ہے۔ (5.11b)

(*hai rahta kareeb ke masjid woh*)

He helped me at **odd** times

اس نے گاڑھے وقت پر میرا ساتھ دیا۔ (5.12a)

(*diya sath mera par waqt gaadhe usne*)

He was dressed in **dark** blue shirt.

اس نے گاڑھے نیلے رنگ کی قمیض پہن رکھی تھی۔ (5.12b)

(*thi rakhi pahan kameez ki rang neele gaadhe usne*)

The text in parentheses is the transliteration of Urdu sentence as it reads from right to left. In English translation, the **bold** words represent the meaning of words قریب and گاڑھے being used in the example sentences.

Similarly, the use of the Hindi word संबंध in the following sentences, corresponds to two different meanings:

मेरा उससे कोई संबंध नही है। (5.13a)

(*Mera usse koi sambandh nahi hai*)

(I do not have any *relation* with him.)

आज कक्षा में परीक्षा के संबंध में जानकारी दी गयी। (5.13b)

(*Aaj kaksha mein pariksha ke sambandh mein jankaari di gayi*)

(Information *about* the examination was given in the class today.)

Figure 5.7 lists different senses of the Hindi word कलम as found in Hindi WordNet.

1. (R) कलम, पेन, लेखनी, अक्षरजननी — स्याही के संयोग से कागज आदि पर लिखने का उपकरण "यह कलम किसी ने मुझे उपहार स्वरूप प्रदान की है"
2. (R) कलम, क्रलम, आँखिया — वह औजार जिससे महीन चीज काटी या खोदी जाए "वह कलम द्वारा संगमरमर पर राम का चित्र बना रहा है"
3. (R) कलम, कलम — पेड़ की यह टहनी जो दूसरी जगह बैठाने या दूसरे पेड़ में पैबंद लगाने के लिए काटी जाए "कलम से तैयार वृक्ष के फल स्वादिष्ट और बड़े होते है"
4. (R) तूलिका, तूलि, तीली, कूँची, कूची, ब्रश, कलम, कलम, अक्षरतूलिका — चित्रकार के रंग भरने की कलम "वह तूलिका से चित्र मे रंग भर रहा है"
5. (R) कलम, कलम — चित्र अंकित करने की किसी विशेष स्थान या परम्परा की शैली "यह राजस्थानी कलम है"
6. (R) कलम, कलम — बही-खाते आदि में लिखा जाने वाला कोई मद "इसमें एक कलम छूट गई है"
7. (R) कलम, कलम — कनपटी के पास का वह स्थान जिस पर गाल की ओर कुछ दूर तक बाल रहते है "बाल बनवाते समय कलम के बाल छोटे करा लेना"
8. (R) कलम, कलम, लेखनी, अक्षरजननी, वर्णिका — लकड़ी आदि का बना वह लेखन उपकरण जिसे स्याही में डुबा-डुबाकर लिखा जाता है "छात्र नरकट की कलम से लिख रहा है"
9. (R) कलम, कलम — सिर के वे बाल जो कनपटी के पास होते है "नाई ने तुम्हारे कलम को ठीक से नहीं काटा है"

Figure 5.7 Different senses of the Hindi word कलम as listed in Hindi WordNet

5.4.1 Syntactic Ambiguity

As discussed in Chapter 4, there are different ways in which a sequence of words can be grammatically structured. Each structuring leads to a different interpretation. The structural ambiguities discussed in the previous chapter were syntactic ambiguities. One frequently quoted example is the sentence, *The man saw the girl with the telescope.* It is unclear—ambiguous— whether the man saw a girl carrying a telescope, or he saw her through his telescope. It is the syntax, not the meaning of the words, which is unclear. The meaning is dependent on whether the preposition 'with' is attached to the girl or the man. This is an example of a specific type of syntactic ambiguity termed PP attachment ambiguity. Consider another example:

$$\text{Stolen painting found by tree.} \tag{5.14}$$

Two alternative syntactic representations make this sentence structurally ambiguous:

1. A tree found a stolen painting.
2. A stolen painting was found near a tree.

Selectional restrictions may help disambiguate this. The verb 'find' usually takes an agent with the property, animate. Co-occurrence statistics between the verb and the preposition on one hand, and between preposition and noun on the other hand can also be used to resolve attachment ambiguity. A simple model based on this information computes log-likelihood ratio.

$$\lambda(\text{v, n, p}) = \log \frac{P(\text{p/v})}{P(\text{p/n})}$$

where $P(\text{p/v})$ is probability of seeing a PP with p after the verb v and $P(\text{p/n})$ is the probability of seeing a PP with p after the noun n.

If $\lambda(\text{v, n, p}) < 0$, then we attach the preposition to the noun and if $\lambda(\text{v, n, p}) > 0$, we attach it to the verb.

5.4.2 Semantic Ambiguity

This occurs when the meaning of the words themselves can be misinterpreted. For example, the meanings of words in a phrase can be combined in different ways, leading to different interpretations.

Iraqi head seeks arms.

The homograph 'head' can be interpreted as a noun meaning either chief or the anatomical head of a body. Likewise, the homograph 'arms' can be interpreted as a plural noun meaning either weapons or body parts.

5.4.3   Pragmatic Ambiguity

Pragmatic ambiguity refers to a situation where the context of a phrase gives it multiple interpretation (Kamsties 2001) as in, *Give it to the kids.* Here 'it' may refer to many things depending on the context. Consider a larger context.

> *Cake is on the table. I have prepared some snacks. Give it to kids.*

It is not clear whether 'it' refers to cake or snacks or both. Perhaps a larger context may help us.

> *Cake is on the table. I have prepared some snacks. Give it to kids.*
> *Kids enjoyed cake and snacks.*

Now, it is clear that 'it' refers to both the snacks and the cake. Resolving these types of ambiguities require discourse processing, which is discussed in the next chapter. In this chapter, we focus mainly on lexical semantic ambiguity. We also discuss how sectional restriction can be used to handle structural ambiguities.

5.5   WORD SENSE DISAMBIGUATION

Having discussed various types of ambiguities we now focus on identifying the correct sense of words in a particular use. The first attempt at automated sense disambiguation was made in the context of machine translation. In his famous *Memorandum,* Weaver (1949) discusses the need for word sense disambiguation (WSD) in machine translation, and outlines an approach to WSD, which underlies all subsequent work on the topic.

> *If one examines the words in a book, one at a time as through an opaque mask with a hole in it one word wide, then it is obviously impossible to determine, one at a time, the meaning of the words. [...] But if one lengthens the slit in the opaque mask, until one can see not only the central word in question but also say N words on either side, then if N is large enough one can unambiguously decide the meaning of the central word. [...]The practical question is: "What minimum value of N will, at least in a tolerable fraction of cases, lead to the correct choice of meaning for the central word ?"*

The gist of this excerpt is that words can be disambiguated (Weaver 1949) only in context, or, equivalently, context contains useful information to disambiguate a word. A number of WSD approaches are found in literature that use the context of an ambiguous word to disambiguate its meaning. Others use selection restriction information to resolve ambiguity, particularly to handle PP attachment ambiguity. In the following lines, we first elaborate use of selectional restriction (or preferences) in WSD and then discuss stand alone WSD approaches.

5.5.1 Selectional Restriction-based Word Sense Disambiguation

We pointed out in Section 5.3 that selectional restrictions or preferences can be used in parsing to eliminate flawed meaning representations. This can be viewed as a form of indirect word sense disambiguation. We now explore this idea. Consider the following sentences:

The institute will *employ* new employees. ('to hire')   (5.15a)

The committee *employed* her proposal. ('to accept')   (5.15b)

One can intuitively differentiate the senses of *employ* in sentences (5.15a) and (5.15b) with the complements of each *employ*. To be more precise, *employ* in (5.15a) restricts its subject and object nouns to those associated with the semantic features human/organization and human, respectively. On the other hand, *employ* in (5.15b) restricts its subject and object nouns to those associated with the semantic features human/organization and idea, respectively. Consequently, given employees as the object, the sense to hire is selected as the interpretation of employ in (5.15a), and the sense *to accept* is ruled out. The same reasoning can be used to select the sense *to accept* as the interpretation of *employ* in (5.15b).

In fact, the disambiguation process can be seen as mutually propagating semantic constraints to each polysemous word through selectional restriction.

PP attachment ambiguity can also be resolved using selectional restriction. For example, in sentence (5.14), *finding* is an activity that some animate being, not a tree, can perform. This selectional restriction information will help eliminate the parse that attempts to assign *tree* as subject of *found*.

Similarly, the PP attachment ambiguity in the sentence *The man saw the bird with the telescope* will be handled correctly using selectional restrictions. However, this approach does not always work as in sentence, *The man saw the girl with the telescope*. Though, selectional restriction helps us identify that seeing is an activity that can be performed using a telescope, it is not unusual to see with naked eye a girl having a telescope.

The selectional restriction-based disambiguation can be integrated into a semantic analyser. We can follow a *rule-to-rule strategy* as discussed in Section 5.2. With this approach, fragments of meaning representations are created and checked for selectional restriction violation as soon as their syntactic constituents are generated. A parse that violates selectional restrictions are blocked from further processing. The selectional restriction about arguments of predicates can be encoded in terms of WordNet synsets.

The selectional preference model proposed by Resnik (discussed in Section 5.3) has also been used for WSD. Recall that we have already discussed how to compute selectional association between a predicate and a class. To apply these notions in disambiguation, we simply select the sense that has highest selectional association between a hypernym and the predicate. As Resnik's algorithm makes use of WordNet in disambiguation, we discuss it along with other knowledge-based algorithms. The approach, however, is applicable only in cases where the predicate is unambiguous and selects the correct sense of the argument. We need some other method to handle cases when both the predicate and argument are ambiguous.

5.5.2 Context-based Word Sense Disambiguation Approaches

Approaches to stand-alone WSD that make use of context of ambiguous word basically fall into one of the following two general categories:
- Knowledge-based
- Corpus-based

However, this strict classification does not hold for a number of algorithms, which attempt to combine the features of both. The algorithm proposed by Yarowsky (1995) and Luk (1995) are two that take a hybrid approach.

Knowledge-based (dictionary-based) approaches utilize information from an explicit lexicon or knowledge base to disambiguate a word. The lexicon may be a machine-readable dictionary, a thesaurus, or ontology. Hand coded knowledge may also be used. The work has been carried out using existing lexical knowledge sources such as WordNet (Aggire and Rigau 1996; Resnik 1995; Voorhees 1995), *LDOCE* (Guthrei et al. 1991) and *Roget's International Thesaurus* (Yarowsky 1992).

A *Corpus-based approach* extracts information on word senses from a large sense-tagged corpus. The information used to annotate an ambiguous word is distributional information, context, and further knowledge that has been annotated in the corpus or added during pre-processing. Distributional information about an ambiguous word refers to the frequency distribution of its senses. Context refers to words found to the right and/or left of a word. Additional knowledge sources that can be used include lemmas, part-of-speech and syntactic annotations, etc.

The data acquisition bottleneck (Gale et al. 1992a) is the major difficulty of a corpus-based approach. In order to be useful for WSD, a corpus has to be annotated with word senses. Creating sense tagged corpus is a labour intensive task. Obtaining sense tagged data in languages other than English is very difficult. Manually sense tagging corpora with the help of a dictionary sense listing or WordNet hierarchy has been utilized. However, it is time consuming. An alternative is to use bootstrapping or unsupervised approaches which are less demanding in terms of tagged data, although tagged data is still needed for evaluation.

There are two possible approaches to *corpus-based* WSD systems:

1. Supervised WSD
2. Unsupervised WSD

Supervised approaches rely on a sense-tagged training corpus for disambiguation. It is an application of the machine learning approach for creating a classifier. It corresponds to the classification task. During training, information about context words and distributional information about the different senses of ambiguous words are collected from the corpus. During testing, the most likely or most similar sense, as computed on the basis of training data, is chosen. The existence of a sense-tagged corpus is a prerequisite for these algorithms.

Unsupervised approaches make use of raw or unlabelled text corpora for training and annotated data for evaluation. A completely unsupervised algorithm cannot do sense tagging. It only does sense discrimination, i.e., it just discriminates among various senses instead of tagging them. Unsupervised WSD algorithms can be considered as a clustering task. We cluster the context of an ambiguous word into a number of groups and discriminate between them without labelling them (Schutz 1998). As in part-of-speech tagging, the performance of unsupervised algorithms is usually on the lower side as compared to supervised algorithms.

Knowledge-based Approaches

A knowledge-based approach utilizes a machine readable dictionary, thesaurus, or ontology to assign the correct sense to an ambiguous word. We begin the discussion with Lesk's work which pioneered the use of dictionary definitions in word sense disambiguation.

Lesk's algorithm: Lesk's method consists in determining the overlap between words in the sense definitions of ambiguous words and the definitions of context words surrounding these ambiguous words in a given text. The most likely sense is the one having relatively large overlap between its definition and the context. Yarowsky (1992) took this idea further by including additional evidences from corpora and training machine learning algorithms.

Suppose that an ambiguous word w has k senses $s_1, s_2, s_3, \ldots, s_k$, and $ds_1,$ $ds_2, ds_3, \ldots, ds_k$ are the definitions of these senses represented as a bag of words occurring in the definition. The context to be disambiguated is c. dw_j is the dictionary definition of a word w_j occurring in context c, where sense distinctions are ignored. if w_j has multiple senses $s_{j1}, s_{j2}, \ldots, s_{jm}$, then $dw_j = \cup_{ji} ds_{ji}$. Lesk's algorithm can be described as shown in Figure 5.8.

for $i = 1$ to k do

$\qquad$ score(s_i) = overlap $(d_{sk}, \cup_{w \in c} dw_j)$

end

choose $\hat{s}$ s.t. $\hat{s} = \text{argmax}_{sk}$ score(s_k)

Figure 5.8 Lesk's algorithm

The overlap can be simply the number of words in common between the sense definition and the definitions of the words w_j in the context. Some of the alternative measures that can be used are shown in Figure 5.9.

Dice Coefficient	$\dfrac{2\lvert X \cap Y \rvert}{\lvert X \rvert + \lvert Y \rvert}$
Jaccard Coefficient	$\dfrac{2\lvert X \cap Y \rvert}{X \cup Y}$
Cosine	$\dfrac{2\lvert X \cap Y \rvert}{\sqrt{\lvert X \rvert \times \lvert Y \rvert}}$

Figure 5.9 Similarity measures

To give a more concrete notion of the algorithm, we consider an example.

Example 5.4 Consider the three senses of noun *ash* in WordNet 2.0 along with their definition.

Sense		Definition
s_1	ash	the residue that remains when something is burned
s_2	ash, ash tree	any of various deciduous pinnate-leaved ornamental or timber trees of the genus Fraxinus
s_3	ash	strong elastic wood of any of various ash trees; used for furniture and tool handles and sporting goods such as baseball bats)

We have to disambiguate the following two contexts:
1. The house was burnt to ashes while the owner returned.
2. This table is made of ash wood.

Using the number of words that the contexts have in common with the sense definition, the scores assigned to these context is

Context		Scores	
	s_1	s_2	s_3
1.	1	0	0
2.	0	0	1

The two contexts are disambiguated correctly using these scores.

Lesk reported disambiguation accuracies of 50–70% when the method was applied on a sample word. Lesk's algorithm is simple to implement and requires no training data. Though the disambiguation accuracy is not impressive, his method served as the basis for most dictionary-based disambiguation work that follows. Lesk's method is highly sensitive to words found in the dictionary. As you can see in Example 5.4, disambiguation was based only on one word that was found to co-occur in the definition of the context word and dictionary sense definition of 'ash'. The presence or absence of a single word may change the result drastically as a direct overlap between contexts and individual sense definition, is needed to disambiguate a word. Lesk acknowledged this problem and mentioned that his algorithm may fail to correctly disambiguate a number of ambiguous words if no words co-occurred between the context and the sense definitions of the word. He suggested the use of dictionaries with larger definitions such as the *Oxford English Dictionary*. The suggestion, however, remains untested.

One way to handle this lack of evidence problem observed by Lesk, is to expand the definition by including words related to, but not appearing in, individual sense definitions. For example, in the sentence, *I went to the bank to deposit cash*, the word *deposit* appears in the context of the ambiguous word *bank*, but the definition of *bank* in the *American Heritage Dictionary* does not contain *deposit*. We can expand the sense definition of *bank* by including deposit. But, knowing that deposit can be used for expansion does not solve our problem. We must know in which sense of *bank* this expansion is to be carried out. One can easily guess that this expansion is to be carried out in the definition corresponding to the financial sense. But how can this be done automatically? Thesauruses and dictionaries that include subject codes or semantic categories in their entries can help us. These subject codes roughly correspond to the conceptual categories of words. The underlying assumption in using semantic categorization for disambiguation is that semantic categories of words in a context determine the semantic category of the context as a whole, and thus identify the word sense being used. For example, *Longman's Dictionary of Contemporary*

English (LDOCE) (Procter 1978), includes the subject code EC (economic) for the financial sense of bank. This subject code helps us in knowing that *deposit* is related to the *financial* sense of bank. This expansion improves the chance of overlap with the correct sense. Roget's thesaurus also provides semantic categories (Roget 1946). Walker (1987) used this idea and proposed a simple algorithm by incorporating subject codes. His algorithm is based on the assumption that the subject codes assigned to a word reflects the sense of the word. If a word has more than one subject code then it will have more than one sense.

Let $t(s_k)$ be the subject code of sense s_k of an ambiguous word w occurring in context c. We disambiguate w by counting the number of words in the context having $subj(s_k)$ as one of their subject code. The sense with the highest count is selected. We can see the algorithm in Figure 5.10.

$$
\begin{aligned}
&\text{for } i = 1 \text{ to } k \text{ do} \\
&\quad score(s_k) = \sum_{w_j \, in \, c} d(\text{subj}(s_k), w_j) \\
&\text{end} \\
&\text{choose } s' \text{ s.t. } s' = \arg\max_{s_k} score(s_k)
\end{aligned}
$$

$d(\text{subj}(s_k), w_j) = 1$ iff $\text{subj}(s_k)$ is one of the subject code of w_j and 0 otherwise. The score is number of words in context whose subject code matches with the subject code of sense s_k.

Figure 5.10   Walker's algorithm for sense disambiguation

A general categorization of words into topics is often inappropriate for a particular domain. For example, *mouse* may be listed both as an animal and an electronic device in a thesaurus; but in a computer magazine it will rarely be in use for the thesaurus category 'animal'. A general topic categorization may also have a problem of coverage.

Wilks et al. (1990) attempted to expand dictionary definition with words that commonly co-occur with that definition; the idea being that commonly co-occurring words are semantically related to those in the definition. He derived this information from definitions in the dictionary. Wilks's work was based on the LDOCE. This vocabulary produces a large number of word co-occurrences. To perform word sense disambiguation, the context in which a target word occurs is also expanded to include words that co-occur with those already in the context. This expansion adds semantically related word to definitions, which increases the chances of correct disambiguation.

Yarowsky (1992) proposed an algorithm to adapt topic classification. He derived classes of words by starting with words in common categories in Roget's thesaurus. Other than word relationships Roget's thesaurus also supplies an explicit concept hierarchy consisting of up to eight increasingly refined levels. Each occurrence of a word under different categories of the thesaurus represents a different sense of word, i.e., the categories correspond roughly to word senses (Yarowsky 1992). A set of words in the same category are semantically related.

We now discuss a WSD algorithm proposed by Resnik (1997) which uses the WordNet class hierarchy to disambiguate noun senses. The sentences in the training corpus need to be parsed in order to extract syntactic relations such as subject-verb, verb-object, adjective-noun, head-modifier, and modifier-head. A syntactic relation involves two words: a noun n to be disambiguated and a verb, adjective or another noun which is involved in some relation with n. Let n has k senses $s_1, ..., s_k$. Suppose the syntactic relation R holds for n and the verb v. For each of these k senses, Resnik's method computes a class C_i as:

$$C_i = \{c | c \text{ is an ancestor of } s_i\}$$

$$a_i = \max_{c \in C_i} (A_R(v, c))$$

We approximate this value using frequency counts:

$$a_i \approx \max_{c \in C_i} \left(\sum_{w \in c} \frac{\text{count}_R(v, w)}{|\text{classes}(w)|} \right)$$

where $\text{count}_R(v, w)$ is the number of times the word w occurs in syntactic relation R with v, and class(w) is the number of classes to which the word w belongs. The algorithm selects the sense s_i for which a_i is greatest.

Let us illustrate the algorithm with the help of an example (adapted from Ng and Zelle (1997). Consider the following sentence:

I would like to drink coffee.

Suppose we want to disambiguate the noun *coffee*. The syntactic relation between *drink* and *coffee* is that of verb-object. The noun *coffee* has four senses in WordNet: coffee as a kind of (1) beverage, (2) tree, (3) seed, and (4) colour. The algorithm searches the parsed training corpus to get all occurrences of *drink* involved in a verb-object relation. Let us assume that the corpus contains only three such occurrences—*drink tea, drink milk,* and *drink cocoa*. The goal is to identify that the nouns *tea, milk,* and *cocoa* are most similar to the beverage sense of *coffee* without requiring that they be manually tagged with the correct sense in the training corpus. As shown in Figure 5.11, each of the four nouns *coffee, tea, milk,* and *cocoa* have multiple senses. For example, the noun *tea* has five senses: as a kind of beverage, meal, leaf, party, or bush. The sense that is most commonly shared by these four nouns is the *beverage* sense, whereas the remaining senses are only pointed to by one individual noun.

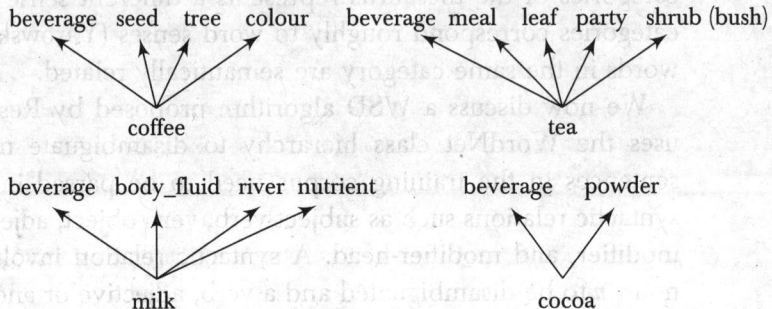

Figure 5.11   Vaious senses of *coffee, tea, milk,* and *cocoa*

In this example, n is the noun *coffee*, with four senses s_1 = beverage, s_2 = tree, s_3 = seed, and s_4 = *colour*. Assume that each of the syntactic relations *drink coffee, drink tea, drink milk,* and *drink cocoa* occur exactly once in our raw corpus. Given the concepts shown in the figure, we calculate the associations $a_1 = 1/4 + 1/5 + 1/4 + 1/2 = 1.2$, $a_2 = a_3 = a_3 = 1/4 = 0.25$. The highest value is a_1, hence, the algorithm selects sense s_1, i.e., the *beverage* sense. See Resnik (1997) for further details.

Supervised Learning of WSD

Most popular approaches to WSD use supervised machine learning methods to train a classifier using a set of labelled instances of the ambiguous word. The labels in the training set are contextually appropriate sense of the ambiguous word w. This classifier is then used to decide the correct sense of unlabelled instances of the ambiguous word. Typically, the learning algorithm requires its training examples, as well as test examples, to be encoded in the form of feature vectors. As discussed earlier, the context of the word provides useful information to disambiguate a word. We map the context into the feature vector. The linguistic features commonly used are collocational and co-occurrence features. Collocational features refer to position specific features, located to the left or right of the target word. Typical features include the word and its part-of-speech. For example, the feature vector consisting of two words to the left and one word to the right of the ambiguous word *silver* in sentence 5.9b is [made VBD a DT silver JJ speech NN]. The co-occurrence features ignore the position information. In this, words themselves are used as features. Usually, a small number of frequently used words are selected as features and the frequency of these words (features) within a fixed size window with the target word at the centre, are used as feature values. The context is represented as vector of these features.

The most widely known supervised algorithms are *Bayesian classification*, *k-nearest neighbour (k-NN) classification*, and *decision lists*. We now discuss the use of *Naive Bayes classification* approach of word sense disambiguation.

Bayesian Classification The specific algorithm we describe here was introduced by Gale (1992). The classifier assumes that we have a corpus in which each occurrence of an ambiguous word is labelled with its correct sense. The words around the ambiguous word are used to define a context window. The classifier treats the context of word w as a bag of words without structure. No feature selection is done. All the words occurring in the context window contribute in deciding which sense of the ambiguous word is likely to be used with it. What we want to find is the most likely sense s' for an input context c of an ambiguous word w. This is obtained as

$$s' = \arg\max_{s_k} P(s_k/c)$$

As it is difficult to collect statistics for this equation, we apply the Bayesian formula to compute it.

$$s' = \arg\max_{s_k} \frac{P(c/s_k)}{P(c)} \times P(s_k)$$

Gale's classifier is an instance of Naive Bayes classifier. Naive Bayes classifier is widely used in machine learning due to its efficiency and its ability to combine evidence from a large number of features. In our case, the features are words w_j that occur in the window defining the context of ambiguous word w. The Naive Bayes assumption is that the features are independent of one another. With this assumption, we can estimate the conditional probability of the context as the product of the probabilities of its individual features given that sense. Thus, we get the following approximation:

$$P(c/s_k) = \prod_{w_j \, \text{inc}} P(w_j/s_k)$$

The Naive Bayes assumption has two consequences.

First, the structure and order of the word within the context is ignored. That is, we treat context as a bag of words.

Second, the presence of one word in the context is independent of the presence of another.

This is clearly not true. For example, *vote* is more likely to occur in a context that contains *election* than in a context that contains *cricket*. The Naive Bayes assumption is therefore inappropriate if there are strong conditional dependencies between attributes. But there is a surprisingly large number of cases in which the assumption works well. With the Naive Bayes assumption, the decision rule for classification becomes

$$s' = \arg\max_{s_k} \frac{\prod\limits_{w_j \text{ inc}} P(w_j/s_k)}{P(c)} \times P(s_k)$$

As the denominator $P(c)$ in the preceding equation is the same for all senses, it does not affect the final ranking of senses. Dropping it, we get

$$s' = \arg\max_{s_k} \prod_{w_j \text{ inc}} P(w_j/s_k) \times P(s_k)$$

We use the log of probabilities to simplify computation. Thus, the sense s' assigned to w is given by the expression

$$s' = \arg\max_{s_k} \left[\log P(s_k) + \sum_{w_j \text{ inc}} \log P(w_j/s_k) \right]$$

$P(s_k)$, the prior probability of sense s_k can be computed as the proportion of the sense s_k in the tagged training corpus.

$$P(s_k) = \frac{C(s_k)}{C(w)}$$

Similarly, $P(w_j/s_k)$ can be calculated as

$$P(w_j/s_k) = \frac{C(w_j, s_k)}{\sum_i C(w_i, s_k)}$$

where $C(s_k)$ is number of occurrences of s_k in the training corpus, $C(w)$ is number of occurrences of word w in the training corpus, and $C(w_i, s_k)$ is number of occurrences of word w_i in a context of sense s_k in the training corpus.

With this discussion, we can state precisely, the task to be performed during training and testing phases of supervised word sense disambiguation. Figure 5.12 summarizes these steps.

Training   Training a Naive Bayes classifier corresponds to collecting statistics of an individual feature with respect to each sense of the target word in a sense-tagged training corpus. In particular, the statistics we need to collect is the prior probability of each sense of a target word and the probability of individual features given that sense. As discussed earlier, these statistics can be calculated by counting the number of occurrences of senses and words in the training corpus.

To ensure that non-observed events are not assumed to have a probability of zero, some smoothing is required.

Testing During testing, we use the classifier to assign the correct sense to unknown instances. The most likely sense is used for this purpose as identified by the last equation in Figure 5.12.

Supervised Word Sense Disambiguation:

Training

Calculate $P(s_k)$ and $P(w_j/s_k)$:

$$P(s_k) = \frac{C(s_k)}{C(w)} \qquad P(w_j/s_k) = \frac{C(w_j, s_k)}{\sum_i C(w_i, s_k)}$$

Testing

Calculate the appropriate sense for a new context:

$$s' = \arg\max_{s_k} \left[\log P(s_k) + \sum_{w_j \text{ inc}} \log P(w_j/s_k) \right]$$

Figure 5.12 Steps in supervised disambiguation

***k*-Nearest Neighbour or Memory-based Learning** An alternative to Bayes classifier is the *k*-nearest neighbour (*k*-NN) learning. A *k*-NN model memorizes all the contexts in the training set by their associated features. This corresponds to creating a vector representation for each context c. During testing, a new context is presented to the classifier. The classifier first selects k contexts in the training set that are closest to the new context. A score is then assigned to each of the *k*-NN context and the sense of the highest scoring context is assigned to the new context. Figure 5.13 depicts the steps in this algorithm.

Training: Calculate context vector ($\vec{c}$) for each context c in training set

Testing: Given input context: $\vec{c}_{new}$

1. Calculate set A of k nearest neighbour context of $\vec{c}_{new}$:

   $$A = \{ \vec{c} \, | \text{sim} \, (\vec{c}_{new}, \vec{c}) | \text{ is maxim}, |A| = k \}$$

2. $\text{score}(c_{new}, s_j) = \sum_{ci \in A} (\text{sim}(\vec{c}_{new}, \vec{c}_i) \times \delta_{ij})$

   where $\delta_{ij} = 1$ if $\vec{c}_i$ has the sense s_j and 0 otherwise.

3. Finally,

   $$s' = \arg\max_j \text{score}(c_{new}, s_j)$$

Figure 5.13 Steps in *k*-NN based learning for disambiguation

Several researchers have used supervised learning from sense tagged corpora. Gale et al. (1992b) reported 90% disambiguation accuracy for six ambiguous nouns in the Hansard corpus: drug, duty, land, language, position, and sentence.

Mooney (1996) reported that a Naive Bayes classifier and a neural network achieved the highest performance, both achieving around 73% disambiguation accuracy in assigning one of six senses to a corpus of example of word *line*.

There are two major obstacles that impede the acquisition of knowledge from large corpora:

1. Difficulty in manually sense tagging a training corpus
2. Data sparseness

The problem of data sparseness which is quite common in corpus-based work is especially severe for WSD. An enormous amount of text is required to capture all senses of an ambiguous word. Even the few available large training corpuses may fail to do so. For example, the Brown Corpus contains only eight occurrences of the relatively common word *ash*, out of which only one occurrence corresponds to its *tree* sense. Similarly, the sense *remains of a cremated body* of *ash,* although quite common, does not appear in LDOCE.

Further, many possible co-occurrences for a given ambiguous word are unlikely to be found in even a very large corpus.

As discussed earlier, smoothing can be used to get around the problem of infrequently occurring events and, in particular, to ensure that non-obscured events are not assigned a probability of zero.

The bootstrapping (Hearst 1991) approach and use of bilingual corpora has been suggested to avoid manual sense tagging of large training corpus and to handle data sparseness problem.

Bootstrapping   The Bayes classifier attempts to combine evidence from all words in the context window to help disambiguation. This requires a large sense tagged training set to collect evidences. Hearst (1991) proposed the bootstrapping approach to eliminate the need for a large training set. The bootstrapping method relies on a relatively small number of instances labelled with senses having a high degree of confidence. This could be accomplished by manually tagging those instances of an ambiguous word for which the sense is clear (Hearst 1991). These labelled instances are used as seeds to train an initial classifier. The classifier is then used to extract more training instances from the remaining untagged corpus. As the process is repeated, the training corpus grows and the number of untagged instances are reduced. The iteration continues until the remaining untagged corpus is empty or no new instance can be annotated.

For example, in the word *bass*, we can begin with *fish* as a reasonable sense for sense *bass*[1], as presented in the WordNet, and *play* as a reasonable sense for *bass*[2] (bass in music). A small number of instances can be labelled with the sense 1 and sense 2. These labelled instances are then used to extract a larger set of labelled instances.

Tatar and Serban (2001) suggest an algorithm which combines bootstrapping with elements of Naive Bayes algorithm. Hearst indicates that an initial set of at least 10 occurrences is needed for the procedure and that 20 or 30 occurrences are necessary for high precision.

Bilingual Corpora   A bilingual corpus consists of two corpora, one of which is a translation of the other. As different senses of an ambiguous word often translate differently in another language, a bilingual corpus can be used for disambiguating word senses. For example, the Hindi word कलम is translated as *pen* in the writing sense and *graft* in the transplant sense. Gale et al. (1992b, 1993) use the bilingual Hansard corpus to avoid manual sense tagging of a corpus. The Handsard corpus consists of transcriptions in French and English of the proceedings of the Canadian parliament. They first automatically aligned the bilingual corpus and then tagged the words of the aligned corpus using the basic assumptions that translations of a word reflect the senses of that word.

However, this method has some limitations, since many of the ambiguities are preserved in the target language (for example, the French word *sours* and English word *mouse*). Further, the available large-scale parallel corpora are domain specific, e.g., the Hansard corpus of Canadian parliamentary debates. Such corpora usually contain many ambiguous words with skewed distributions, i.e., one sense is used in most occurrences.

Dagan et al. (1991) and Dagan and Itai (1994) propose the use of two monolingual corpora and a bilingual dictionary. Monolingual corpora from diverse genre and domain are much easier to obtain than parallel corpora. The disambiguation method used was as follows.

Given a source language sentence containing ambiguous words, the disambiguator generates every combination of translated words. For each combination, it then examines the target language corpus counting how often the combination occurs in the corpus. The most frequent combination is assumed as the correct translation and the senses corresponding to this translation are selected.

Dagan et al. (1991) reported a disambiguation accuracy of 75% and 92% for German/English and Hebrew/English respectively on 105 (73 Hebrew, 32 German) words.

Unsupervised Methods of WSD

Unsupervised methods of WSD eliminate the need for sense tagged training data. Instead, these approaches take feature-value representations of unlabelled contexts (instances) and group them into clusters. Each cluster can be assumed to represent one sense of an ambiguous word. These clusters can be represented as the average of their constituent feature vectors. Unknown instances are classified as having the sense of the cluster to which they are closest according to the similarity measure.

Strictly speaking, using a completely unsupervised sense disambiguation task, we can only discriminate word senses. That is, we can group together instances of a word used in different senses without knowing what those senses are. However, Yarowsky (1995) proposed an unsupervised algorithm that can accurately disambiguate word senses in a large completely untagged corpus. He exploited two powerful properties of human language in an iterative bootstrapping setup to avoid the need of manually tagged training data (adapted from Yarowsky 1995):

1. One sense per discourse: The sense of a target word is highly consistent within any given document or discourse.
2. One sense per collocation: Nearby words provide strong and consistent clues to the sense of a target word, conditional on relative distance, order and syntactic relationship.

We now briefly discuss steps in Yarowsky's algorithm.

1. Identify all examples of the given polysemous words and store a certain width of their context in a table. This table forms an untrained training set.
2. For each possible sense, identify some training examples representative of that sense. This is accomplished by identifying a number of collocations that are reliable indicators of that sense. We identify these collocations, termed *seed collocations*, either manually or with the help of a dictionary. These collocations are then used to classify the context. The labelled contexts are called *seed set*. The remainder of the examples constitute an untagged residual.
3a. Next, supervised disambiguation algorithm on seed set identifies collocations within the user-specified window that reliably partition the seed training data. These collocations are ranked by the purity of their distribution, which is specified by the log-likelihood ratio. The

log-likelihood ratio between a sense a pair of senses and collocation$_i$ is defined as

$$\log \frac{P(\text{sense}_A/\text{collocation}_i)}{P(\text{sense}_B/\text{collocation}_i)}$$

Consider only those collocations that have a likelihood ratio higher than a set threshold. The list of collocations identified in this step is called a *decision list.*

3b. Apply the resulting classifier to the entire sample set. Identify those members in the residual that are tagged as sense, with probability values above a certain threshold, and add those examples to the growing seed sets. These additions will contain new collocations that are reliable indicators of the previously tagged seed set. This process is iterated to train the classifier and reapplied on the previously tagged training set to get improvement.

3c. The one-sense-per-discourse constraint can be applied to either tag new contexts or to correct wrongly tagged ones.

3d. Repeat Steps 3a–3c iteratively. The seed set will grow, and the residual, shrink. Stop when the algorithm converges into a stable residual set.

4. The classification procedure, learned from the final supervised training step, may now be applied to new data and used to annotate the original untagged corpus with sense tags and probabilities.

5.5.3 Knowledge Sources in WSD

A variety of information, including syntactic (part-of-speech, grammatical structure), semantic (selection restriction) and pragmatic (topics) information as well as dictionary (definitions), and corpus (collocation) specific information, can be utilized as a knowledge source in WSD. Here is a list of some of the information sources deemed useful in disambiguation.

Context of a word The context of a word can be regarded as the words surrounding the ambiguous word. A word only can be disambiguated in its context. The context is therefore useful in determining the meaning of a word in a particular usage.

Frequency of a sense This information is generally used in statistical approaches to measure the likelihood of each possible sense. Usually, this statistics is gathered over some sense-tagged corpus.

Part-of-speech Part-of-speech information can reduce the number of possible senses a word can have. For example, in WordNet 2.0 *bitter* has

3 senses as noun, 7 senses as adjective, and one sense as a verb. The use of *bitter* as a verb does not lead to ambiguity.

Collocations   These may provide useful information about the sense of a word. For instance, the noun *match* has 9 senses listed in WordNet but only one of these applies to football match.

Selectional preferences   Semantic restrictions that predicates place on their argument can be used for disambiguation. For instance, *eat* in the *have a meal* sense prefers humans as subjects. This knowledge is similar to the argument-head relation, but selectional preferences are given in terms of semantic classes, instead of plain words.

Domain   In a particular domain, only one sense of a word is likely to be used. Thus, information about domain furnishes useful information for disambiguation. For example, in the domain of sports, the *cricket bat* sense of *bat* is preferred.

Besides these, thematic role of a word (subject or object), sentence structure, semantic word properties, and pragmatic information may also be utilized in sense disambiguation. All this information can be used together with general knowledge about the situation to rule out impossible readings.

5.5.4 Applications of WSD

Word sense disambiguation (WSD) is only an intermediate task in NLP, like POS tagging or parsing. Accurate WSD is important for many applications, e.g., machine translation and information retrieval.

One of the first applications of WSD was machine translation, for which, disambiguating the sense of a source language word is crucial for accurately selecting its translation equivalent in the target language. The Hindi word फल, for example, can either have the sense of the English word *fruit*, or the sense of परिणाम (result). In order to correctly translate a text containing फल, we need to know which sense is intended.

Word sense disambiguation is also useful in information retrieval (see Chapters 9 and 10). Here, WSD helps rule out documents that are irrelevant and uses keywords that match the query. For example, a query such as 'easy chairs' would match all documents that contain the words 'easy' and 'chair'. Some documents would contain information about furniture while others would contain information about people. In principle, sense-tagged documents could help rule out the unwanted subset of documents. Sense-tagged documents could also improve information retrieval because different documents use different words to refer to similar contents.

Some evidence that WSD improves this task has been offered by Schütze and Pedersen (1995), and Jing and Tzoukermann (1999). However, there is disagreement over the usefulness of WSD for information retrieval.

Parsing is another task where lexical semantics is helpful. Consider the sentence, *I baked the chicken in the oven.* The semantic knowledge about *bake* helps parser correctly attach the preposition and eliminate erroneous parses. However, little empirical research has been carried out on the usefulness of word sense disambiguation for parsing.

5.5.5   WSD Evaluation

Evaluation is important in all NLP tasks. It has always been a problem in disambiguation research, as the only way to judge the performance of a disambiguator is to manually check its output. Manual checking is time consuming and because of this, most disambiguators have been evaluated only on a small number of words. The SENSEVAL initiatives have simplified the evaluation task. The basic metric used for evaluating word sense disambiguation algorithm is precision and recall. Precision measures the fraction of correctly tagged instances in the total set. This requires access to an annotated corpus. Two such corpuses are now available: the SEMCOR (Landes et al. 1998) corpus and SENSEVAL (Kilgariff and Rosenzweig 2000) corpus. These metrics fail to give any credit to an algorithm that makes only broad distinctions between senses, as they consider sense match to be exact. Some metrics have been proposed to give partial credit to instances where a broader sense is selected.

Yet another completely automatic method for evaluating disambiguators is the pseudo word method. In this technique, ambiguity is artificially introduced in the corpora by taking all occurrences of two words and replacing them with artificial or pseudo words, for example *people* and *language* by a new word *peolang*. These artificial words are then considered ambiguous, which can be disambiguated to any of the original words. However, the ambiguity introduced does not match the ambiguity found in real situations.

SUMMARY

This chapter explored important concepts related to semantic processing.
- Semantic analysis is concerned with meaning representation of linguistic inputs.
- A meaning representation bridges the gap between linguistic and commonsense knowledge.

- A meaning representation language must be verifiable and unambiguous. It should support the use of variables and inferencing and must be expressive enough to handle the wide variety of content found in natural languages.
- Predicate argument is a fundamental structure used to convey meaning in human languages. A meaning representation language should support the representation of this structure.
- The principle of compositionality says that meaning of a sentence can be built using meaning of its parts.
- Syntax-driven semantic analysis uses the syntactic constituents of a sentence to build its meaning representation. Semantic attachments are used to specify how the meaning of a construction is composed from meaning of its constituents.
- Semantic grammars provide an alternative way for creating meaning representation.
- Lexical semantics is an important part of semantic analysis. It deals with meaning of words, internal structure of words, and relationships between words.
- A word can have multiple senses; only one of these applies in a particular use. Identifying the correct sense is important in the semantic analysis of a sentence.
- Word sense disambiguation is concerned with identifying the correct sense of a word.
- Selectional restrictions can be used to disambiguate word senses.
- The context of a word provides useful information about word sense. Context-based disambiguation algorithms can be broadly classified into knowledge-based and corpus-based approaches.
- Corpus-based approaches use either supervised or unsupervised learning. Supervised methods require tagged data whereas unsupervised methods eliminate the need of tagged data but usually perform only word sense discrimination.
- The knowledge sources used by word sense disambiguation algorithms include context of word, sense frequency, selectional preferences, collocation, and domain.

REFERENCES

Agirre, E. and G. Rigau, 1996, 'Word sense disambiguation using conceptual density,' *Proceedings of COLING-96*, Copenhagen, pp. 16–22.

Cowie, J., L. Guthrie, and J. Guthrie, 1992, 'Lexical disambiguation using simulated annealing,' *Proceedings of the 14th International Conference on Computational Linguistics (COLING-92),* Nantes, France, pp. 359–65.

Dagan, I'do, Alan Itai, and U. Schwall 1991, 'Two languages are more informative than one,' *Proceedings of the ACL,* 29, California, USA, pp.130–37.

Dagan, Ido and Alon Itai, 1994, 'Word sense disambiguation using a second language monolingual corpus,' *Computational Linguistics,* 20(4), MIT Press, Cambridge, MA, pp. 563–96.

Fillmore, C. J., 1968, 'The case for case,' *Universals in Linguistic Theory,* E. W. Bach and R.T. Harms (Eds.), Holt, Rinehart and Winson, New York, pp. 1–88.

Gale, William A., Kenneth W. Church, and David Yarowsky, 1992a, 'Estimating upper and lower bounds on the performance of word-sense disambiguation programs,' *Proceedings of the ACL,* 30, pp. 249–56.

_____1992b, 'Using bilingual materials to develop word sense disambiguation methods,' *Proceedings of the International Conference on Theoretical and Methodological Issues in Machine Translation,* pp. 101–12.

_____1993, 'A method for disambiguating word senses in a large corpus,' *Computers and the Humanities,* 26, pp. 415–39.

Gruber, J.S., 1965, 'Stuides in lexical relations,' *PhD Thesis,* MIT, Cambridge, MA.

Guthrie J., L. Guthrie, Y. Wilks, and H. Aidinejad, 1991, 'Subject-dependent co-occurrence and word sense disambiguation,' *ACL-91,* pp. 146–52.

Hearst, M.A., 1991, 'Noun homograph disambiguation using local context in large text corpora,' *Proceedings of the 7th Conference of the UW Centre for the New OED and Text Research Using Corpora.*

Jing, H. and E. Tzoukermann, 1999, 'Information retrieval based on context distance and morphology,' *Proceedings of the 22nd Annual international ACM SIGIR Conference on Research and Development in Information Retrieval (SIGIR -99),* Seattle, WA, pp. 90–96.

Joshi, A.K., 1987, 'An introduction to tree adjoining grammars,' *Mathematics of Language,* A. Manaster-Ramer (Ed.), Amsterdam/Philadelphia, John Benjamins, pp. 87–114.

Kaplan, Abraham, 1950, 'An experimental study of ambiguity and context,' Mimeographed, 1950. [Published as: Kaplan, Abraham, 1955, 'An experimental study of ambiguity and context,' *Mechanical Translation,* 2(2), November, pp. 39–46.

Kearns, Kate, 2000, *Semantics,* St. Martin's Press, New York.

Kamsties, E., 2001, 'Surfacing ambiguity in natural language requirements,' *PhD Thesis,* Fraunhofer-Institue für Experimentelles Software Engineering, Kaiserslautern, Germany.

Keenan, E.L. and L.M. Faltz, 1985, *Boolean Semantics for Natural Language,* D. Reidel Publ. Comp.

Kilgarriff, Adam and Joseph Rosenzweig, 2000, 'Framework and results for English SENSEVAL,' *Computers and the Humanities,* 34(1–2), pp. 15–48.

Landes, S., C. Leacock, and R.I. Tengi, 1998 'Building semantic concordances,' *WordNet: An electronic lexical database,* C. Fellbaum (Ed.), MIT Press, Cambridge, MA, pp. 199–216.

Procter, P., 1978, *Longman Dictionary of Contemporary English,* Longman Group Essex, England.

Lesk, Michael, 1986, 'Automated sense disambiguation using machine-readable dictionaries: how to tell a pine cone from an ice cream cone,' *Proceedings of the 1986 SIGDOC Conference,* Toronto, Canada, pp. 24–26.

Luk, A.K., 1995, 'Statistical sense disambiguation with relatively small corpora using dictionary definitions,' *Proceedings of the 33rd Annual Meeting of the Association for Computational Linguistics,* Somerset, NJ, pp. 181–88.

Ng, Hwee Tou and John Zelle, 1997, 'Corpus-based approaches to semantic interpretation in natural language processing,' *American Association for Artificial Intelligence,* pp. 45–64.

Resnik, P., 1995, 'Using information content to evaluate semantic similarity in a taxonomy,' *Proceedings of IJCAI.*

———1996, 'Selectional constraints: An information-theoretic model and its computational realization', *Cognition,* 61, pp. 127–59.

———1997, 'Selectional preference and sense disambiguation,' *Proceedings of the ACL SIGLEX Workshop on Tagging Text with Lexical Semantics:Why, What, and How?,* Association for Computational Linguistics, Somerset, NJ, pp. 52–57.

Roget, P.M., 1946, *Roget's International Thesarus,* Thomas Y Crowell, New York.

Schütze, Hinrich, 1998, 'Automatic word sense discrimination,' *Computational Linguistics,* 24(1).

Schütze, Hinrich and Pedersen, 1995, 'Information retrieval based on word senses,' *Proceedings of SDAIR-95,* Las Vegas, Nevada.

Tatar, D. and G. Serban, 2001, 'A new algorithm for WSD,' *Studia Univ._Babes-Bolyai, Informatica,* 2, pp. 99–108.

Voorhees, E.M., 1993, 'Using WordNet to disambiguate word senses for text retrieval,' *Proceedings of the 16th Annual International ACM SIGIR Conference on Research and Development in Information Retrieval,* Pittsburgh, Pennsylvania, pp.171–80.

Walker, Donald, 1987, 'Knowledge resource tools for accessing large text files,' *Machine Translation: Theoretical and Methodological Issues,* Sergei Nirenberg (Ed.), Cambridge University Press, Cambridge, England.

Weaver, Warren, 1949, *Translation,* Mimeographed, Reprinted in Locke, William N. and Booth, A. Donald, 1955, (Eds.), *Machine Translation of Languages,* John Wiley & Sons, New York, pp. 15–23.

Wilks, Y., D. Fass, C. Guo, J. E. McDonald, T. Plate, and B.M. Slator, 1990, 'Providing machine-tractable dictionary tools,' *Machine Translation,* 5(2), pp. 99–154.

Yarowsky, D., 1992, 'Word-sense disambiguation using statistical models of Roget's categories trained on large corpora,' *Proceedings of the 14th International Conference on Computational Linguistics (COLING-92),* Nantes, France, pp. 454–60.

———1995, 'Unsupervised word sense disambiguation rivaling supervised methods,' *Proceedings of the 33rd Annual Meeting of the Association for Computational Linguistics,* Cambridge, MA, pp. 189–96.

EXERCISES

1. Give two examples each of cases where the principle of compositionality does not hold and where it holds only partially.
2. Give the semantic attachment for any two of the evidences identified in Question 1.
3. What is the use of the lambda expression in syntax-drive semantic analysis?
4. What are the advantages and disadvantages of using semantic grammar?
5. For each of the following verbs give all the selectional restrictions possible: think, eat, bark, fly.
6. Evaluate the one-sense-per-discourse constraint on a corpus. Find sections with multiple uses of an ambiguous word and determine how many of them refer to a different meaning.
7. Find out the number of senses of 'still', 'bat', and 'cricket' in WordNet.

8. Create a small corpus from a newspaper. Consider articles from specific domains (e.g., sports, whether, finance, politics, etc.). Using WordNet, determine how many senses there are for each of the open class words in each sentence.

9. Tag correctly the ambiguous words found in Question 8. Which word has the maximum number of tags?

10. Consider the verbs appearing in the corpus prepared in Question 8 and give selection restrictions that can be imposed on their argument. Also mention the cases that can be disambiguated correctly using the selectional restriction information.

LAB EXERCISES

1. Install WordNet 2.1 on your machine. Find out how WordNet is accessed using command line.

2. WordNet can also be accessed from Java using an API. Find and install an API. Now write a small program in Java to demonstrate how WordNet is made to get lexical information.

3. Write a program to extract a five-word window for a target word in the corpus created in Question 8.

4. Identify 10 co-occurring words in the definition of any ambiguous word of your choice in the corpus created in Question 8. Use these words as features. Write a program to create context vector of a target vector. Consider a window size of 10 words with target word in the middle.

5. Using a dictionary, simulate Lesk algorithm.

6. Write a Java or Perl program that
   (i) Extracts the word from a line
   (ii) Finds the senses of the word in WordNet
   (iii) Prints out the most frequent sense (the first)

CHAPTER 6

DISCOURSE PROCESSING

CHAPTER OVERVIEW

This chapter is concerned with discourse and world knowledge. A discourse comprises a sequence of sentences that must be interpreted with respect to the context. World knowledge is needed to find the connection between two sentences, to resolve ambiguities, and to infer new information to get a coherent view of the information being communicated through spoken or written text. Discourse and world knowledge is especially important for interpreting pronouns and the temporal aspects of the information conveyed, resolving ambiguities, and understanding metaphors and ellipses. The chapter introduces these concepts and discusses anaphor resolution methods. There are various patterns of discourse. Identifying the discourse relation between two or more sentences requires knowledge of how the discourse is structured. We, therefore, present a theoretical framework for discourse analysis.

6.1 INTRODUCTION

A discourse is most commonly described as the language above the sentence level or as 'language in use'. An arbitrary collection of sentences does not make sense to us. In order to make sense, a text must consists of sentences that are related to each other. This means that there exists a structure above the sentence that is needed for interpretation of text. This structure is known as the discourse structure, and the collection of interrelated sentences is a discourse.

However, Widdowson (1995) argued that a discourse unit can be smaller than a sentence. For instance, we easily interpret the word 'Ladies' written beside a seat on a bus. We do not use the dictionary definition of ladies to understand this. Instead, our interpretation is based on the context; it uses real world knowledge, which is not there in the text itself. Discourse analysis thus makes use of contextual knowledge.

Discourse analysis deals with the intended meaning of textual units. The following text illustrates this idea:

Excuse me. You are standing on my foot.

This sentence is not just a plain assertion; it is a request to someone to get off your foot. Again, the intended meaning is not present in the text itself. However, we understand the meaning because that is the way language is used generally.

The preceding discussion makes it clear what is meant by 'language above the sentence level' or 'language in use'. Discourse analysis involves the study of the relationship between language and contextual background. Contextual knowledge that is needed to interpret a sentence includes situational context, background knowledge, and co-textual context. Situational context is the knowledge about physical situations existing in the surroundings at the time of utterance. Background knowledge includes cultural knowledge and interpersonal knowledge. Co-textual context is the knowledge of what has been said earlier.

There are many types of discourse, including written, spoken, and signed discourse as well as monologue and dialogue. In this chapter, we focus on the monologue type of discourse. A monologue involves a speaker (writer) and a hearer (reader). The communication is unidirectional from speaker to hearer (whereas in a dialogue, the role of the participant alternates periodically). Many of the discourse problems are shared by the various types of discourse. However, they need different techniques to process them. Throughout this chapter, we use the term sentence and utterance interchangeably.

The phenomena that operate at discourse level include cohesion and coherence. Cohesion is a textual phenomenon, whereas coherence is a mental phenomenon. A text is cohesive if its elements link together, and coherent if it makes sense. Cohesion studies how words are linked to each other in the text. This linking can be forwards or backwards and meshes the text together. Language make use of cohesive devices like references, ellipsis, repetitions, and conjunctions to achieve this linking. We discuss these cohesive devices in Section 6.2. Resolving pronominal references is essential in applications like information extraction and text summarization. An information extraction system often needs to fill slots that correspond to named entities. In order to make this possible, pronouns referring to those named entities need to be identified first. Similarly, if a sentence in a text uses a pronoun that refers to an entity in a previous sentence, which is not included in the text, that sentence will be unreadable.

How we resolve these references is discussed in Section 6.3. In this chapter, we discuss these phenomena and the methods used to resolve them. What we speak or write appears unified. Coherence refers to this property of being meaningful and unified. We elaborate on this concept, explain various coherence relations, and discuss Hobbs's (1985) abductive framework for determining local coherence, in Section 6.4.

6.2 COHESION

Cohesion bounds text together. Consider the following piece of text:

> Yesterday, my friend invited me to her house. When I reached, my friend was preparing coffee. Her father was cleaning dishes. Her mother was busy writing a book.

Each occurrence of *her* in the preceding refers to the noun phrase *my friend's*. Here is the same text with *my friend's* substituted at each place where *her* appears.

> Yesterday, my friend invited me to my friend's house. When I reached, my friend was preparing coffee. My friend's father was cleaning dishes. My friend's mother was busy writing a book.

To most of us, this repetition is undesirable. This type of over-specification is avoided in the earlier text through the use of *her*. We say that *her* is cohesive with *my friend's*. This is just a simple example of cohesion with pronominal references. Any communication in a speaker-hearer environment assumes the presence of some shared knowledge. The encoding of this message by the speaker and its subsequent interpretation take place within the context of this shared knowledge. The speaker avoids encoding information that is obvious or known to the hearer.

Pronominal reference is just one type of reference. There are many others including ellipses, which we discuss next.

6.2.1 Reference

Reference is a means to link a referring expression to another referring expression in the surrounding text, as in the following example.

> Suha bought a printer. It cost her Rs. 20,000.   (6.1)

Here, 'her' refers to a person named 'Suha' and 'it' represents an entity named 'printer'. Both 'it' and 'her' in sentence (6.1) refer to entities that have been previously introduced into the discourse. Such a reference is called an anaphoric reference. There are primarily five types of references: indefinite, definite, pronominal, demonstrative, and ordinal.

Indefinite Reference

An indefinite reference introduces a new object to the discourse context. The most commonly used form of indefinite reference involves the use of the determiner 'a' (or 'an') as in the following sentence.

> I bought *a printer* today. (6.2)

Other markers of indefinite reference are the quantifier 'some' [sentence (6.3a)] and the determiner 'this' as in sentence (6.3b).

> *Some* printers make noise while printing. (6.3a)
>
> I met *this* girl earlier in a conference. (6.3b)

These references are also termed non-anaphoric references.

Definite Reference

A definite reference refers to an object that already exists in the discourse context, as illustrated in the following example.

> I bought a printer today. The printer didn't work properly. (6.4)

The first sentence in this example introduces a new entity, which is referred to by the definite reference in the second sentence.

In most cases, the determiner structure of the noun phrase helps make the distinction between definite and indefinite referents.

Pronominal Reference

Pronominal references are the references that use a pronoun to refer to some entity. Sentence (6.5) illustrates this type of reference.

> I bought a printer today. On installation, it didn't work properly.
> (6.5a)
>
> Zuha forgets her pen drive in lab. (6.5b)

In (6.5a), 'it' refers to printer. In (6.5b), 'her' refers to Zuha.

However, the use of 'it' as a referent should be clearly distinguished from its pleonastic use, as in "It is raining" or in "It is okay", where 'it' refers to a state and not to an entity. Denber (1998) and Mitkov (1999) classify the use of 'it' as pleonastic in sentences like "It seems more likely that…" or "It is obvious that…". Though it appears that the phrase following 'that' can be regarded as a referent in these cases.

Pronominal referents put stricter constraints on the distance between the entities being referred to and their introduction. Usually, a pronominal reference refers to an object introduced in the previous one or two sentences, whereas definite noun phrases can refer to objects further back.

A pronominal reference can refer to an entity before it is actually introduced in the discourse context, as in the following example.

> Having installed it, I found that the printer was not
> working properly. (6.6)

This type of pronominal reference is called cataphoric reference.

A pronominal reference can be a part of a quantified context as in sentence (6.7).

> All students should sign their project report. (6.7)

A reference need not always refer to a noun phrase. Sentence 6.8 illustrates a situation in which the pronominal referent 'it' refers to the event introduced in the previous sentence.

> The printer I bought today doesn't work properly.
> *It* surprised me. (6.8)

Here, 'it' refers to the event of printer not working properly.

Demonstrative Reference

> I bought a printer today. I had bought one earlier in 2004. *This*
> one cost me Rs. 6,000 whereas that one cost me Rs. 12,000.
>
> (6.9)

Here, 'this' refers to the printer I bought today and 'that' refers to the printer I bought in 2004.

Quantifier or Ordinal Reference

An ordinal reference uses an ordinal, such as 'first', 'one', etc. Some authors call these referents as one-anaphora.

> I visited a computer shop to buy a printer. I have seen many
> and now I have to select *one*. (6.10)

Here, 'one' refers to a printer I have seen.

Thus, it introduces a new entity into the discourse context. The use of 'one' in sentence (6.10) is different from the formal, non-specific use, as in the sentence, 'One should be confident while facing interviews', and its use as the numeral one, as in, 'She got her trousers shortened by one inch'.

Inferables

Inferable referents refer to entities that can be inferred from other entities explicitly evoked in the text, as illustrated in sentence (6.11).

> I bought a printer today. On opening the package, I found
> *the paper tray* broken. (6.11)

Here, the paper tray does not introduce a new object in the discourse context; instead it refers to the paper tray of the printer introduced in first sentence.

Generic Reference

A generic reference refers to a whole class instead of an individual or specific entity.

> I saw two laser printers in a shop. *They* were the fastest printers available. (6.12)

Here, 'they' refers to laser printers in general and not to ones of those I saw.

6.2.2   Ellipsis

Ellipsis is a form of grammatical cohesion. It refers to the phenomenon where a part of a sentence (utterance) is omitted or left unpronounced. The reader uses the surrounding text to recover the omitted text. Ellipsis, like references, are cohesive devices that avoid repetition. Consider the following sentences:

> Do you take fish?
> Yes I do.

'Yes I do' is an example of an ellipsis in which the verb phrase has been deleted. Instead of saying 'Yes I take fish', the speaker omits the text 'take fish' because it is not necessary. The reader (hearer) retrieves the missing text from the previous sentence. Here is another example:

> I know that lady. Do you?

Here 'Do you' is an ellipsis, which stands for 'Do you know that lady'. Here, the verb phrase 'know that lady' is left out. The elided verb phrase is understood from the previous sentence.

The clause containing the elided verb phrase is called the target clause, and the clause from which the verb phrase is understood (taken) is called the source clause. The presence of references in the source clause may lead to ambiguities in target clause, as in the following example:

> Seema loves her mother and Suha does too.

This sentence may mean that Suha likes Seema's mother or that Suha likes her own mother. The first reading is often called a *strict* reading whereas the second reading is called a the *sloppy* reading. In a strict reading, the pronoun in the target clause refers to the same entity as the pronoun in the source clause.

6.2.3 Lexical Cohesion

References and ellipsis are forms of grammatical cohesion that avoid repetition of clauses. There are a number of other lexical phenomena that languages use to achieve cohesion, which come under the category of lexical cohesion. Unlike grammatical cohesion, lexical cohesion exploits repetition, either explicitly or implicitly, to introduce stylistic effect. Lexical cohesion devices include repetition, synonymy, and hypernymy. The use of repetition for style is illustrated in the following examples:

Ba ba black sheep,
Have you any wool?
Yes sir, yes sir,
Three bags full.

Instead of repeating the same word, synonymy uses a word that means the same in the given context; hypernymy uses a super-ordinate.

6.3 REFERENCE RESOLUTION

We now discuss the factors and methods that help resolve various types of referents.

6.3.1 Constraints and Preferences

Constraints that rule out certain referents need to be checked. This includes the constraints of number, gender, and case agreement. Apart from these strict constraints, there are a number of preferences and semantic constraints that can be used to identify a preferred referent. Let us have a look on these constraints and preferences.

Person Agreement

The referent and the referring expressions must agree in person. Consider the following examples:

Zuha and I bought a camera. *We* like capturing nature scenes.

(We = I and Zuha)   (6.13a)

Zuha and Prabha bought a camera. *We* like capturing nature scenes.

(6.13b)

Resolving 'We' into 'Zuha and Prabha' is incorrect in (6.13b).

Case Agreement

The position where a pronoun is used constraints its form. For example, in the object position, we use the accusative case of a pronoun (e.g., him, her, them), whereas in the subject position, we use the nominative form of a pronoun.

Gender Agreement

English distinguishes between male, female, and non-personal genders in the use of third person pronouns. These constraints need to be checked when resolving pronominal references.

> Zuha bought a printer. She is printing now.
>
> $\qquad$ (She = Zuha, not the printer) $\qquad\qquad$ (6.14a)
>
> Zuha bought a printer. It is printing now.
>
> $\qquad$ (It = the printer, not Zuha) $\qquad\qquad$ (6.14b)

Selectional Restrictions

Selectional restriction, placed by verbs on their arguments, can also help in resolving references.

> Zuha put an apple on the table. Suha is eating *it*.

The pronoun 'it' has two possible referents: apple and table. However, the verb eat requires some edible entity as its object. As the table is not edible, the correct referent is the apple.

As discussed in Chapter 5, violations to these restrictions are sometimes acceptable. That is why we usually call them preferences rather than restrictions. Quite often, these violations refer to metaphoric uses, as in the following sentence.

> A range of shopping malls are opening in the city. The local vendors fear that *they* will swallow their livelihood.

Here, 'they' refers to shopping malls. Shopping malls are not something that can swallow, but its use is not semantically odd to us and is quite acceptable. Resolving these types of references requires semantic knowledge.

Recently Introduced References

While resolving references, the entities introduced more recently are considered of greater importance than those introduced further back in the text.

Grammatical Role

Grammatical roles played by an entity in a sentence, provide useful clues on their salience. For example, an entity in the subject position can be considered more important than one in the object position.

Parallelism

The structural parallelism that exists in a sentence can be used to resolve references, as in the following example:

Zuha went with Suha to the computer shop. Danish went with her to a computer institute.

The parallelism effect suggests that 'her' refers to Suha and not to Zuha.

Repeated Mention

This refers to the idea that entities that are focussed on in prior discourses are more likely to continue to be focussed on in subsequent discourses. Hence, it is more likely that they are referred to by the pronoun, as in the following excerpts from *SPAN* magazine:

> *Lucid was the first among the six women to join the astronaut program. A veteran of five space flights, logging 223 days in space, Lucid holds the international record for the most flight hours in orbit by any American, and any women in the world. She spent 180 days on the Russian space station Mir in 1996.*
>
> *In 1998 **she** wrote in* The Scientific American *that **she** viewed the Mir mission as the perfect opportunity to combine two of her passions: flying airplanes and working in laboratories.*

Here, the pronoun 'she' in the second paragraph refers to Lucid, who is the focus in the prior discourse.

Intra-sentential Syntactic Constraints

The reflexive use of pronouns is constrained by the syntactic relationship between the referential expression and the antecedent noun phrase. Usually, it co-refers with the subject of the innermost clause that includes it. The following examples clarify this constraint:

Preghna bought herself a laptop. [herself = Preghna]   (6.15a)

Preghna bought her a laptop. [her ≠ Preghna]   (6.15b)

6.3.2 Reference Resolution Algorithms

An algorithm that uses all these constraints and preferences requires a vast amount of knowledge, which is difficult to achieve. Till date, none of the existing algorithms for pronouns incorporates all of them. However, most use syntactic constraints, either directly or indirectly. In the following lines, we discuss pronoun resolution algorithms introduced by Lappin and Leass (1994), Grosz et al. (1995), Mitkov (1998), and Lappin (2003).

Resolution of Anaphora Procedure

Lappin and Leass (1994) proposed an algorithm that uses some of the constraints and preferences discussed in the previous section, to resolve pronominal anaphora. They called it RAP—Resolution of Anaphora Procedure. It uses a salience value, derived from the syntactic structure, to rank a filtered set of NP candidates. It uses no semantic information. The algorithm uses a set of filters to identify pleonastic pronominal references and to eliminate candidate NPs that (i) do not agree in number, gender, or person, and (ii) violate syntactic co-reference constraints. The salience factors used by Lappin and Leass are listed in Table 6.1 along with their initial weights.

Table 6.1 Salience factors and their weights (Lappin and Leass 1994)

Sentence recency	100
Subject emphasis	80
Head noun emphasis	80
Existential emphasis	70
Accusative emphasis	50
Non-adverbial emphasis	50
Indirect object and oblique complement emphasis	40

The non-adverbial emphasis rewards NPs not occurring in the adverbial PPs. The salience factors considered by RAP, implements the following hierarchy of grammatical roles (adapted from Jurafsky and Martin 2000):

Subject > existential predicate nominal > direct object> indirect object | prepositional object > demarcated adverbial PP

The following are a few examples explaining these grammatical roles:

Suha bought a laser printer.	(subject, head noun)	(6.16a)
There are only *two printers* working properly in the lab.		
	(existential predicate)	(6.16b)
The engineer repaired the *printer*.	(object)	(6.16c)
Suha got her *printer* repaired today.	(indirect object)	(6.16d)
Suha placed the printer on the *table*.	(prepositional object)	(6.16e)
In *her new printer*, a scratch was found.		
	(demarcated adverbial NP)	(6.16f)

In sentence (6.16a), Suha receives 80 points for subject and 80 points for being denoted as a head noun. This is also true for Suha in sentences (6.16d) and (6.16e).

The salience value assigned to a candidate referent is the sum of the weights of salience factors associated with it. The referents in the current sentence are assigned initial weights as mentioned in Table 6.1. A sentence recency weight of 100 is assigned to all discourse referents introduced in the current sentence. Weights assigned by all salience factors to a referent before the current sentence are degraded by a factor of two. The steps in the pronoun resolution algorithm are as follows:

1. Create a list of potential referents.
2. Filter out potential referents that do not agree in number, person, and gender with the pronoun.
3. Remove referents that do not pass intra-sentential syntactic co-reference constraints.
4. Calculate the total salience value for each potential referent. Modify the salience value to account for role parallelism between the pronoun, the referent, and the cataphoric reference. In a cataphoric reference, a referent appears after the pronoun. Such a reference is penalized by assigning a negative weight of −175. Grammatical role parallelism refers to a situation in which the pronoun and the referring entity play the same role. A positive weight of 35 is assigned to such entities.
5. Select the referent with the highest salience value.

We now explain these ideas with the help of an example.

Example 6.1 Consider the following text:

Suha saw a laptop in the shop. She enquired about it.
She bought it. (6.17)

We begin with the first sentence and assign weights to the references, as shown in the following table:

	Recency	Subject	Existential	Object	Ind-obj	Non-adv	Head Noun	Total
Suha	100	80	—	—	—	50	80	310
Laptop	100	—	—	50	—	50	80	280
shop	100	—	—	—	—	50	80	230

The first sentence does not contain any pronominal reference. Now consider the next sentence. As mentioned earlier, we first degrade the salience value of each referent by a factor of two, as shown in the following table:

Referent	Phrases	Salience value
Suha	{Suha}	155
Laptop	{a laptop}	140
shop	{the shop}	115

The second sentence contains two pronouns: 'she' and 'it'. The gender agreement rules out 'laptop' and 'shop' from the list of potential referents for 'she' leaving only 'Suha'. So, the algorithm returns 'Suha' as the referent of 'she'. We now update the salience values by adding 'she' in the equivalence class for Suha and adding its salience score (=310) to Suha. Since 'she' is in the current sentence, it receives a recency score of

100. Other factors that contribute to its score are subject position (=80), not in adverbial position (=50), and head noun (=80), giving a total of 310. The updated values are listed in the following table:

Referent	Phrases	Salience value
Suha	{Suha, she}	465
Laptop	{a laptop}	140
shop	{the shop}	115

The next NP in the current sentence is 'it', which can refer to either 'shop' or 'laptop'. We first update the values by adding 35 to the salience score of 'laptop' to incorporate parallelism preference as both 'it' and 'laptop' are in the object position. As the weight of 'laptop' is more than that of 'shop', we select it as the referent.

The next table shows the updated discourse model after processing the second sentence. The score of 'it' is 100 (recency) + 50 (obj) + 50 (non-adv) + 80 (head-noun) = 280.

Referent	Phrases	Salience value
Suha	{Suha, she}	465
Laptop	{a laptop, it}	455
shop	{the shop}	115

We now move on to the last sentence. Before considering references in this sentence, we first reduce the values by half.

Referent	Phrases	Salience value
Suha	{Suha, she}	232.5
Laptop	{a laptop, it}	227.5
shop	{the shop}	57.5

The first noun phrase in the last sentence is 'she'. Step 2 of the algorithm suggests that the possible referent is Suha. So, we stop here and consider Suha as the referent. The following table shows the updated discourse model after this step. To distinguish 'she' in the current sentence from the one mentioned in first sentence, we represent it as she_1.

Referent	Phrases	Salience value
Suha	{Suha, she, she_1}	597.5
Laptop	{a laptop, it}	227.5
shop	{the shop}	57.5

The next noun phrase in the current sentence is 'it'. The possible referents are 'shop' and 'laptop'. Both 'it' and 'laptop' occur in the object position, so we add 35 for the parallelism preference to 'laptop', yielding a value of 245. As the salience score of laptop is high, 'it' will resolve to 'shop'.

Centering Algorithm

Centering theory (Grosz et al. 1995) uses a discourse model representation consisting of a forward-looking centre (C_f) and a backward-looking centre (C_b). The algorithm presumes the existence of a single entity 'centred' on a given point in the discourse. If U_{n-1} and U_n are two adjacent utterances, then the backward-looking centre of U_n, $C_b(U_n)$, represents the most prominent entity in the discourse after U_n has been interpreted. The forward-looking centre of U_n, $C_f(U_n)$, is an ordered list of entities mentioned in U_n. The centre $C_b(U_n)$, is the highest ranked entity of $C_f(U_{n-1})$. The elements of $C_f(U_n)$ are ordered based on their grammatical role in the utterance. The highest rank is given to a subject followed by the direct object, the indirect object, the oblique, and the adjuncts. This ranking is similar to that of the grammatical role hierarchy used by Lappin and Leass. However, no numerical weight is assigned to the entities on the list. The algorithm computes inter-sentential relationship between the pair of utterances U_{n-1} and U_n, as shown in Table 6.2, to find the preferred referents of the pronoun.

Table 6.2 Inter-sentential relationship

	$C_b(U_n) = C_b(U_{n-1})$ or undefined $C_b(U_{n-1})$	$C_b(U_n) \neq C_b(U_{n-1})$
$C_b(U_n) = C_p(U_n)$	Continuing	Smooth shift
$C_b(U_n) \neq C_p(U_n)$	Retaining	Rough shift

where $C_p(U_n)$ is the highest ranked, forward-looking centre in U_n, called preferred centre. The algorithm uses the following rules:
1. If some elements of $C_f(U_{n-1})$ are realized as the pronoun, then so is $C_b(U_n)$.
2. The transitions between utterances are ordered as follows:

    Continuing > Retaining > Smooth shift > Rough shift

The transition types define the referent assignment. The referent that results in the most preferred relation in Rule 2 is assigned to the pronoun. The algorithm uses the ordering of the element in the previous C_f list to break the tie. The steps in the algorithm are summarized as follows:
1. For each potential referent, generate C_b–C_f combination.

2. Filter out candidate referents that violate the co-reference constraints (number, gender, person, selectional preferences, etc.).
3. Rank referents by transition ordering.

Example 6.3   Consider sentences (6.17), which are reproduced here.

Suha saw a laptop in the shop. She enquired about it. She bought it.

$C_f(U_1)$: {Suha, laptop, shop}

$C_p(U_1)$: Suha

$C_b(U_1)$: undefined

U_2 contains two pronouns: *she* and *it*. *She* is compatible with *Suha* and *it* is compatible with *laptop* and *shop*.

$$C_b(U_2) = \text{highest ranked element of } C_f(U_1) = \{\text{Suha}\}$$

It has two possible referents. Assuming that *it* refers to the laptop, the assignments would be

$C_f(U_2)$: {Suha, laptop}

$C_p(U_2)$: Suha

$C_b(U_2)$: Suha

Result: continue

Assuming that *it* refers to the shop, the assignments would be

$C_f(U_2)$: {Suha, shop}

$C_p(U_2)$: Suha

$C_b(U_2)$: Suha

Result: continue

Since both assignments result in a 'continue' transition, the ordering in $C_f(U_1)$ is used to break the tie. Since *laptop* precedes *shop* in C_f list of U_1, *laptop* is assigned as the referent.

In sentence U_3, *she* is compatible with Suha and *it* is compatible with *laptop*. Resolving them is trivial.

Mitkov's Pronoun Resolution Algorithm

Mitkov (1998) proposed a robust, knowledge-poor approach for resolving anaphors. His algorithm uses a noun phrase extractor and a filtering module to identify a list of potential candidates for the antecedent. The input to the algorithm is part-of-speech tagged text. The noun phrase extractor, extracts NPs up to two sentences back from the anaphor. A referential filter eliminates pleonastic (semantically null) uses of pronouns, as in 'It is important to note that...'. The filtering module filters out NPs that do not agree in number, person, or gender, with the pronoun. The remaining NPs constitute the list of potential referents. The algorithm applies a list of antecedent indicators on them. These indicators are drawn empirically and implement syntactic salience or lexical preferences, some of which are domain dependent. The NP with the maximum aggregate score is proposed as the referent. To break the tie, a priority of preferences is used. The highest priority is given to candidate with higher score for immediate reference, followed by best collocation pattern score. If it does not work, the lexical (verb) preference score is used. If this also fails, the algorithm simply selects the most recent NPs from the candidate list as the referent. The antecedent indicators used by the algorithm are derived empirically. We now discuss these indicators.

Definiteness   Definite NPs are more important for resolving anaphoric references than indefinite ones. The algorithm penalizes indefinite NPs by assigning them a negative score (–1). If no definite article, possessive or demonstrative pronouns, appear in the paragraph, the rule is ignored.

Givenness   The NPs representing themes are considered good candidates for referents and are awarded a score of 1, the remaining NPs receive a score of zero. The theme appears first and has fair chances of having a co-referential link with the previous text.

Indicating Verbs   Mitkov considered a set of verbs as good indicators for identifying salient NPs. The verb set considered by Mitkov is

> {discuss, present, illustrate, identify, summarize, examine, describe, show, check, develop, review, report, outline, consider, investigate, explore, assess, analyse, synthesis, study, survey, deal, cover}

Mitkov considered the NPs following these verbs as preferred antecedents and assigned them a score of 1.

Lexical Repetition   This indicator gives preference to lexically reiterated candidates. NPs that are repeated two or more times in a paragraph are assigned a score of 2. Sequences of NPs that are synonyms, or have the same head, are considered the same when counting their occurrences, e.g., printer, laser printer, and fast laser printer, are regarded the same.

Section Heading Preferences   A candidate noun phrase that matches a noun phrase appearing in a section heading, gets a score of 1.

Non-prepositional Noun Phrases   Non-prepositional noun phrases are preferred over noun phrases that are part of prepositional phrases. This preference is weakened by penalizing (–1) a prepositional noun phrase, for example,

Load the printer with paper before restarting it. [it = printer]

Paper is penalized for being part of the prepositional phrase 'with paper'. This helps resolve the pronoun 'it' correctly. NPs which are part of prepositional phrases are usually indirect object. This preference is thus in accordance with the following ranking used by the centering theory:

Subject > direct object > indirect object

Collocation Pattern Preference An NP whose collocation pattern is identical with that of a pronoun is assigned a weight of 2. The collocation pattern considered for this preference is NP|Pronoun Verb and Verb NP| Pronoun.

Immediate Reference The heuristic used to implement this preference is described by the following pattern. (You) $Verb_1$ NP...(con) (you) $Verb_2$ it (con (you) $Verb_3$ it), where con $\in$ { and /or / before/ after/...}. The noun phrase immediately after $Verb_1$ is a more likely candidate for the antecedent of the pronoun 'it' immediately following $verb_2$ and is therefore given a preference by assigning it a score of 2.

Referential Distance The referential distance indicator suggests that noun phrases occurring recently are more likely candidates than those introduced further back. This preference is realized in a complex sentence using the following preference hierarchy.

Noun phrases in the previous clause (score = 2) > noun phrase in the previous sentences (score =1) > noun phrases occurring two sentences back (score = 0) > noun phrases occurring three sentences back (score = –1).

For a simple sentence, noun phrases that occurred in the previous sentence are the best candidates for referents, followed by noun phrases situated two sentences back, followed by noun phrases situated three sentences further back (score 1, 0, –1 respectively).

Term Preference NPs representing terms in the field are deemed better than non-terms, and are therefore assigned a score of 1.

Mitkov pointed out that the antecedent indicators are just preferences, not absolute factors. In certain cases, one or more indicators do not point to the correct antecedent. Sometimes, the erroneous clues given by some indicators are counterbalanced by preferences assigned by other indicators. The steps in the algorithm are summarized in Figure 6.1. To give a concrete idea, we illustrate the algorithm with an example.

1. Extract noun phrases in the current sentence and previous two sentences (if available). Consider only the noun phrases appearing to the left of the anaphora (pleonastic references are dropped).
2. Filter out NPs which do not agree in number, gender, or person, The remaining NPs are potential candidates.
3. Apply antecedent indicators to each potential candidate. The candidate with the maximum aggregate score is accepted as the referent of the pronoun. Break the tie using the collocation pattern score. If this does not work, use the lexical (verb) preference score, otherwise select the most recent NPs from the candidate list as the referent.

Figure 6.1 Steps in Mitkov's pronoun resolution algorithm (adapted from Mitkov 1998)

Example 6.2 Consider sentences (6.17), which are reproduced here.

Suha saw a laptop in the shop. She enquired about it. She bought it.

Noun phrases in the first sentence are: {Suha, laptop, shop}.

Now consider the pronoun 'she' in the second sentence. As mentioned in Step 1 of the algorithm, the potential candidates are NPs to the left of anaphor, i.e., NPs identified in sentence 1. Step 2 eliminates 'laptop' and 'shop' from this list, due to non-agreement in the gender, leaving only one potential candidate. Hence, we accept 'Suha' as the antecedent of 'she'. Next, the pronoun to be resolved is 'it'. The candidate referents for 'it' are {Suha, laptop, shop}. Step 2 eliminates Suha from this list. Now, we apply the antecedent indicators to each of the remaining candidates. As 'shop' is a prepositional NP, a negative score (-1) is assigned to it. The total score for shop is -1 + definiteness 1 + referential distance 1 + indicating verbs 0 + term preference 1 + section heading 0 + collocation $0 = 2$; and for 'laptop' the score is definiteness 1 + referential distance 1 + indicating verbs 0 + term preference 1 + non-prepositional noun phrase 0 + section heading 0 + collocation 0 + givenness $1 = 4$. So, we correctly resolve 'it' into laptop. Identifying the antecedent of pronouns in the third sentence is left as an exercise to readers.

Mitkov reported a success rate of 89.7% for the genre of technical manuals, which is better than any other success rate existing on the same genre. In particular, any knowledge-poor algorithm will have difficulties with sentences that have more complex syntactic structures. This is because the algorithm does not use any syntactic knowledge. The approach can be easily adapted for other languages as well.

Sequenced Model

Lappin (2003) proposed an integrated sequenced model of anaphora and ellipsis resolution. Figure 6.2 outlines the architecture of this model. The sequenced model combines relatively inexpensive and robust syntactic salience and recency-based approaches, with lexical preference model and adductive inference model, to achieve higher accuracy. These models are invoked one after the other in sequence. First, the syntactic salience and recency-based methods are applied to identify referring expressions for a pronoun. If this method fails to reliably resolve some references, then the lexical preference model is invoked. The unresolved cases include references for which the difference in the salience scores of top two candidates is less than a certain threshold value. The lexical preference model exploits semantic and real-world knowledge to resolve these references. The computational cost incurred in this approach is higher than in the first model. If all the references are resolved successfully, the process terminates, otherwise an even more expensive abductive inference model (Kehler 2000, 2002) is called.

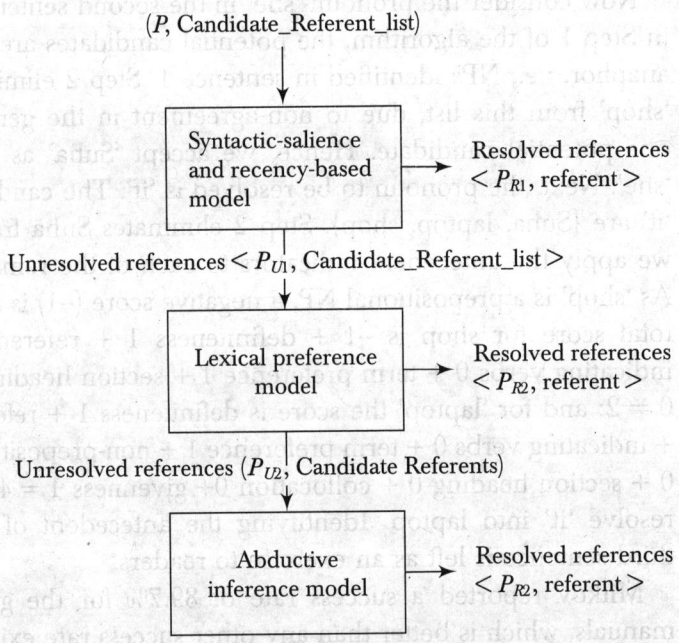

$(P,$ Candidate_Referent_list$)$

Syntactic-salience and recency-based model $\rightarrow$ Resolved references $< P_{R1},$ referent $>$

Unresolved references $< P_{U1},$ Candidate_Referent_list $>$

Lexical preference model $\rightarrow$ Resolved references $< P_{R2},$ referent $>$

Unresolved references $(P_{U2},$ Candidate Referents$)$

Abductive inference model $\rightarrow$ Resolved references $< P_{R2},$ referent $>$

Figure 6.2 Architecture of a sequenced model

6.4 DISCOURSE COHERENCE AND STRUCTURE

Consider the following text:

> *Biomass is emerging as a viable source of power for rural electrification in India. At first glance, Kirgavalu may look like a typical village in southern Karnataka.*

Both sentences in the passage are well formed and independently interpretable. But the passage seems a bit odd. The reason is that we try to establish a connection between the first and the second sentence. We raise questions such as how the Kirgavalu village is connected with biomass. In this case, we find it difficult to understand the connection. By raising such questions, we point out that the discourse is not coherent. In order to make the text coherent, we might build an understanding that perhaps the Kirguvalu village has a biomass plant. The attempt on the part of hearer (reader) to establish a connection between a pair of sentences suggests that merely grouping well-formed, independently interpretable sentences, does not yield meaningful passages; coherence is required to produce a meaningful composition. It is also needed for discourse comprehension. Coherence is different from cohesion. *Cohesion* refers to the grammatical relationship between words, referring forwards or backwards to other words, or substituting words or phrases, within the text.

There are a number of relations that connect utterances (sentences). Consider the following text:

Section 5.2 deals with sentence level meaning representation.

(6.18a)

In particular, we discuss the general characteristics of meaning representation languages (Section 5.2.1), (6.18b')
and computational approaches to semantic analysis (syntax-driven semantic analysis and semantic grammars in Section 5.2.3).

(6.18b")

Next, we discuss the internal structure of words, their relationships, and their meanings in Section 5.3. (6.18c)

Sentence (6.18a) introduces a topic. The next sentence elaborate on that by breaking it into subtopics, clauses (6.18b') and (6.18b"). A temporal relation exists between (6.18b) and (6.18c) indicated by 'next', which links the topics we intend to discuss. Figure 6.3 illustrates these relations.

A number of researchers have pointed out that such relations exist and have proposed various instances. Joseph Grimes in *Thread of Discourse* (1975) includes alternation, specification, equivalence, attributions, and explanations. Grimes called these relations rhetorical predicates,

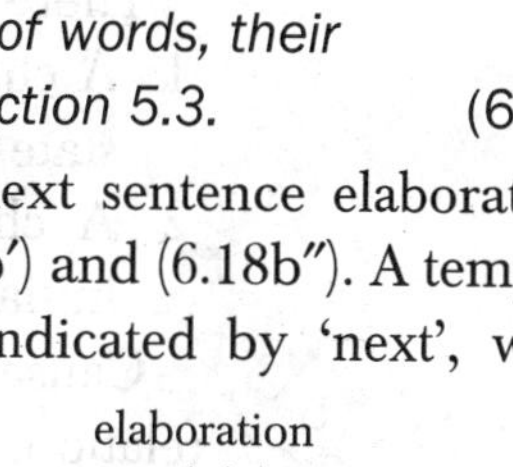

Figure 6.3 Structure of passage (6.18)

whereas Hobbs called them coherence relations. Robert Longacre (1976) include conjunction, contrast, comparison, alternation, temporal overlap and succession, implications, and causation. The list of coherence relations proposed by Hobbs (1979) includes result, explanation, occasion, parallel, and elaboration.

6.4.1 Coherence Relations

Hobbs (1985) described the process of interpreting discourse as 'a process of using knowledge acquired in past to construct a theory of what is happening in the present.' Understanding discourse requires identification of the coherence relations in the discourse. We now illustrate some coherence relations. In all of these examples, we assume that S_0 and S_1 are two consecutive utterances (sentences). Cue phrases, such as 'so', 'because of', 'but', 'while', and 'therefore', are good indicators of relations between discourse segments.

Occasion

Consider the following passage:

> At 9:00 a.m., the train arrived at Allahabad. The conference
> was inaugurated at 10 a.m.   (6.19)

One way to read the passage to make it coherent is by assuming that someone who wants to attend the conference was on the train. That is, the first event sets up the occasion for the second. This relation is different from the causality relation. There is nothing special about the train that causes the conference to be inaugurated.

There are two cases that define occasion relation:

1. A change of state can be inferred from the assertion of S_0, whose final state can be inferred from S_1.
2. A change of state can be inferred from the assertion of S_1, whose initial state can be inferred from S_0.

Cause and enablement can be regarded as special cases of the occasion relation. Here is another illustration of occasion relation (adapted from Hobbs 1985):

> Go out of this door.   (6.20a)
> Turn right.   (6.20b)
> Go to the second room.   (6.20c)

Sentence (6.20a) describes a change of location and assumes an orientation. The final state of the location holds during the event described in (6.20b). The initial state of the change in location described in (6.20c) can be inferred from (6.20b). Similarly, the orientation assumed in (6.20a) is the initial state for the change in state described in 6.20b, and its final state is assumed in 6.20c. Figure 6.4 shows the inferences that need to be drawn to satisfy the definition. It is possible to find more than one relation between a pair of sentences, provided they do not involve inconsistent assumptions.

Type	6.20a	6.20b	6.20c
1	loc 1 → loc 2	loc2	
2		loc 2	loc 2 → loc 3
2	angle 1	angle 1 → angle 2	
1		angle 1 → angle 2	angle 2

Figure 6.4   Occasion relation in sentences (6.20)

Here is another example of occasion relation:

> Increment the counter by one.
> If it is 100, reset it to zero. (6.21)

Here, the value of the counter is changed, which is presupposed in the second sentence.

Explanation

The segment S_1 is an explanation of S_0 if S_1 describes an event or state that could cause the state or event described in S_0. Explanation is a relation that relates a segment of discourse to the listener's prior knowledge. It is formally defined as follows:

Infer that the state or event asserted by S_1 causes, or could cause, the state or event asserted by S_0.

> Suha ate all the rice in bowl. She was very hungry. (6.22)

In this passage, S_1 explains the event asserted by S_0.

Causality may sometimes be explicitly stated as in the following statements:

> Suha ate all the rice in the bowl because she was very hungry.
> I get late because of the procession on the roads.

Elaboration

The definition of the elaboration relation involves identical entities. It is defined formally as follows:

> Infer the same proposition from the assertions of S_0 and S_1.

A simple example of elaboration relation is

> Saif scored an unbeaten century today. He was in full
> swing and made 108 not out on 87 balls. (6.23)

From the first sentence, and from what we know about an unbeaten century, we infer that Saif made more than 100 runs and he remains not out. By assuming that 'he' refers to Saif, we infer the same proposition from S_1 and thus establish the elaboration relation.

Parallel

The parallel relation is based on the similarity of entities. In this context, we say that two entities are similar if they share some property. More formally, the parallel relation is defined as follows:

Infer $p(x_1, x_2,...)$ from the assertion of S_0 and $p(y_1, y_2, ...)$ from the assertion of S_1, where x_1 and y_2 are similar, for all i.

Suha likes reading novels. Zuha enjoys reading science fiction.

$$(6.24)$$

For each of the segments, we infer that a person likes reading books. Zuha and Suha are similar in that they are both people, reading novels, and reading science fiction. The predicate in this case may be hobby.

Contrast

There are two cases that define contrast relations:

1. Infer $p(x)$ from the assertion of S_0 and $\neg\, p(y)$ from the assertion of S_1, where x and y are similar.
2. Infer $p(x)$ from the assertion of S_0 and $\neg\, p(y)$ from the assertion of S_1, where there is some property q such that $q(x)$ and $\neg\, q(y)$.

Here is a simple example of the first case:

Suha does not like cricket. But she likes cricket more
than any other game.

$$(6.25)$$

Exemplification

Infer $p\,(X)$ from the assertion of S_0 and infer $p(x)$ from the assertion of S_1, where x is a member or subset of X. This relation is illustrated as follows:

Suha bought a printer today. It is a laser printer.

$$(6.26)$$

6.4.2 Discourse Interpretation

Hobbs (1985) suggested that the problem of discourse interpretation can be solved by decomposing it into six sub-problems.

1. Logical Notation or Knowledge Representation

The first sub-problem deals with the problem of representation. In order to interpret discourse, a logical representation of natural language sentences is required. First order predicate logic representation is one such representation that has been used to translate natural language representation into logical representation, and supports reasoning based on that representation.

2. Syntax and Semantics

This is concerned with the translation of text, sentence by sentence, into logical notation or representation. This problem has been researched heavily in linguistics and computational linguistics (Woods 1970; Montague 1974), and is considered to be solved to a large extent for common syntactic constructions.

3. Knowledge Encoding

This deals with the representation of the world and the language in the knowledge base. This knowledge is required for interpreting discourse. However, the task of encoding world knowledge is not trivial. We must decide what knowledge to represent, how to represent it, and whether the new knowledge being added is consistent with what is already in existence. A lot of research is focused on this problem. We make some general assumptions about how knowledge is encoded and assume the existence of specific facts in the knowledge base, so that we may continue to the discussion of the more important problem of 'how to use this knowledge in interpretation'. For example, we can have the following fact in knowledge base:

$$(\forall x)(\exists y) \text{ printer } (x) \rightarrow \text{ cartridge } (y, x)$$

4. Deductive Mechanism

In order to use stored facts, we must have some deductive mechanism. One such rule of inference is modus ponens, which permits us to infer cartridge (y, x) from:

$$(\exists x) \text{ printer } (x)$$

and $\quad (\forall x) (\exists y) \text{ printer } (x) \rightarrow \text{ cartridge } (y, x)$

5. Discourse Operations or Specifications of Possible Interpretation

There are certain problems in discourse such as co-reference resolution. These problems need to be resolved first to interpret text. This requires the identification of these problems and a specification of what it means to solve them. For example, a specification might state that the existence of an entity described by the definite noun phrase, can be inferred from the previous text and knowledge base.

6. Specification of the Best Interpretation

Discourse operations may yield many solutions to a discourse problem. This sub-theory deals with identifying the most economic interpretation for a sentence. The factors that govern cost of the solution include complexity of proof, salience of axiom used, and redundancy in the interpretation. Let us take a close look at what the discourse problems are. Discourse problems can be divided into those problems that can be solved using information within the sentence, and those that involve the relation of the sentence to its context.

The within-sentence problems include problems of co-reference resolution, e.g., resolving pronouns, definite noun phrases, and missing arguments; identifying intended predicates where predicate is non-specific; satisfaction of selectional constraints; and determining the internal coherence of the discourse. The internal coherence problem deals with inferring relationship, such as causality, between sentences.

The second type of discourse problem considers the relationship between the sentence and the world.

6.4.3 Abductive Interpretation of Local Coherence

Hobbs (1985) presented an abductive framework for determining local coherence of the utterance. 'Abductive' means that assumptions are allowed at various costs. The method seeks the most economic interpretation of a sentence, such as an explanation that uses a small number of assumptions or one that uses the most specific properties of the input. In abductive inference, we make assumptions that need not be provable. Hobbs used an etc. predicate to represent all other properties that must be true for an axiom, but which were too vague to be stated explicitly. These predicates are assumed at a certain cost, not proved. A predicate with a low assumption cost will be preferred to one with high assumption cost. We now explain how the coherence of a segment is established with the help of an example.

Example 6.3 Consider the following text:

The local administration stopped the trade union
from meeting. They feared violence.   (6.27)

We need to establish local coherence in this segment. One way to prove that there is a coherence relation between the sentences is to prove that there is an explanation relation, i.e.,

Explanation $(e1, e2)$

This relation will hold if there is a causal relation between them:

Cause $(e2, e1)$

The logical form of the sentences and the hypothesized causal relation between them is given by the following expression.

$(\exists \ s, \ l, \ m, \ u, \ f, \ y, \ v)$ stops $(s, \ l, \ m) \wedge$ meeting $(m, \ u) \wedge$ cuase $(f, \ s) \wedge$ fear $(f, \ y, \ v) \wedge$ violent $(v, \ z)$

To prove this expression, we require axioms representing world knowledge in addition to axioms about coherence relation. The world knowledge needed for establishing coherence in this example is as follows:

1. If there is a fear event f imposed by someone, say y, of violence v, it means that y does not want violence (v).
2. A meeting m, by trade union u, causes violence.
3. If someone y, does not want (diswant) event v, and v is caused by m, then that will also cause y to diswant m.
4. If the local administration does not want something, then they will stop it.
5. And finally, cause is transitive, i.e., if $e1$ causes $e2$ and $e2$ causes $e3$, then $e1$ causes $e3$.

The axioms representing these sentences are as follows:

$(\forall f, y, v)$ fear $(f, y, v) \rightarrow \exists\, d)$ diswant $(d, y, v) \wedge$ cause (f, d)

$(\forall m, u)$ meeting $(m, u) \rightarrow (\exists\, v, z)$ cause $(m, v) \wedge$ violent (v, z)

$(\forall m, v, d, y)$ cause $(m, v) \wedge$ diswant $(d, y, v) \rightarrow (\exists\, d1)$ diswant $(d1, y, m) \wedge$ cause $(d, d1)$

$(\forall d1, l, m)$ diswant $(d1, l, m) \wedge$ localadministration $(l) \rightarrow (\exists s)$ stop $(s, l, m) \wedge$ cause $(d1, s)$

$(\forall e1, e2, e3)$ cause $(e1, e2) \wedge$ cause $(e2, e3) \rightarrow$ cause $(e1, e3)$

The derivation is shown in Figure 6.5. During the derivation, we also unify y with l.

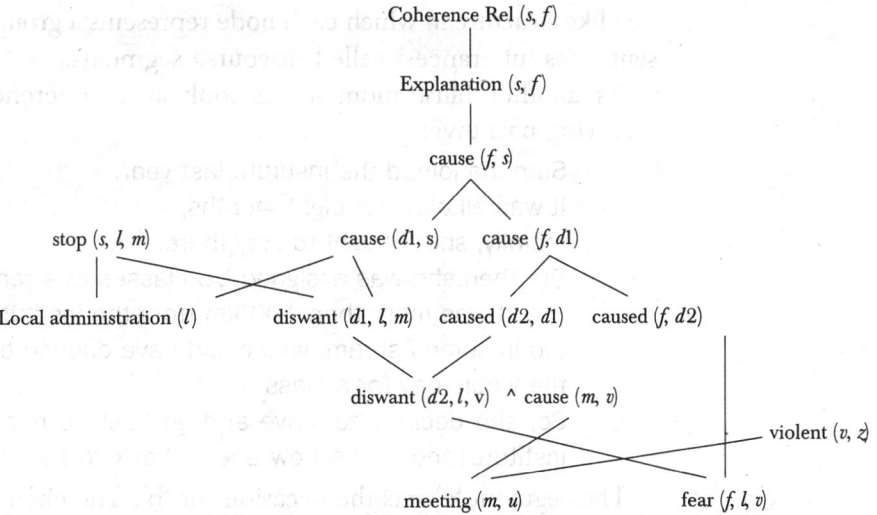

Figure 6.5 Interpretation of sentences (6.27)

Sometimes, the speaker uses a connective, which makes coherent relations explicit. For example, the use of 'because' to join the two sentences in segment (6.27), resulting in *The local administration stopped trade union from meeting because they feared violence,* would make the coherence relation explicit. The literal cause (f, s) becomes part of the logical form of the sentence and we need not to assume it.

We can extend this framework to establish the coherence of larger discourse.

6.4.4 Discourse Structure

So far we have discussed only the relations between a pair of sentences. In fact, it is possible to establish such relations in longer discourse. A discourse has a structure. For example, sentences (6.18b′) and (6.18b″) are related by an occasion relation. They combine to give a segment, which is linked with (6.18c) by an occasion relation. The resulting composed segment is related to (6.18a) by an elaboration relation. The coherence relationships between these sentences assign a 'coherence structure' to the discourse, as shown in Figure 6.6. It is a tree-like structure in which each node represents a group of locally coherent sentences (utterances) called discourse segments.

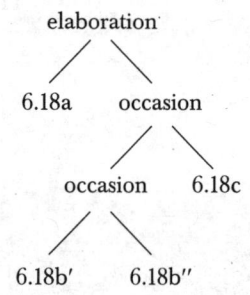

Figure 6.6 Coherence structure of text (6.18)

As another illustration, let us look at a coherence structure of the following narrative:

Sumitha joined the institute last year.	(6.28a)
It was all okay for eight months,	(6.28b)
Initially, she thought to stay there.	(6.28c)
But then she was assigned UG classes at a remote centre.	(6.28d)
Um, it was more than 100 km from the institute and that too in some Ashram, who could have enough time to waste the whole day for a class.	(6.28e)
So, she decided to leave and go back to her parent	(6.28f)
institute|and that's how she is back to the university.	(6.28g)

The segment 'a' sets the occasion for 'b'. The circumstance of segment 'd'–'e' causes and thus occasions the events of 'f'. The segments 'c' and 'f' are contrasting relations. 'a'–'e' and 'f' are related by a set of events and its outcome. Figure 6.7 illustrates the structure.

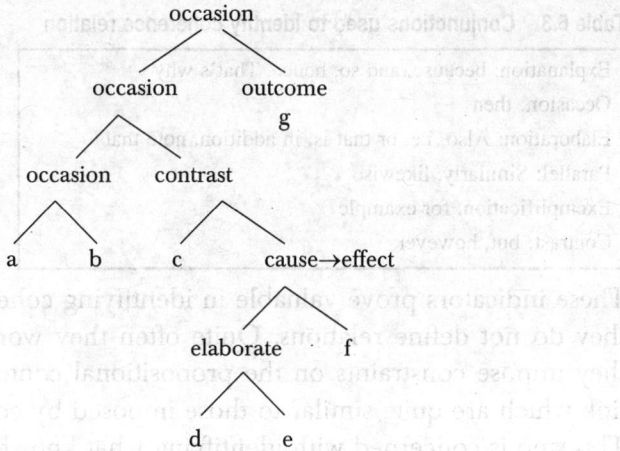

Figure 6.7 Structure of text (6.28)

Such a structure of discourse is useful in explaining classical problems of 'topic', 'genre', and 'coherence drift' that occur in ordinary conversation.

Hobbs proposed a four-step procedure for analysing discourse. As we move from one step to another, difficulty increases. The steps involved in the procedure are discussed below.

1. Identify one or two major breaks in the text and divide the text into two or three segments. This division corresponds to most natural division that one can carry out intuitively. This process is repeated for each segments obtained at first place. The iteration continues until we get single clause. The output of this step is a tree structure of the text. For example, in text (6.18), the major break comes between sentences (6.18a) and (6.18c). Within sentence (6.19b), there is a break between the first clause and the second clause of the sentence. This gives the tree structure of Figure 6.5.

2. In step 2, the non-terminal nodes of the tree are labelled with coherence relations. We follow a bottom-up approach to get an understanding of what is being representing by the composed segment. Thus, in Figure 6.5, the node linking (6.18b′) and (6.18b″) is labelled with the occasion relation. The node linking the resulting segment and (6.18c) is labelled with occasion relation and son. This step requires an understanding of different types of relations. Some simple heuristics based on what conjunctions need to be inserted might help identifying coherence relation. For example, if we can insert 'then' between S0 and S1 and reversing the order of the segment changes the sense, then the occasion relation is quite likely. If 'because' seems to be appropriate between S0 and S1, then explanation relation is preferred candidate. Table 6.3 lists useful conjuncts to identify coherence relation.

Table 6.3   Conjunctions used to identify coherence relation

Explanation: because, and so, hence, That's why
Occasion: then
Elaboration: Also, i.e. or that is, in addition, note that
Parallel: Similarly, likewise
Exemplification: for example
Contrast: but, however

These indicators prove valuable in identifying coherence relation but they do not define relations. Quite often they work because usually they impose constraints on the propositional content of clauses they link which are quite similar to those imposed by coherence relations.

3. This step is concerned with identifying what knowledge underlies the (discourse) the composed segment. We need to specify the knowledge or beliefs that support assignment of coherence relation to the nodes.

4. Step 4 is concerned with validation of hypotheses made in step 3. This requires consideration of longer corpus and construction of a knowledge base that would support the analysis of al of the text in the corpus.

SUMMARY

This chapter introduces discourse concepts and presents methods for analysing discourse.

- Discourse is concerned with linguistic and extra-linguistic phenomena that give text a unified and meaningful form.
- Discourse can be defined as 'language above the sentence level.
- Another way, we can describe discourse is by saying that it is 'language in use'.
- Discourse processing uses situational context, and cultural, social, interpersonal, and linguistic knowledge.
- Language uses cohesive devices like references, ellipsis, conjunctions, repetitions, etc., to achieve bound text.
- Ellipsis refers to the phenomenon where part of a sentence (utterances) is omitted or left unpronounced.
- Reference resolution techniques use various constraints and preferences to identify the preferred referent.
- Cohesion relates words to other words within the text; coherence relates sentences.
- The coherence relationships between sentences assign a 'coherence structure' to text.

REFERENCES

Brennan, S., M. Friedman, and C. Pollard, 1987, 'A centreing approach to pronouns,' *ACL-87*, pp. 155–62.

Denber, Michel, 1998, 'Automatic resolution of anaphora in English,' Technical report, Eastman Kodak Co., http://www.wlv.ac.uk/~le1825/anaphora_resolutionpapers/denber.ps.

Grimes, Joseph, 1975, *The Thread of Discourse*, Mouton and Company, The Hague, The Netherlands.

Grosz, B. J., A.K. Joshi, and S. Weinstein, 1995, 'Centering: a framework for modelling the local coherence of discourse,' *Computational Linguistics*, 21(2), pp. 175–204.

Hobbs, Jerry, 1979, 'Coherence and coreference,' *Cognitive Science*, 3, pp. 67–90.

_____1985, 'On the coherence and structure of discourse,' Technical Report 85-37, Center for the Study of Language and Information (CSLI), Stanford, CA.

Kehler, A., 2000, 'Pragmatics,' Chapter 18 in D. Jurafsky and J. Martin (Eds.), *Speech and Language Processing*, Prentice Hall, Upper Saddle River, NJ.

_____2002, *Coherence, Reference, and the Theory of Grammar*, Stanford, CSLI, CA.

Lappin, S., 2003, A 'sequenced model of anaphora and ellipsis resolution,' *Anaphora Processing: Linguistic, Cognitive, and Computational Modelling*, A. Branco, A., McEnery, and R. Mitkov (Eds.), John Benjamins, Amsterdam, pp. 3–16.

Lappin, Shalom and Herbert J. Leass, 1994, 'An algorithm for pronominal anaphora resolution,' *Computational Linguistics*, 20(4), pp. 535–61.

Longacre, Robert, 1976, *An Anatomy of Speech Notions*, The Peter Rider Press, Ghent, Belgium.

Mann, William and Sandra Thompson, 1986, 'Relational propositions in discourse,' *Discourse processes*, 9(1), pp. 57–90.

Mitkov, R., 1998, 'Robust pronoun resolution with limited knowledge,' *Proceedings of the 18th International Conference on Computational Linguistics (COLING'98)/ACL'98 Conference*, Montreal, Canada, pp. 869–75.

_____1999, 'Anaphora resolution: The state of the art, 1999,' Paper based on the COLING'98/ACL'98 tutorial on anaphora resolution, University of Wolverhampton.

Widdowson, H.G., 1995, 'Discourse analysis—a critical view,' *Language and Literature*, 4(3), pp. 157–72.

EXERCISES

1. Which aspects of discourse are language problems and which are general AI/knowledge representation problems?
2. Differentiate between 'cohesion' and 'coherence'. Do you think that a non-cohesive text can be coherent? Is it possible for text to be cohesive but not coherent? Give appropriate examples.
3. What are lexical cohesion devices? Discuss any two of them.
4. Given the sentence, *The shopkeeper showed a Dell laptop to Suha and she liked it very much*, find possible referents for the pronouns 'she' and 'it', and give the score of these referents for the following antecedent indicators: definiteness, referential distance, indicating verbs, term preference, section heading, non-prepositional noun phrase, collocation.
5. Differentiate between elaboration and explanation relations with the help of suitable examples.
6. Identify the coherence relation between the following sentences:
   (a) There is a train on Platform 6.
   (b) Its destination is New Delhi.
   (c) There is another train on Platform 7.
   (d) Its destination is Varanasi.
7. Give an abductive interpretation of the following text:
   The farmer was worried. There was little rain this year.
   List all the assumption and world knowledge used in this process.

LAB EXERCISES

1. Write a program that takes a text file as input and outputs all the sentences involving ellipsis.
2. Write a program which takes tagged text as input and list candidate referents for each pronoun appearing in it. Consider anaphors only and look only in the sentence and its preceding sentence for referents.
3. Write a program that checks gender agreement in the list of candidate referents in the output of the previous program.
4. Take a document in your native language. Identify all the pronouns and their referents occurring in it. Try to find constraints and preferences discussed in this chapter that might be useful in resolving the pronouns.
5. Use the constraints and preferences identified in Exercise 1 in a program that resolves pronominal references automatically.

CHAPTER 7

NATURAL LANGUAGE GENERATION

CHAPTER OVERVIEW

This chapter is concerned with automated natural language generation. The goal of language generation is to produce natural language text from its computer-based representation. The chapter begins with an introduction to natural language generation (NLG). Next, it describes various architectures of an NLG system. A general framework for NLG is then presented. This includes a discussion on various approaches to sentence generation. Finally, some of the NLG applications have been listed.

7.1 INTRODUCTION

Natural language generation (NLG) is one of the oldest applications in computational linguistics. One of the most widely cited initial works in this field is ELIZA (Weizenbaum 1966), a keyword-based computer-generated conversation system. This simple system is more like a child reproducing a memorized sentence when asked to make sentences from given words, as illustrated in the following conversation with ELIZA:

Eliza: Hello. I am ELIZA. How may I help you?
You: I am feeling a little bit sleepy.
Eliza: How long have you been feeling a little bit sleepy?
You: For almost half an hour.
Eliza: Please go on.

The inflexibility of such a system is self-evident. The NLG modules were used mainly in the interlingua approach of machine translation (MT). In this case, the content and form of the sentence to be generated is fully determined by the source sentence. In fact, the match between the two serves as a yardstick of quality for MT systems.

In this chapter, we discuss the automatic generation of a set of natural language sentences for given inputs, which are generally non-linguistic (Jurafsky and Martin 2000). The text is generated sentence by sentence, but the structural relationship between the sentences or clauses is represented and maintained through various mechanisms and structures. The theme of the discourse is also maintained by the system to generate only coherent sentences.

To generate the text, we have to answer two questions:

(a) What to write?

(b) How to write it?

The first answer will give the content of the text and should be the input to the system. The issue here is, how elaborate should this input be. If it is not provided in sufficient detail, the expansion of the idea and formation of the sentence is an open problem and may have infinite solutions.

The second answer determines the language and the structure of the text. The choice of appropriate words, called lexical words, is also varied. The way the system should interact, e.g., with a command, or an interrogative, or a negative sentence, etc, is to be determined. The linguistic knowledge, i.e., the grammar of a language, argument structure or sub-categorization, rules for morphological modifications, etc., is to be represented and organized, and a mechanism has to be evolved to apply this knowledge on inputs from the previous stage so that an actual, intended and valid sentence can be generated. An NLG system can thus be viewed as the implementation of these tasks. A large number of NLG systems have been designed and implemented for specific applications. However, such heuristic methods are not likely to succeed in the development of a generic system.

As the number of choices to be applied in NLG is quite large, the process of NLG can be viewed as a network which applies several constraints at various levels. Representations of meaning, structural and linguistic knowledge and their modifications for domain independent sentence generation are still under research and do not fall within the scope of this chapter. In this chapter, we present a basic sketch of the NLG process. In Section 7.2, we discuss the types of architectures of NLG systems and in Section 7.3, various useful representations are introduced.

7.2   ARCHITECTURES OF NLG SYSTEMS

Several tasks need to be implemented and aggregated into a framework to create a generation system. The framework will define the constraints on the system. We discuss three types of architectures relevant to NLG systems, namely the pipelined, interleaved, and integrated systems. Although, a large number of generators have been developed with different goals and it is very difficult to fit them in one model or the other, our purpose here is to help to understand the organization of various tasks in an NLG system.

Three of these tasks are obvious. Taking a top-down approach, the first task is the overall planning of the discourse. The second is to plan the texts and their structure, and the third task is to actually realize the plan and generate the sentences. The three corresponding modules are (a) discourse planner, (b) text planner, and (c) surface realizer.

The discourse planner decides the ordering and structure of the text. The output of this module is usually represented as a tree. The text planner decides which words and phrases are used to express the concepts and relationship specified by the discourse planner. The tasks performed by this module include sentence aggregation, lexicalization, and referring-expression generation. The surface realizer takes the output of the text planner module and generates individual sentences. These tasks are not monolithic tasks, but combines many sub-tasks. The decision to include a sub-task either in discourse planning, text planning or surface realization depends on the application. For example, lexicalization (or choice of words) is performed both during discourse planning and surface realization in DRAFTER (Paris et al. 1995) and DRAFTER 2 (Scott, Power, and Evans 1998), while in HEALTHDOC (Hirst et al. 1997), it is implemented in the text planning stage and in KUNET (Teich and Bateman 1994), it is implemented in the surface realization stage (Mellish et al. 2004). The continuation of tasks over various stages, and the designing of structures to pass information from one stage to another, yields the following three architectures.

As can be seen in Figure 7.1(a), the pipelined architecture, the most common architecture (Rags Mellish et al. 2004), divides NLG tasks into three distinct groups and passes the processed information from one stage to another in a pipelined fashion. This serial method of processing is obvious and simple. But it requires that each stage must complete its tasks before the next stage starts its own. In NLG systems, information at the discourse planner level is vague but becomes more specific as more and more constraints are met on the way to the surface realizer level.

For example:

He danced with Jaya and she got angry. (7.1a)

He forced Jaya to dance and she got angry. (7.1b)

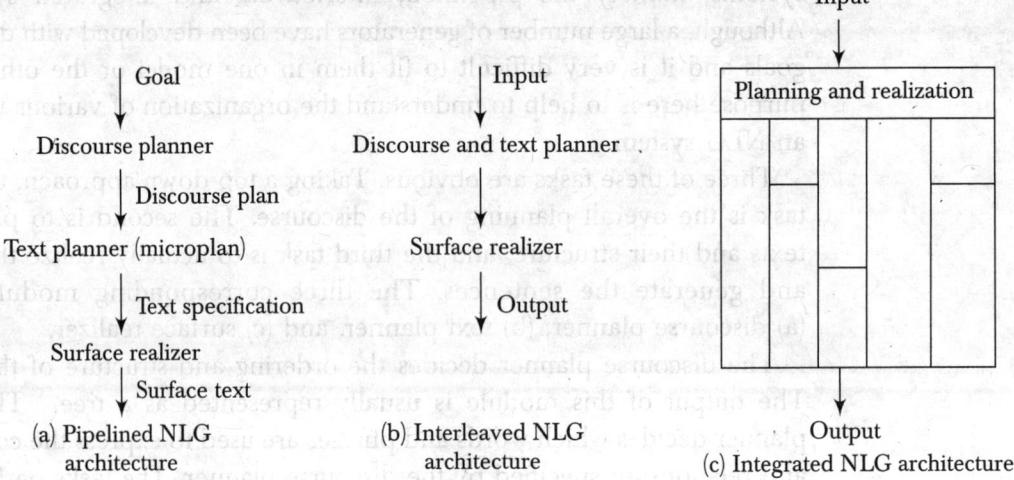

Figure 7.1 Natural language generation architecture

In sentences (7.1a) and (7.1b), the choice of 'dance' or 'forced to dance' can change the referent of 'she'. Hence, if the two parts of the sentences are generated independently and then aggregated, then depending on the desired meaning, one may change the choices afterwards. Such a process is not commensurate with the pipelined architecture. Commitments are made at each stage and the flow goes in a predetermined path. To support such a rethinking in generation, the interleaved NLG architecture, as denoted in Figure 7.1(b), has been proposed by Hovy (1985).

In the interleaved architecture, the discourse and text planner stages are merged into a single stage, to indicate that control and information can be passed back to any task in the two stages. As the demarcation between discourse planning and text planning is thin, one can separate out the higher order functions in discourse planning and pass only the control back to text planning. The name interleaved indicates that the task of generation is completed in the 'planning → realization → planning →...' sequence of stages. Hence, unlike pipelined architecture, only the appropriate commitments are made at text planning stage and, in turn, at surface realization stage in each cycle. Hovy used this approach to generate pragmatic effects in text in PAULINE (Hovy 1988). Some of the other systems which use this architecture are MUMBLE (McDonald 1980; McDonald and Pustejovsky 1985), IPG (Smedt and Kemper 1987) and PROTEUS (Davey 1979). Although it can be theoretically argued that the sentences generated by pipeline and interleaved architectures can be similar, the flexibility levels and the complexity and performance in different contexts will be different for the two architectures.

A different architecture, called *integrated architecture*, has been discussed by Kantrowitz and Bates (1992). Generation of sentences in an interactive environment necessitates a kind of 'blackboard' architecture, which can trigger rules or operators, and transform or combine structures, objects, actions, and events appropriate to the situation. GLINDA, an integrated NLG, was introduced to allow interaction between computer simulated agents and human beings. The two architectures discussed earlier are more suitable for discourse generation in offline situations.

It can be argued that if we divide each task into sub-tasks at each of the three stages, many of these fine-grained sub-tasks will be similar. These orthogonal sub-tasks can be implemented directly as and when required. This breaks the boundaries of stages and all three stages are integrated into one. For example, a common operation called serialization can fulfil the task of goal-ordering at discourse or text planning stage or the task of implementing precedence relation grammar at the surface realization stage (Kartrowitz and Bates 1992). Similarly, the aggregation of sentences into groups, or words and phrases into a clause, can be abstracted in one operation termed transformation. Hence, in Figure 7.1(c), all stages are fused into one, but the figure shows some sub-tasks which may be interleaved. In this chapter, we discuss tasks and representational issues using one example generated by pipelined architecture.

7.3 GENERATION TASKS AND REPRESENTATIONS

In this section, we present an overview of the common tasks involved in the generation process and the representations employed in it. Many new systems have defined variants of these tasks and representations, but they are not within the scope of this book.

As discussed in the previous section, NLG can be defined as a single task that maps the non-linguistic input to a linguistic output. This task can be divided into three sub-tasks, as shown in Figure 7.2.

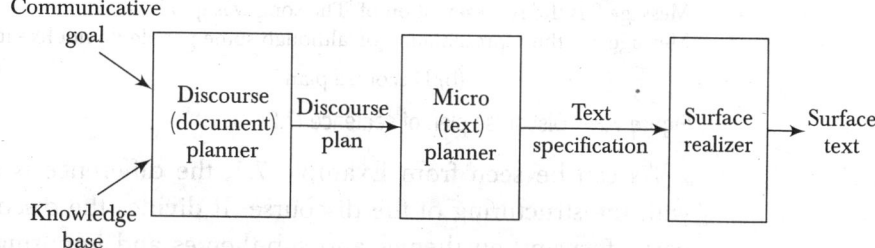

Figure 7.2 Input and output of three NLG tasks

7.3.1   Discourse Planner

Given a communicative goal—such as the description of an event or explaining a procedure to a new user—and a knowledge base representing the content in a non-linguistic manner—such as the weather data for WeatherReporter (Essers and Dale 1998), a parse tree from a machine translator, a propositional logic based, or KONE-styled knowledge base— the goal is to produce a discourse plan that represents the content and structure of the discourse. For example, let us consider the generation of the following discourse.

Example 7.1

Savitha sang a song. The song was good although some people do not like it.   (7.2)

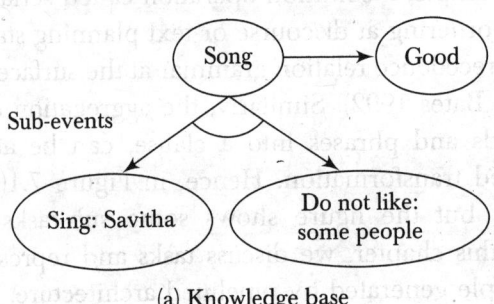

Sub-events

(a) Knowledge base

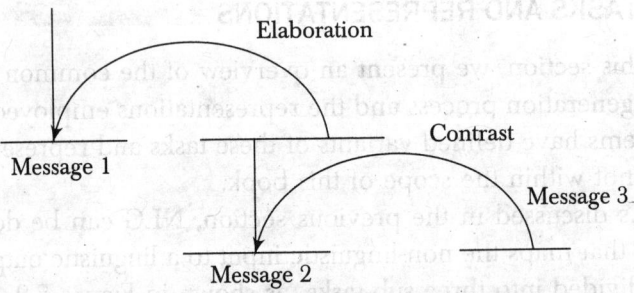

Message 1 is the representation of 'Savitha sang a song'.
Message 2 is the representation of 'The song was good'.
Message 3 is the representation of 'although some people do not like it'.

(b) Discourse plan

Figure 7.3   Discourse plan of sentence (7.2)

As can be seen from Example 7.1, the difference is more concerned with the structuring of the discourse. It divides the discourse into logical parts, focusing on themes and sub-themes and imposing order on those parts. The relationships between the sentences or parts of sentences are determined with the help of valid operations on the computationally represented structures. These structures give a definite framework to discourse generation. The structure, once decided, works as a limitation to flexibility.

Several representations have been employed in practice and no standard exists at present. One of the representations, called text schemata, involves structure representation using augmented transition network (ATN). Each node in ATN represents a state of the activity of discourse structure generation and each arc represents the optional transition to another state when one sentence part is added to the existing structure. Detailed discussion of ATN is available in textbooks and we will not discuss it here. But this kind of structural pattern representation fails as soon as the available information increases beyond a threshold limit of possibilities. We discuss another widely employed representation called rhetorical structure here. It is to be noted that the relationship of the sentence parts may have a hierarchy. It forces us to consider the representation in a different manner.

Rhetorical structure is the structure imposed on text based on the relationships that hold between parts of the text (Mann and Thompson 1987). These relationships make the text coherent. Mann and Thompson (1987) suggested a list of 23 rhetorical relations to describe the structure of text. In general, this list is not standard and different researchers have used different subsets of these relations in NLG. We have already discussed a subset of these relations in Chapter 6 (Section 6.4.1). In rhetorical structure theory (RST), the central segment of text taking part in a rhetorical relation is termed nucleus whereas the peripheral segment is termed satellite. For example, in Example 7.1, the sentence, 'Savitha sang a song' is the nucleus and 'The song was good although some people do not like it' is a satellite.

Most rhetorical relations are asymmetric as we interpret them in terms of nucleus and not vice versa. Figure 7.4 is the graphical representation of sentence (6.23) of Chapter 6. In this figure, the text is shown at the bottom and rhetorical relations are built upon them. The discourse plan in Figure 7.3 shows a hierarchical representation of rhetorical structure in which the satellite consists of two clauses related by contrast relationship. These clauses are represented by Message 2 and Message 3 in Figure 7.3, and as a pair they are related to Message 1 by elaboration relation.

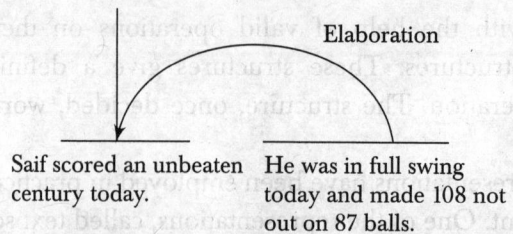

Figure 7.4   Graphical representation of sentence (6.23)

In NLG, RST has been most commonly used to describe discourse plan as in Figure 7.3.

7.3.2   Micro-planning

The micro-planning or text planner is concerned with tasks such as how information should be grouped into sentence-sized chunks, what lexical items should be used and how entities in the domain should be referred to. The micro-planner takes as input the high level structure, the macro plan, produced by the discourse planner and carries out detail planning. This detail planning is called micro-planning. It produces a text specification usually expressed in the form of a tree. Each leaf node in this tree structure is a sentence plan which is fed to surface realizer. The sentence plan contains sufficient information for generating the sentence. We now describe each of the three tasks discussed.

Sentence Aggregation

The sentence aggregation task decides how messages in the discourse plan are grouped together into sentences. For example, Message 2 and Message 3 in the discourse plan of Figure 7.3 can be combined into a single sentence:

> The song was good, although some people did not like it.

Or each message can be expressed as a single:

> Savitha sang a song. The song was good. Some people did not like the song.

The sentence aggregation system may also decide to combine all the three messages:

> Savitha sang a good song, although some people did not like it.

Note that in each case, the content remains the same. Aggregation is needed usually to enhance readability and fluency of the text to reader.

Lexicalization

Lexicalization is the process of choosing appropriate words or phrases to realize concepts that appear in the message. For example, we might specify that the phrase 'did not like' should be used to express Message 3. Another choice may be 'dislike'.

Referring Expression Generation

Referring expression generation is the task of determining appropriate words or phrases to identify domain entities. This task requires the consideration of contextual factors, in particular the content of previous communication with the user. For example, whether or not 'she' can be used to refer to 'Savitha' depends on what other entities have been mentioned in the previous sentences in the text.

All these three tasks must be performed regardless of the approach used by the surface realizer. A wide variety of representations have been proposed to specify sentence plans. Template-based representations and the use of abstract representation language are common. Abstract representation languages specify the words (nouns, verbs, adjectives, etc.) of a sentence and their relationship. One of the most popular abstract languages is sentence planning language (SPL) (Kasper 1989). As the exact form of the input accepted by the surface realizer is dependent on the approach used, we defer the detail representation of the sentence plan using a specific notation. An abstract representation for the sentence 'Savitha sang a song' may be given as

```
(S1
: subject    (Savitha)
:process    (sing)
:object     (song)
:tense      (past)
)
```

7.3.3 Surface Realization

The surface realization component takes the sentence specification produced by the micro planner and generates individual sentences. The two widely used approaches for this task are based on systemic grammar and functional specification grammar. We use these approaches to generate the following text:

Savitha sang a song.

There is no consensus on the level of representation used by the surface realizer. Some approaches specify only the propositional content while others specify the grammatical form, e.g., past tense and future tense and lexical items ('Savitha', 'sang', and 'song').

Approaches based on both systemic grammar and functional unification grammar work using functional specifications. Most NLG systems do not use syntactic specification because such systems start with meaning representations, with which it is more natural to specify functions rather than syntactic forms.

Approaches Based on Systemic Grammar

These approaches are motivated by systemic functional linguistics (SFL) (Halliday 1985), which is concerned with the function of language. Systemic grammar is a part of SFL, describing mapping from function into surface form.

Systemic grammar is represented using a directed, acyclic and/or graph, called a system network. Figure 7.5 shows a simple system network. The curly braces in the network represents 'and', and the vertical lines represent

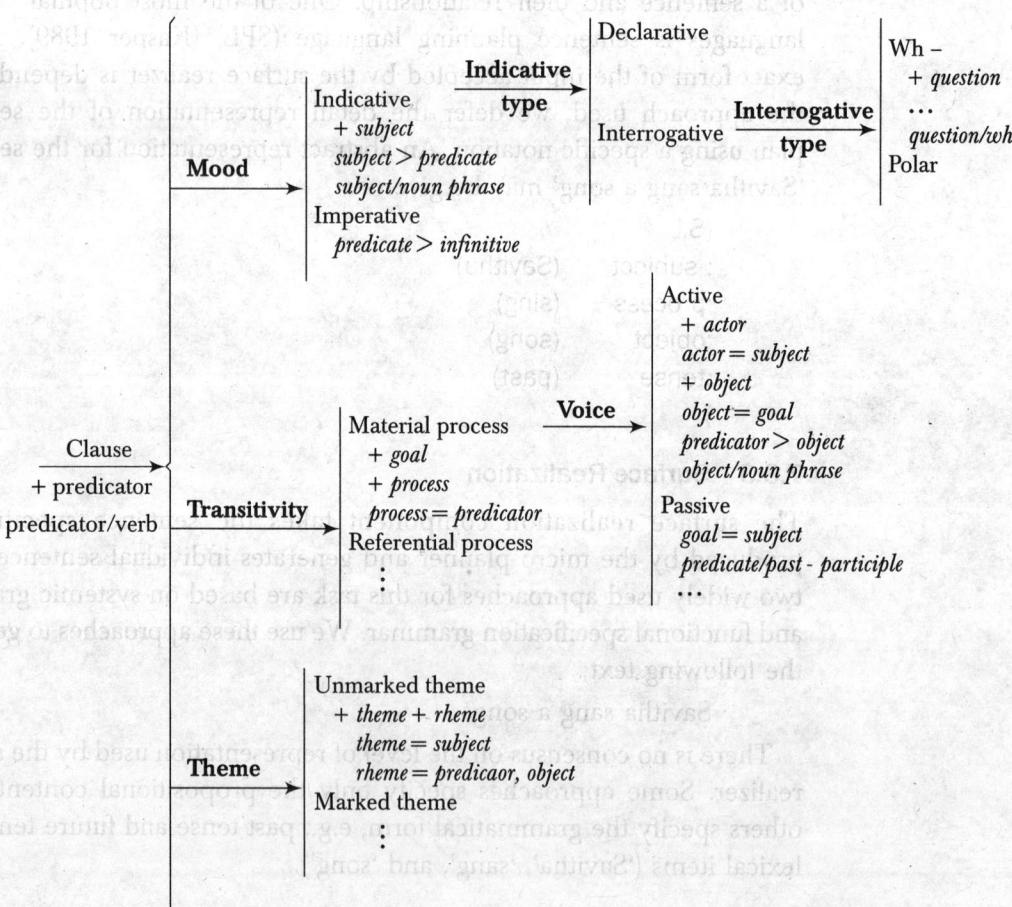

Figure 7.5   A simple systemic grammar

'or'. This means every clause (shown on extreme left) can have features for mood, transitivity, and theme. Mood can be either indicative or imperative, but not both. Similarly, transitivity can be a material process or a referential process. With this grammar specification, our example sentence is an indicative, declarative clause representing a material process with an unmarked theme.

The italicized statements written below the grammar features (e.g., indicative and declarative), map the features into syntactic form and are called realization statements. Realization statements associated with a feature specify constraints on the final form of the expression. The operators used in this specification are defined as follows.

> +X—Insert the function X. For example, the grammar in Figure 7.5 specifies that all clauses will have a predicator and each indicative sentence will have a subject and finite, etc.
>
> X=Y—Conflate the functions X and Y. Using this operator, the grammar assigns different functions to the same portion of expression. For example, active clauses conflate the actor with the subject and object with the goal.
>
> X > Y—This operator is used to impose ordering between two functions, i.e., order X before Y. For example, the indicative sentences place subject before predicator and finite after predicator.
>
> X/A—Classify the function X with grammatical or lexical feature A. Classifying X with these features leads to a recursive pass through the grammar at a lower level. For example, the active clauses insert an object which must be a noun phrase, the specification of which requires another pass through the grammar.

A systemic sentence analysis yields a three layer representation. For example, the analysis of the sentence, 'Savitha sang a song':

Mood	subject	predicator	object
Transitivity	actor	process	goal
Theme	theme	rheme	

The first layer is the mood layer, which specifies a simple declarative structure with subject, predicator (verb) object. The transitivity layer tells us that 'Savitha' is the actor of the process of 'singing' and the goal, the object acted upon, of this process is 'song'. Finally, the third layer, the theme layer, indicates that 'Savitha' is the theme, the main topic of discussion, of the sentence. These three layers describe three different sets of functions, called meta-functions. These meta-functions are known as interpersonal meta-function, ideational meta-function, and textual meta-function.

The interpersonal meta-function refers to set of functions responsible for maintaining the interaction between the writer and the reader.

The ideational meta-function deals with propositional content of the expression. It corresponds to the transitivity layer which captures semantics in terms of the nature of the process being expressed and various case roles.

The textual meta-function is concerned with issues pertaining to discourse, e.g., thematic and referential issues. For the example under consideration, this objective is fulfilled by explicitly marking 'Savitha' as the theme of the sentence.

Using a fully specified system network, we can generate sentences using the following steps:

1. Traverse the system network to choose the appropriate features and retrieve associated realization statements.
2. Build an intermediate expression that is compatible with the constraints imposed by realization statements.
3. Make a recursive pass of the grammar at a low level to complete the specification of any incompletely specified function.

The generation process can be viewed as a choice-making process (Reiter and Dale 2000). A pass through the grammar makes a choice regarding lexical and grammatical features. This choice is based upon input specification and information from the knowledge base. The successive recursive passes make a series of increasingly fine-grained choices. Taken together, these choices describe the syntactic characteristics of the sentence.

To illustrate this process, we traverse the grammar to generate the sentence, 'Savitha sang a song'. We begin with the following input specification:

```
(
  : process      sang-1
  : actor        Savitha
  :goal          song
  :speechact     assertion
  :tense         past
)
```

Here, we assume that objects are instances for which lexical entries exist in the knowledge base. Thus, sang-1 is an instance identifying the process involved in the expression, while Savitha-1 is the actor and song-1 is the object. The input also specifies that the intended expression is an assertion in past tense.

The traversal starts from leftmost clause feature in Figure 7.5. The associated realization statements cause the insertion of predicator and classify it as verb. It then enters mood system which chooses indicative and declarative features because input is an assertion. The realization statement associated with it inserts subject function and orders it as subject then predicator. The resulting structure becomes

Mood	subject	predicator

Next, we traverse the transitivity system. This system consults the knowledge base to get information regarding sang-1 action. We assume that sang-1 action is marked as the material process in the knowledge base. Thus, the transitivity system selects material process, inserts goal and process function and conflates process with the predicator. As the input specification and the knowledge base do not specify that the sentence is passive, the system next chooses the active feature. The realization statements associated with it insert the actor and the object, and the system conflates them with the subject and the goal respectively. It also orders the object after predicator. This gives the following structure:

Mood	subject	predicator	object
Transitivity	actor	process	goal

Finally, the theme network chooses an unmarked theme, inserts theme and rheme and conflates them with subject and predicator/object pair respectively. This yields the following full function structure (as discussed earlier):

Mood	subject	predicator	object
Transitivity	actor	process	goal
Theme	theme	rheme	

The generation process makes recursive passes to fully specify phrases, lexical items, etc. For example, it enters a noun phrase network to produce 'Savitha' and 'a song'. The lexical 'sang', 'Savitha', and 'song', are retrieved from the knowledge base. We assume here that the lexical items are associated with object instances.

Approach Based on Functional Unification Grammar

Functional unification grammar (FUG) uses unification to manipulate feature structures (as discussed in Chapter 5). In this section, we illustrate the use of FUG in NLG with the help of an example. We generate the sentence once again using a simple functional unification formalism (FUF) (shown in Figure 7.6). The FUF is an implementation of FUG developed by Elhadad (1995).

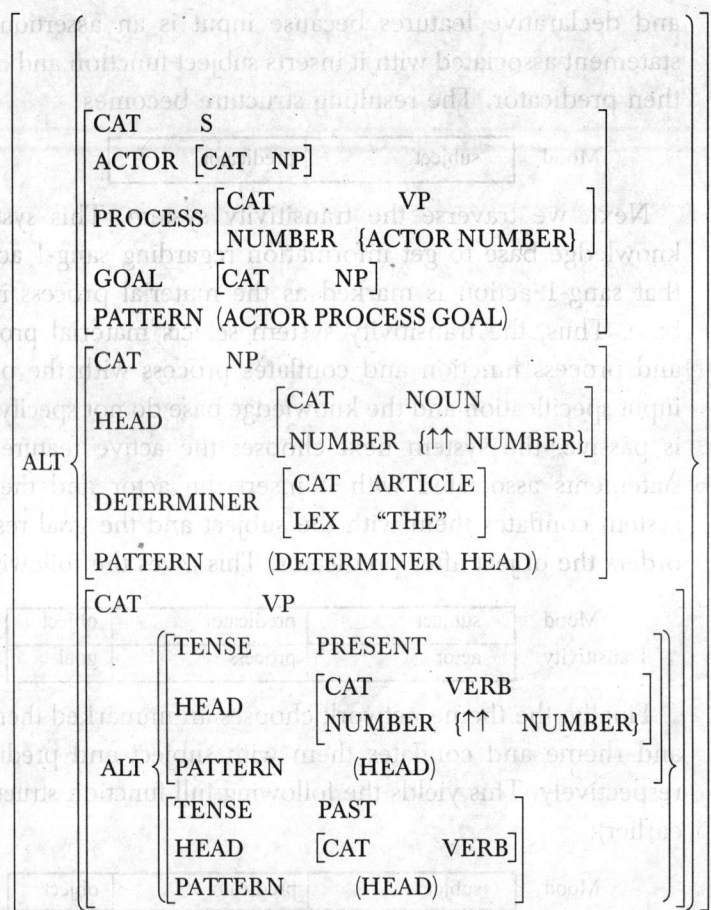

Figure 7.6 A simple FUF grammar

The leftmost ALT feature in the grammar describes a top level alternative, e.g., sentences (cat S), noun phrases (cat NP), and verb phrases (cat VP). The curly brace signifies that any of these alternative categories may be followed. The feature structure of each category also specifies a pattern that tells us the order of the features in that category, e.g., for the category sentence (S), the ordering is actor, process, and goal. At the sentence level, the grammar supports actor, process, and goal features with the categories NP, VP, and NP respectively. The process feature also enforces subject-verb agreement on number using the number feature. Here, the agreement tells us that the number of the process must unify with the path {actor number}, i.e., with the number of the actor.

At the noun level, the grammar supports head and determiner features. The agreement enforced by the head feature of the noun phrase is that the number of the head of the NP must agree with the number of the feature two levels up.

The VP level has its own alternation between present and past tenses. In the case of present tense, the head will be single verb.

The input feature structure, called functional description (FD), provides the details of intended sentence. An FD for the example sentence is as follows:

$$
\begin{bmatrix}
\text{CAT} & \text{S} \\
\text{ACTOR} & \begin{bmatrix} \text{HEAD} & \begin{bmatrix} \text{LEX} & \text{SAVITHA} \end{bmatrix} \end{bmatrix} \\
\text{PROCESS} & \begin{bmatrix} \text{HEAD} & \begin{bmatrix} \text{LEX} & \text{SING} \end{bmatrix} \\ \text{TENSE} & \text{PAST} \end{bmatrix} \\
\text{GOAL} & \begin{bmatrix} \text{HEAD} & \begin{bmatrix} \text{LEX} & \text{SONG} \end{bmatrix} \end{bmatrix}
\end{bmatrix}
$$

The input specification consists of an actor 'Savitha', a goal 'song', and the process of singing a song. The input also specifies the lexical items (nouns and verbs) and tense. We now unify the input with the grammar shown in Figure 7.6.

The first pass through the grammar unifies the input features with the features at the top level of the grammar. For example, the ACTOR feature combines the lexical item 'Savitha' from input FD and category 'NP' from the grammar.

The generation process makes a recursive pass through the grammar to complete the specification of NP and VP. In particular, it enters the NP level twice to complete the specification of actor and goal; and the VP level once to complete the specification of process. Figure 7.7 shows the resulting FD. In this FD, each complex constituent has a pattern specification, while each single constituent has a lexical specification. The pattern specification is used to specify the ordering of words. Using this pattern specification, the system generates the output: 'Savitha sang a song.'

7.4 APPLICATIONS OF NLG

The NLG systems have been used to provide natural language interfaces

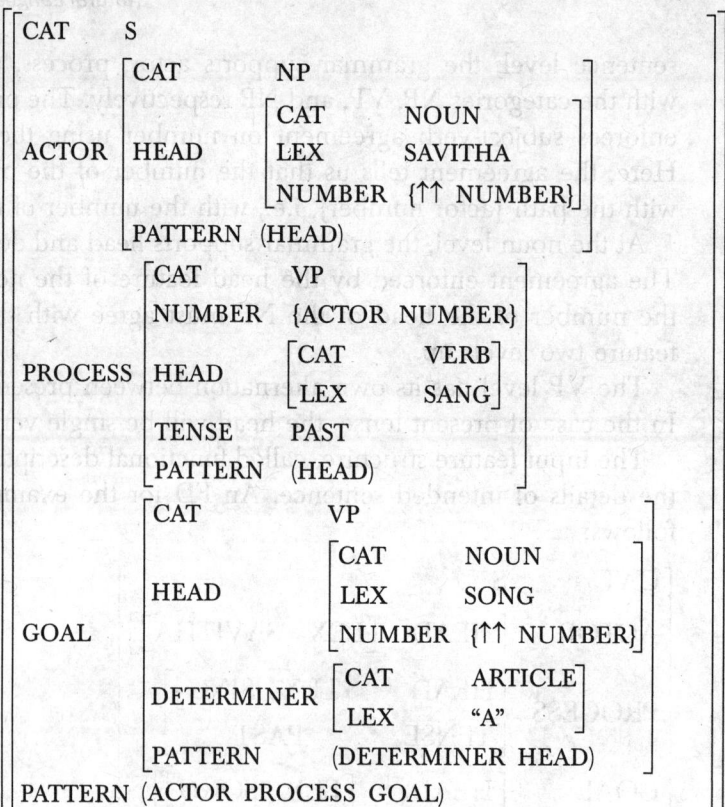

Figure 7.7 FD with fully specified S, NP, and VP

to many databases, such as airline schedule database, accounting databases or spreadsheets, expert system knowledge bases, etc. The internal representations used by these systems cannot be understood easily by a non-expert user. NLG technology has been fruitfully utilized to present such data in a human-readable form. Its use is not just restricted to databases and spreadsheets; it can be used to summarize graphical and speech data. NLG is also needed for generating the abstractive summary of a textual document. Given here is a list of some of the previous work involving application of the NLG technique.

1. The NLG is used to summarize statistical data extracted from database or spreadsheet (Iordanskaja et al. 1992).

2. It is used in the WeatherReporter system to create multi-sentence weather summaries from the database of numerical meteorological data, collected automatically by weather service as mentioned in Essers and Dale.

3. Maybury (1995) used NLG to create summaries from event data.
4. The NLG produces answers to questions about an object described in a knowledge base (Reiter et al. 1995).

SUMMARY

- Natural language generation (NLG) deals with the automatic generation of a natural language sentence using non-linguistic inputs.
- In order to generate text we have to identify (a) what to write and (b) how to write it.
- The three main tasks in NLG are discourse planning, text planning, and surface realization.
- These tasks are usually broken into many sub-tasks and may be organized in a number of ways, leading to pipeline, interleaved or integrated types of architecture.
- In the interleaved architecture, the discourse and text planning stages are merged to indicate that the control and information can be passed back to any task in the two stages.
- The discourse planner decides ordering and structure of the text that is to be generated. Rhetorical relationships are often used to express the structure of the text. The output is usually represented in the form of a tree.
- The text (micro) planner decides words and phrases to express the concepts and structures present in a discourse plan. The tasks performed by it are sentence aggregation, lexicalization and referring-expressions generation.
- The surface realizer takes sentence plans produced by the text planner and generates individual sentences.
- Two widely-used approaches for the surface realization task are based on systemic grammar and functional unification grammar.

REFERENCES

Caldwell, D. and T. Korelsky, 1994, 'Bilingual generation of job descriptions from quasi-conceptual forms,' *Proceedings of the Fourth Conference on Applied Natural Language Processing*, Association for Computational Linguistics, pp. 1–6.

Coch, J., 1996, 'Evaluating and comparing three text production techniques,' *Proceedings of COLING-1996*.

Essers, Victor and Robert Dale, 1998, 'Choosing a surface realizer: exploring the differences in using KPML/NIGEL and FUF/SURGE,' *Proceedings of PRICAI 98*, Singapore, September.

Halliday, M., 1985, *An Introduction to Functional Grammar*, Edward Arnold, London.

Hirst, G., C. DiMarco, E. Hovy, and K. Parsons, 'Authoring and generating health-education documents that are tailored to the needs of the individual patient,' *Proceedings of the Sixth International Conference on User Modelling*, Italy.

Hovy, Eduard, 1985, 'Integrating text planning and production in generation,' *Proceedings of the Ninth IJCAI*, Los Angeles, CA.

_____1988, *Generating Natural Language under Pragmatic Constraints*, Lawrence Erlbaum Associates, Hillsdale, New Jersey.

Iordanskaja, L., Kim, M., Kitteredge, R., Lavoie, B., and Polguere, A., 1992, 'Generation of extended bilingual statistical reports,' *Proceedings of Coling-92*, Nantes, August, pp. 1019–23.

Jurafsky, Daniel and James H. Martin, 2000, *Speech and Language Processing. An Introduction to Natural Language Processing, Computational Linguistics and Speech Recognition*, Prentice Hall, NJ.

Kantrowitz, Mark and Joseph Bates, 1992, 'Integrated natural language generation system,' *LNAI*, 587, Springer-Verlag.

Kasper, R., 1989, 'A flexible interface for linking applications to Penman's sentence generator,' *Proceedings of the 1989 DARPA Speech and Natural Language Workshop*, pp. 153–58.

Maybury, M.T., 1995, 'Generating summaries from event data,' *Information Processing and Management*, 31(5).

McKeown, K., K. Kukich, and J. Shaw, 1994, 'Practical issues in automatic document generation,' *Proceedings of the Fourth Conference on Applied Natural Language Processing*, pp. 7–14.

Mellish, Chris, Mike Reape, Donia Scott, Lynne Cahill, Roger Evans, and Daniel Paiva, 2004, 'A reference architecture for generation systems,' *Natural Language Engineering*, 10, 3(4), Cambridge University Press, UK, pp. 227–60.

Paris, C., K. Vander Linden, M. Fischer, A. Hartley, L. Pemberton, R. Power, and D. Scott, 1995, 'A support tool for writing multilingual instructions,' *Proceedings of 14th International Joint Conference on Artificial Intelligence*, pp. 1398–1404.

Reiter, Ehud and Robert Dale, 2000, 'Building applied natural language generation systems,' *Natural Language Engineering*, 1(1), Cambridge University Press.

Scott, D., R. Power, and R. Evans, 1998, 'Generation as a solution to its own problem,' *Proceedings of the Ninth International Workshop on Natural Language Generation* (INLG 1998), Niagara-on-the-lake, Canada, pp. 256–65.

Springer, S., P. Buta and T. Wolf, 1991, 'Automatic letter composition for customer 'service,' *Proceedings of the Innovative Applications of Artificial Intelligence Conference (CAIA-1991)*, pp. 67–83.

Weizenbaum, R., 1966, 'ELIZA—A computer program for the study of natural language communication between man and machine,' *Communications of ACM*, 9(1).

EXERCISES

1. In what ways does NLG differ from the earlier applications discussed in this book?
2. Discuss the pipelined architecture of the NLG system.
3. How is the interleaved architecture different from the integrated architecture?
4. What is the role of micro-planner in NLG?
5. Use the systemic grammar given in the chapter to analyse the following sentences:
   (a) Savitha will sing a song.
   (b) Savitha sings a song.
6. Extend the systemic grammar in the chapter to handle the following sentence:
   Savitha danced at the party.
7. Use the FUF grammar given in the chapter to build a fully unified FD for the following sentences:
   (a) Savitha will sing a song.
   (b) Savitha sings a song.
8. Extend the FUF grammar in the chapter to handle the following sentences:
   (a) Savitha danced in the party.
   (b) Will Savitha sing a song?
   (c) Sing a song.

LAB EXERCISES

1. Write a program that takes as an input a paragraph consisting of 5 to 10 sentences without any pronominal references, and introduce appropriate references in it.
2. Write a program to combine two sentences using conjunctions such as 'but', 'although', 'and', 'which', etc.
3. Write a program to generate simple sentences from an intermediate representation of your choice.

CHAPTER 8

MACHINE TRANSLATION

CHAPTER OVERVIEW

Machine translation (MT) facilitates translation of text from one language to another. This chapter discusses various types of MT approaches such as direct, rule-based (transfer and interlingua), corpus-based, and knowledge-based. It also highlights characteristics of Indian languages and translation strategies.

8.1 INTRODUCTION

Machine translation (MT), perhaps the earliest NLP application, is the translation of text units from one language into another, using computers. It is fascinating to think of an MT system that can translate literary works from any language into our native language. So we need not to know Urdu to read the stories of Manto; just feed it to an MT system and get it translated. Such MT systems are able to break the barriers of language by making available rich sources of literature to people across the world.

Machine translation can also help overcome technological barriers. The IT revolution has led to a plethora of information. But the flood of this information is available only in a very small subset of languages and beyond the reach of a significant portion of society. This has led to yet another division in society called the digital divide. MT can be of great help in removing this divide. Multilingual countries, like India where only 3% of the population can understand English (Sinha and Jain 2003), are more particularly in need of MT systems to translate information from English into local languages.

Thus, the dream of MT is not just to unite the world intellectually and culturally, but also to unite it through technology. Achieving this goal seems to be a Herculean task. As Jurafsky and Martin (2000) rightly pointed out,

Translation, in its full generality, is a difficult, fascinating, and intensely human endeavour, as rich as any other area of human creativity.

The goal of achieving error-free translation that reads fluently in target languages is still far off; limited success has been achieved within a restricted domain. For example, the well-known METEO system automatically translates hundreds of Canadian weather bulletins every day with 95% accuracy (Chandioux 1988, Macklovitch and Isabelle 1990). Existing state-of-the art systems in MT do not permit completely automatic, high quality, general-purpose translations—one of the three facets has to be compromised. We can have an automatic, high quality system for sub-languages, or an automatic general-purpose system yielding rough translation, or an interactive general-purpose system with pre- or post-editing.

Achieving high quality translation is difficult. This is partly because a text must be thoroughly understood before it can be translated. In the previous chapters, we discussed various types of ambiguities that make it difficult to understand a text. We provide a quick review of such issues that make machine translation hard in Section 8.2. In Section 8.3, we discuss important characteristics of Indian languages. The diverse approaches to MT are discussed in Section 8.4. Section 8.5 is about existing machine translation systems, focussing on Indian languages. Finally, the last section summarizes the main points.

8.2 PROBLEMS IN MACHINE TRANSLATION

There are many structural and stylistic differences among languages, which make automatic translation a difficult task.

8.2.1 Word Order

The arrangement of words in a sentence varies across languages. For example, in English words are usually arranged in the order subject, verb and object; whereas in Indian languages, object usually precedes verb. This makes a word by word translation impractical.

8.2.2 Word Sense

The sense of a word in one language may translate into a different sense with the words of another language. This creates problem in target language word selection.

8.2.3 Pronoun Resolution

Resolving pronominal references is important for machine translation. Unresolved references may lead to incorrect translation.

8.2.4 Idioms

A sentence involving idiomatic expressions is difficult to translate as idioms are composed of words that do not directly contribute to their meaning. Replacing words constituting an idiom with words from the target language can lead to funny and nonsensical translations.

For example, consider the sentence: 'The old man finally kicked the bucket.' If the system does not recognize the idiom 'kicked the bucket', the translation, say in Hindi, will end up as: '*Boodhe aadmi ne ant-ta balti mein laat maari*'.

8.2.5 Ambiguity

Certain languages do not permit certain types of ambiguities. For example, consider the PP-ambiguity in the English sentence: 'The man saw the girl with a telescope.' In order to translate this sentence into Hindi, its PP-ambiguity must first be resolved.

8.3 CHARACTERISTICS OF INDIAN LANGUAGES

A wide variety of languages are used in India. In this section, we take a look at the important characteristics of Indian languages. We categorize Indian languages in the following four broad families:

- Indo-Aryan (e.g., Hindi, Bangla, Asamiya, Punjabi, Marathi, Gujarati, and Oriya)
- Dravidian (Tamil, Telugu, Kannada, and Malayalam)
- Austro-Asian
- Tibetan-Burmese

Languages within each group are structurally similar to each other, to a great extent. We have already pointed out some of the characteristics of Indian languages in Chapters 1, 3, and 4. Considering their use in machine translation, we summarize common characteristics here.

1. Indian languages have SOV (Subject-Object-Verb) as the default sentence structure.
2. Indian languages are free word order, i.e., words can be moved freely within a sentence without changing the meaning of the sentence. The relationship between various components of the sentence is shown by inflecting the components. For example, in Hindi, the following two sentences mean the same thing:

    Surabhi ne Poonam ko ek kitab di.
    Poonam ko Surabhi ne ek kitaab di.

3. Indian languages have a relatively rich set of morphological variants. Unlike English, Indian-language adjectives undergo morphological changes depending upon number and gender.
4. Indian languages make extensive and productive use of complex predicates (CPs). As discussed in Chapter 4, a complex predicate

combines a light verb with a verb, noun or adjective to produce a new verb, e.g., सबा आ गयी (Saba *aa gayi.*) which means 'Saba arrived', not 'Saba came' (see Section 4.6).

5. Indian languages make use of post-position case markers (Karaks) instead of prepositions. In some languages, inflections get attached to objects, to handle English prepositions, e.g., *bazaar-e* (market-to) in Bengali.

6. Indian languages make use of verb complexes consisting of sequences of verbs, e.g., *ga raha hai* and *khel rahi hai.* The auxiliary verbs in this sequence provide information about tense, aspect, and modality. Gender information is also contained in verb group.

7. Most Indian languages have only two genders—masculine and feminine.

8. Sometimes, adjectives are also modified to agree with gender, e.g., *achcha ladka* vs *achchi ladki.*

9. Unlike English, Indian-language pronouns have no associated gender information.

So far, we have made no specific mention of languages in the Dravidian family. Let us now focus on these languages and point out some of the important differences between them and the Indo-Aryan family of languages. We consider Tamil for this discussion.

Tamil is an agglutinative language. Tamil words are formed by combining roots with other grammatical information. These accumulations can go up to a maximum of six in number and include *pakuthy* (prime-stem), *sandhi* (junction), *viha:ram* (variation), *idainilai* (middle part), *sa:riyai* (enunciater), and *vikuthy* (terminator), in that order (Mohanrajah and Veer Singhe). Regardless of the length, type and complexity of Tamil words, we can reduce each word to its monosyllabic root by removing successive accumulations.

Unlike Hindi, Tamil and Telugu do not make use of explicit Karakas. Instead, morphological variations are used to reflect the intended relationship. For example, '*ammayi*' in Telugu means 'girl' and '*ammayiki*' means 'to girl'. Similarly, the Tamil word '*ponnu*' means 'girl' and '*ponnunko*' means 'to girl'.

Tamil singular and plural nouns have the same form of case-terminators. In Tamil, the verb carries information about tense, aspect, modality, and gender. This is unlike languages in the Indo-Aryan family in which auxiliary verbs quite often provide this form of information. In Tamil, all this information is attached to root verb in a pre-defined order.

8.4 MACHINE TRANSLATION APPROACHES

Machine translation approaches can be broadly classified into the following four categories (Figure 8.1):

1. Direct machine translation
2. Rule-based translation
3. Corpus-based translation
4. Knowledge-based translation

We discuss these approaches in the following sections. Unless otherwise stated, we consider the case of English to Hindi translation.

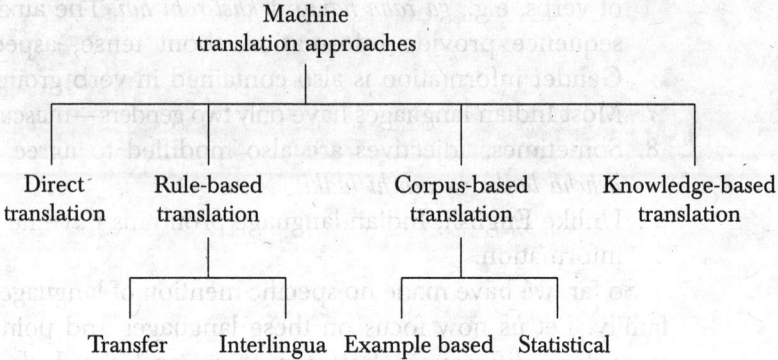

Figure 8.1   Machine translation approaches

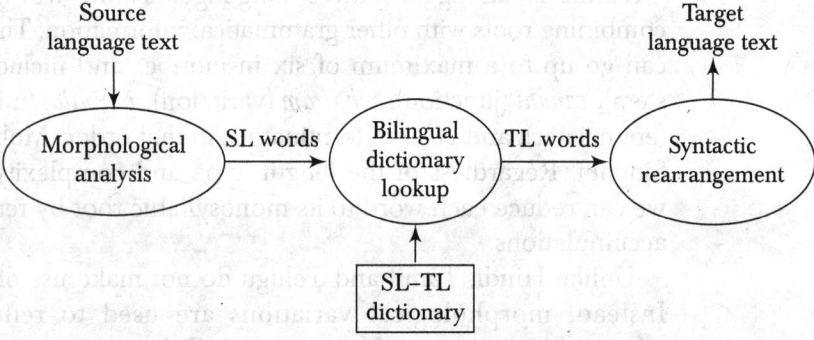

Figure 8.2   Direct machine translation systems

8.5   DIRECT MACHINE TRANSLATION

As the name suggests, direct machine translation systems provide direct translation, i.e., no intermediate representation is used. A direct translation system carries out word-by-word translation with the help of a bilingual dictionary, usually followed by some syntactic rearrangement. These systems are based on the principle that an *MT system should do as little work as possible* (Jurafsky and Martin 2000). Direct translation systems take a monolithic approach towards development, i.e., they consider all the details of one language pair. They involve little analysis of the source text, no parsing, and rely mainly on a large bilingual dictionary. Besides dictionary translation, the analysis performed in this approach includes morphological analysis, preposition handling, syntactic arrangement (so as to reflect correct word order in the target language) and morphological generation. The general procedure for direct translation (English to Hindi) systems can be summarized in the following three steps:

1. Remove morphological inflections from the words to get the root form of the source-language (English) words.
2. Look up a bilingual dictionary to get the target-language words corresponding to the source-language words.
3. Change the word order to that which best matches the word order of the target language, e.g., in a English–Hindi translation system, this would involve changing prepositions to post-positions and changing the subject-verb-object structure to subject-object-verb.

Consider the following English sentence:

> *Khushbu slept in the garden.*   (8.1)

To translate this sentence into Hindi, a direct translation system will first look up a dictionary to get target words for each word appearing in the source-language sentences. Then, the words are re-ordered to match the default sentence structure (SOV) of Hindi. The output of these steps is as follows.

Word translation:

> खुशबू सोयी में बाग
> *Khushbu soyi mein baag*

Syntactic rearrangement:

> खुशबू बाग में सोयी
> *Khushbu baag mein soyi*

Besides word ordering and preposition handling, suffix handling is also needed in order to make the translation acceptable. For example, while translating sentence (8.2), we need to change the Hindi word *ladka* to *ladke* simply to match it to the way Hindi is naturally used. This is termed *idiomatization*.

English sentence:

> *The boy gave the girl a book.*   (8.2)

Word translation:

> *Ladka dee ladki ek kitaab*
> लड़का दी लड़की एक किताब

Syntactic rearrangement:

Ladka ladki ek kitaab dee

लड़का लड़की एक किताब दी

Karaka handling and idiomatization:

Ladke ne ladki ko ek kitaab di

लड़के ने लड़की को एक किताब दी

Other changes include modifying verbs and adjective according to the gender of the subject. Here is an illustration.

She saw stars in the sky. (8.3)

वो देखा तारे में आसमान

Wo dekha tare mein aasman

वो आसमान में तारे देखी

Wo aasman mein tare dekhi

Karak handling and idiomatization:

उसने आसमान में तारे देखे

Usne aasman mein taare dekhe

Translation steps in Telugu and Tamil

Telugu

Khushbu   slept   in   the garden. (8.1)

ఖుష్బు నిద్రపోతుంది లో ఉద్యానవనము

Khushbu   *nidrapotundi*   *lo*   *udyanavanam*

Morphological handling and syntactic rearrangement yields the following translation:

ఖుష్బు ఉద్యానవనములో నిద్రపోతుంది.

Khushbu   *udyanavanamlo*   *nidrapotundi*

The boy   gave   a book   to the girl. (8.2)

అబ్బాయి | ఇచ్చాడు | పుస్తకం | అమ్మాయికి

Abbai   *ichadu*   *pustakam*   *ammayiki*

Reordering the words to get reflect the structure of Tamil sentence, we get

అబ్బాయి అమ్మాయికి పుస్తకం ఇచ్చాడు.

Abbai   *ammayiki*   *pustakam*   *ichadu*

She   saw   stars   in   the sky.   (8.3)

ఆమె   చూసింది   నక్షత్ర   లో   ఆకాశం

Aame   choosindi   nakshatra   lo   aakasham

After morphological variations and syntactic rearrangement, we get

Aame   aakashamlo   nakshatralu   choosindi

ఆమె   ఆకాశంలో   నక్షత్రాలు   చూసింది

Tamil

Khushbu   slept   in the garden   (8.1)

Khushbu   thongivittal   pungavil

Syntactic rearrangement yields the following translation:

Khushbu   pungavil   thongivittal

The   boy   gave   a   book   to   the   girl

Antha   payan   kuduthan   oru   pustakam   ku   antha   ponnun

Syntactic rearrangement and morphological handling yields:

Antha   payan   antha   ponnunku   oru   pustakam kuduthan

She   saw   stars   in the sky.   (8.3)

Aval   parthal   nakshatram   agayathil

After syntactic rearrangement:

Aval   agayathil   nakshatram   parthal

When translating between Indian languages, the task of translation is simpler, as most Indian languages have the same default sentence structure. Hence, a simple word-by-word translation, with a little idiomatization, may lead to an acceptable translation in many cases. For example, a simple word-by-word substitution may suffice in most of the cases when the translation is taking place between Hindi and Urdu sentences.

आज मुख्यमंत्री ने बेगम हजरत महल पार्क में एक सभा
को सम्बोदित करते हुए ये घोषणा की   (8.4)

Aaj mukhyamantri ne begum hazrat mahal park mein ek sabha ko sambodhit karte hue yeh ghoshna ki

A word-by-word translation in Urdu results in the following sentence:

آج وزیراعلیٰ نے بیگم حضرت محل پارک میں ایک جلسہ کو خطاب کرتے ہوئے یہ اعلان کیا۔

Aaj wazir-e-aala ne begum hazrat mahal park mein ek intekhabi jalse ko khitab karte hue yae elaan kiya

The preceding sentence can be understood easily by a Urdu speaking person. With a little idiomitazaion, it becomes acceptable translation.

آج وزیرِ اعلیٰ نے بیگم حضرت محل پارک میں ایک انتخابی جلسہ کو خطاب کرتے ہوئے یہ اعلان کیا۔

Aaj wazir-e-aala ne begum hazrat mahal park mein ek jalse ko khitab karte hue yae elaan kiya.

For Hindi to English translation, such a word-by-word replacement may sometimes be difficult to understand. For example, consider a direct translation of sentence (8.4):

Chief Minister one electoral meeting addressing doing this announcement made

Selection of the correct target language word is another problem in direct translation systems. Consider the case following English to Hindi translation.

Book a ticket for me.   (8.5)

A word-by-word translation does not make it clear whether 'book' is being used as a noun or verb. This makes it difficult to get at the correct meaning.

A direct MT system involves only lexical analysis. It does not consider the structure and relationships between words. It does not attempt to disambiguate words. Hence, the quality of output is often not very good. A direct MT system is developed for a specific language pair and cannot be adapted for a different pair. Being monolithic, it is not possible to take advantage of any of the analysis performed in one MT system to provide translation capabilities to a new language pair. In a multilingual scenario, this may be quite expensive. For example, in order to provide translation capability for n number of languages, we need to develop $n(n-1)$ MT systems.

8.6   RULE-BASED MACHINE TRANSLATION

Rule-based MT systems parse the source text and produce an intermediate representation, which may be a parse tree or some abstract representation. The target language text is generated from the intermediate representation. These systems rely on specification of rules for morphology, syntax, lexical selection and transfer, semantic analysis and generation, and are hence called rule-based systems. These systems make use of extensive lexicons equipped with morphological, syntactic and semantic information, and a large set of rules to map the input text to intermediate representation. Examples of rule-based include the Ariane and SUSY systems. Depending on the intermediate representation used, these systems are further categorized as follows:

1. Transfer-based machine translation
2. Interlingua machine translation

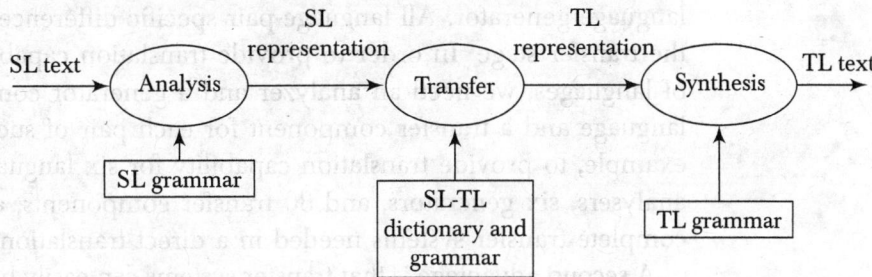

Figure 8.3 Schematic diagram of the transfer-based model

8.6.1 Transfer-Based Translation

As discussed in the previous section, a direct translation model maps directly between source and target language without producing an intermediate representation. An alternative strategy is to use transfer-based translation models. These models transform the structure of the input to produce a representation that matches the rules of the target language. This transformation requires an understanding of the differences between the source and target language. In order to get the structure of the input, some form of parse is needed. Therefore, transfer-based translation involves parsing of the source text. The source language parse structure is then transferred into the target language structure. The generation of the target language text is facilitated by a generation module. Thus, a transfer-based machine translation system has the following three components:

1. Analysis—To produce source language structure
2. Transfer—To transfer the source language representation to a target level representation
3. Generation—To generate target language text using target level structure

The first stage analyses the source text and produces a structure confirming the rules of source language. Details of the analysis vary from system to system. It may involve morphological, syntactic, and semantic analyses. As this stage involves parsing of source text, syntactic ambiguities and lexical ambiguities are better resolved in this approach than in the direct translation approach. The second stage transfers source language representation into target language representation. All the language-pair specific peculiarities are handled by the transfer component. The third stage is responsible for generating the actual target language text.

The main advantage of this approach is its modular structure. The analysis of source language text (i.e., the parser) is independent of target language generator. All language-pair specific differences are captured in the transfer stage. In order to provide translation capability among a set of languages, we need an analyzer and a generator component for each language and a transfer component for each pair of such languages. For example, to provide translation capability for six languages, we need six analysers, six generators, and 30 transfer components, as opposed to 30 complete transfer systems needed in a direct translation approach.

A second advantage is that transfer systems can easily handle ambiguities that carry over from one language to another. For example, we need not manually resolve the PP-attachment ambiguity in the sentence, *The girl plucked a flower with stick*, while translating it into French. Likewise, the lexical ambiguity in the Hindi sentence, *Mujhe sona achchha lagta hai*, need not be resolved while translating it into Urdu.

Transfer systems also help in resolving certain lexical ambiguities. Consider sentence (8.5). The parse tree generated by the transfer system makes it clear that 'book' is used as verb in this sentence and not as a noun. This information prevents 'book' from being translated into '*kitaab*' in Hindi.

A further advancement in the transfer-based approach is the use of reversible representations and processing rules. Feature-structure unification, introduced in Chapter 4, was basically motivated by the need for reversible transformation in MT systems (Jurafsky and Martin 2000). Use of unification and constraint-based grammar formalisms simplify the rules of analysis, transformation, and generation (Hutchins 1995). Instead of a multi-level representation and a large set of rules (for mapping morphological representation to syntactic tree, syntactic tree to semantic tree, and source language representation into an equivalent target language representation), which apply in specific circumstances and to specific representations, there are monostratal representations and an abstract set of rules, with conditions and constraints incorporated into specific lexical entries. The reversible nature of grammar components has the advantage that for a given language the same grammar can be used both for analysis as well as for generation.

The essential idea of the structural transfer algorithm is that SOV in English becomes SVO in Hindi, and post modifiers in English become pre modifiers in Hindi (Rao et al. 1998). Figure 8.4 shows an example of the structural transfer for sentence (8.1).

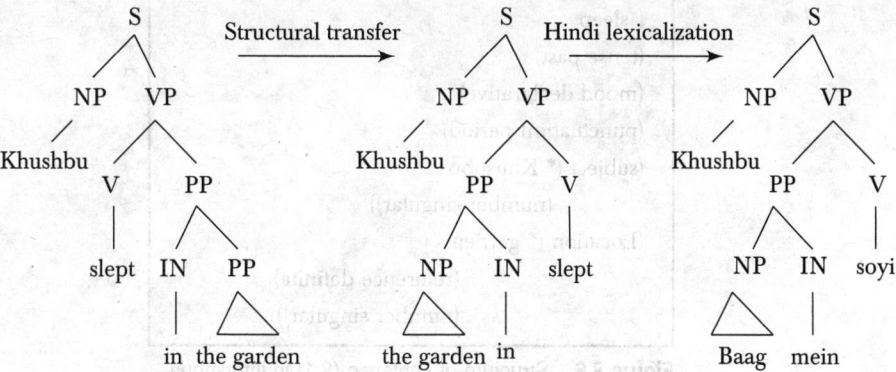

Figure 8.4 Structural transfer of sentence (8.1) from Hindi to English

Transfer systems are perhaps more realistic, flexible, and adaptable in meeting the needs of different levels and 'depths' of syntactic and semantic analysis. One of the advantages of the transfer system is that it can be extended to language pairs in a multilingual environment.

8.6.2 Interlingua-based Machine Translation

The inter-lingual approach was inspired by Chomsky's claim that regardless of varying 'surface' syntactic structures, languages share a common 'deep structure' (Chomsky 1965). In interlingua-based MT approach, the source language text is converted into a language independent meaning representation called 'interlingua'. *An interlingua represents all sentences that mean the same thing in the same way regardless of the source language they happen to be in* (Jurafsky and Martin 2000). From interlingual representation, texts are generated into other languages. Translation is thus a two-stage process, analysis and synthesis, as shown in Figure 8.5. Structure of sentence (8.1) in interlingua is illustrated in Figure 8.6.

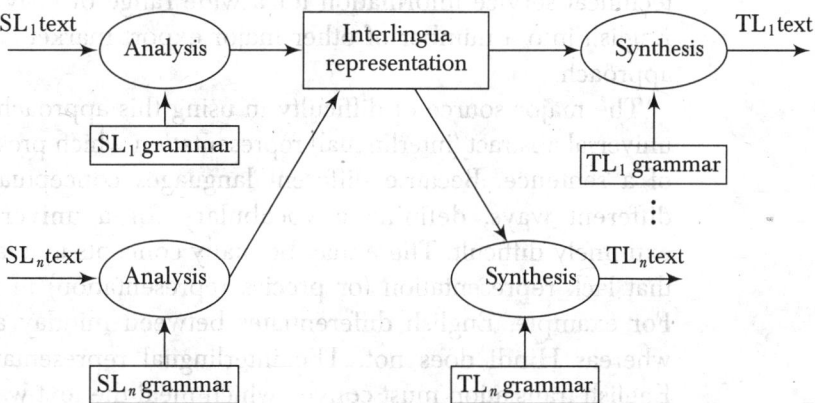

Figure 8.5 Interlingua-based translation model

```
(*sleep
(tense past)
(mood declarative)
(punctuation period)
(subject (* Khushbu
          (number singular))
(Location (* garden
              (reference definite)
              (number singular)))
```

Figure 8.6 Structure of sentence (8.1) in interlingua

In the first stage, source language text is represented in interlingua; in the second stage, target language text is generated. The analysis phase is specific to the source language text. Similarly, the synthesis phase is specific to the target language. This makes it convenient to use in the multilingual environment. The same analysis component may be used for more than one target language. This means that in order to build a multilingual translation capability among *n* number of languages, we need only *n* analysis and n generation components, as opposed to $n(n-1)$ complete MT systems needed in direct translation approach.

The amount of analysis needed in an interlingua system is much more than in a transfer system. An interlingua system has to resolve all ambiguities so that translation to any language can take place from interlingual representation. It therefore requires a semantic analyser as well as a syntactic parser. Another advantage of interlingua is that it is a meaning-based representation and can be used in applications like information retrieval. The KANT (knowledge-based, accurate natural-language translation) system (Nyberg and Mitamura 1992), which translates technical service information for a wide range of heavy machinery from English into a number of other major export market languages, uses this approach.

The major source of difficulty in using this approach lies in defining a universal abstract (interlingual) representation which preserves the meaning of a sentence. Because different languages conceptualize the world in different ways, defining a vocabulary for a universal interlingua is extremely difficult. There may be many concepts in a language or culture that lack representation (or precise representation) in another language. For example, English differentiates between midday and evening meal whereas Hindi does not. The interlingual representation for Hindi to English translation must convey which meal the text was referring to. We need to add additional information into the interlingual representation to make this differentiation. Likewise, in the language of Eskimos there are many different shades of white, and representing their meaning in other languages is difficult. It is also extremely difficult to define an interlingua for a general MT system that can take care of all sorts of ambiguities. However, for restricted domains, the interlingua approach can be used successfully. This is due to specific characteristics of restricted domain (discussed in Chapter 10, Section 10.5), which simplifies the task of defining interlingual representation.

8.7 CORPUS-BASED MACHINE TRANSLATION

Corpus-based MT systems have gained much interest in recent years. The advantages of these systems are that they are fully automatic and require significantly less human labour than traditional rule-based approaches. However, they require sentence-aligned parallel text for each language pair and cannot be used for language pairs for which such corpora do not exist. The corpus-based approach is further classified into statistical and example-based machine translation approaches.

8.7.1 Statistical Machine Translation

Statistical machine translation systems are inspired by the noisy channel model used in speech recognition systems. A noisy channel introduces noise that makes it difficult to recognize the input word. A recognition system based on it builds a model of the channel to identify how it modifies the input and recover the original word. Figure 8.7 shows a direct analogy of the noisy channel model for an English to Hindi MT task. The input is considered as the distorted version of the target language sentence; the task is to find the most likely source language sentence, given the translation. This system models a target language sentence T, given a source language sentence S, as the product of translation probability $P(S/T)$ and target language probability $P(T)$. The translation probability $P(S/T)$ accounts for the adequacy of translation contents, whereas $P(T)$ accounts for fluency of target construction.

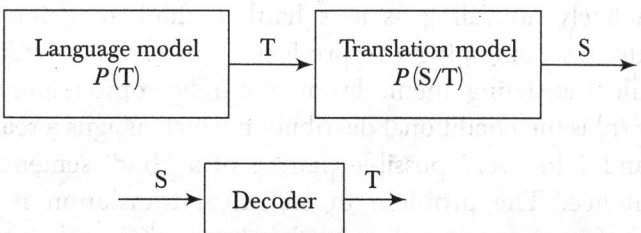

Figure 8.7 Noisy channel model for English to Hindi machine translation

The statistical machine translation system is based on the view that every sentence in a language has a possible translation in another language; a sentence can be translated from one language to another in many possible ways. The choice among these possible translations is a matter of translator's preference. Statistical translation approaches take the view that every sentence in the target language is a possible translation of the input sentence in source language. Let us take the case of English to Hindi translations. This means that every Hindi sentence, h, is a possible translation of an English sentence, e. The probability that '*Ghoda ghaas khata hai*' is a translation of 'Murthy eats apple' is low as compared to the probability of 'Ravi *khana khata hai*' being the English translation of the sentence. We assign to every pair of sentences (e, h) a probability, $P(h/e)$, which is the probability that a translator when presented with an English sentence e, will produce h as its Hindi translation. We also assume that when native speakers of Hindi speak English, they have in mind some Hindi sentence which they translate mentally. The goal of the translation system is to find the sentence h that the native speaker has in mind when he produces e. We minimize our chance of error by selecting that Hindi sentence $\hat{h}$ for which $P(h/e)$ is maximum. Using Bayes's theorem, we write

$$P(h/e) = \frac{P(e, h)}{P(e)} = \frac{P(h) \times P(e/h)}{P(e)}$$

where $\hat{h}$ can be calculated as

$$\hat{h} = \arg\max_{h} P(e/h) = \arg\max_{h} \frac{P(h) \times P(e/h)}{P(e)}$$

Since the denominator is independent of h, we can write the preceding expression as

$$\hat{h} = \arg\max_{h} P(h) \times P(e/h) \qquad \text{(i)}$$

where $P(h)$ is prior probability of Hindi text and $P(e/h)$ is conditional probability of the English text given the Hindi text. The idea is not intuitively appealing as it is hard to think of going through a list of sentences computing the product of prior and conditional probability while translating them. From a purely computational point of view, $P(e/h)$ is the conditional distribution which assigns a real number between 0 and 1 to every possible pairing of a Hindi sentence and an English sentence. The problem of statistical translation is thus reduced to identifying approximation to the distributions $P(h)$ and $P(h/e)$ that are

good enough to achieve an acceptable quality of translation. As suggested by Equation (i) above, this process involves three computational challenges:
1. Estimating the language model probability, $P(h)$
2. Estimating the translation model probability, $P(e/h)$
3. Devising an efficient search for the Hindi text that maximizes their product

These challenges correspond to three problems namely, language modelling problem, translation modelling problem, and search problem. The search problem is also known as decoding.

Language Model

A language model gives the probability of a sentence. Chapter 2 discusses how this probability is computed using n-gram model. So, we proceed with the discussion of translation modelling problem.

Translation Model

The translation model helps compute the conditional probability $P(e/h)$. It is trained from a parallel corpus of English/Hindi pairs. As no corpus is large enough to allow the computation of translation model probabilities at sentence level (Figure 8.8), we break this process into smaller units, e.g., words or phrases, and learn their probability. We think of the Hindi translation of an English sentence as being generated from the English sentence word by word.

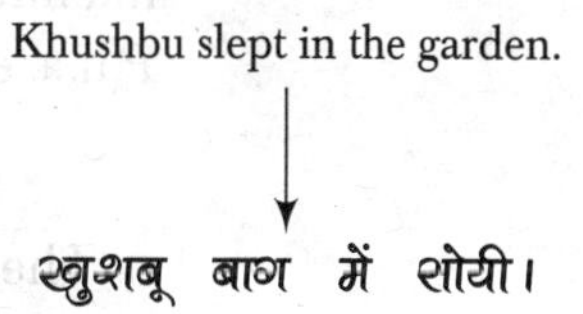

Khushbu slept in the garden.

खुशबू बाग में सोयी।

Figure 8.8 Sentence alignment

We use the notation (t/s) to represent an input sentence s and its translation t. Using this notation, (*Woh baag mein so gayi*/ She slept in the garden) means that '*Wo baag mein so gayi*' is a translation of 'She slept in the garden'. Figure 8.9(a) shows one possible alignment for English sentence (8.1). The alignment shown in Figure 8.9(a) can be represented by this notation (woh baag mein soyi| She(1) slept(4) in(3) the (3) garden(2)). An alignment-mark shows the origin of each English word in the Hindi sentence. Similarly, the alignment in Figure 8.9(b) is represented by (*woh baag mein soyi*| She(1) slept(4) in(3) the garden(2)). In this case, 'the' is not aligned to any Hindi word. If we consider '*woh baag mein so gayi*' as the Hindi translation, then one possible alignment can be [*woh bag mein so gayi*| She(1) slept(4, 5) in(3) the garden(2)]. Initially, we assume that all connections are equally likely. After one iteration, the model learns that the connection between 'she' and 'woh' is more likely. After another iteration, it becomes clear that a connection between 'garden' and '*baag*' is more likely than a connection between 'garden' and '*soyi*'.

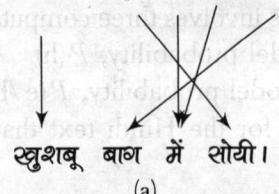
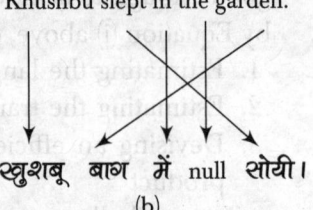

Figure 8.9 Two possible alignments between sentence (8.1) and its Hindi translation

The preceding discussion makes it clear that for any given source–target sentence pair, a number of alignments is possible. Some of them will be more probable. For the sake of simplicity, we consider a translation model based on word-by-word alignment. We denote the set of alignment by A(e, h). If the length of h is l and that of e is m, then there are lm different connections possible, because each of the m Hindi words can be connected to any of the English words. Here, we discuss the translation model in which all connections for each Hindi position are equally likely. Therefore, the order of the words in h and e does not affect P(h/e). The likelihood of (h/e) can be defined in terms of the conditional probability P(h,a/e) as

$$P(e/h) = \sum_a P(e, a/h)$$

The sum is over the elements of A(e, h). We restrict ourselves to the case where each English word has exactly one connection. A Hindi word can be aligned with more than one English word. This kind of alignment can be represented by a series 'a' of length m:

$$\{a_1, ..., a_m\}$$

where each element a_j takes a value between 0 and l. If the word in position j of the English string is connected to the word in position i of Hindi string, then $a_j = i$, and if it is not connected to any Hindi word, then $a_j = 0$. We assume that individual translations are independent and multiply the m translation probabilities for a particular alignment. For the alignment (*woh baag mein soyi* | She(1) slept(4) in(3) the garden(2)), we can compute P(woh baag mein soyi | She slept in the garden) by multiplying the translation probabilities T(*woh*/she), T(*baag*/garden), T(*mein*/in), T(null/the), and T(*soyi*/slept).

To generate a Hindi sentence from an English sentence, we have to folllow the steps as given below:

(i) Select the length of e with probability L, where $L = P[\text{length(e)} = m]$ is a constant, i.e., we assume that all lengths are equally likely with probability L.

(ii) Select an alignment 'a' with probability $P(a/e)$. There are $(1 + 1)^m$ possible alignments. Assuming all possible alignments are equally likely, the probability of an alignment a, $P(a/e)$, is

$$P(a/e) = L \times \frac{1}{(l+1)^m}$$

(iii) Select the jth English word with probability

$$P(e/a, h) = \prod_{j=1}^{m} T(e_j|h_{aj})$$

The joint likelihood of a Hindi string and an alignment given an English string is

$$P(e, a/h) = P(a/h) \times P(e/a, h)$$

$$= \frac{L}{(l+1)^m} \prod_{j=1}^{m} T(e_j|h_{aj})$$

where T is the probability of seeing e_j in an English sentence, given that we see h_{aj} in a Hindi sentence and L as the normalization constant. The alignment is determined by specifying the values of a_j for j from 1 to m, each of which can take any value from 0 to l. Therefore,

$$P(e, a/h) = \frac{(l+1)^m}{L} \sum_{a_a=0}^{l} \cdots \sum_{a_m=0}^{l} \prod_{j=1}^{m} T(e_j|h_{aj})$$

Search

We have to search for the sentence h that maximizes $P(e, h)$:

$$\hat{h} = \arg\max_h P(e, h) = \arg\max_h P(h)P(e/h)$$

This creates a problem as the search space is infinite. Brown, et al. (1993) suggested the use of a variant for stacked search. In stack search, we maintain a list of partial alignment hypotheses. We start with a null hypothesis. which means that the target sentence is obtained from a sequence of source words that we do not know. We represent this entry as (*woh baag mein soyi*|*), where * is a place holder for an unknown sequence of source words. As the search proceeds, it extends entries in the list by adding one or more additional words to its hypothesis. For example, we can extend the initial entry to one or more of the following entries:

(woh baag mein soyi|she(1)*),
(woh baag mein soyi|* garden(2)*),
(woh baag mein soyi|*slept(4))

The search terminates when there is a complete alignment in the list that is more promising than any of the incomplete alignments.

In order to estimate parameters of language model we need a corpus of English text only. However, parameter estimation of translation models requires an aligned bilingual corpus. Translation probabilities, $T(w_1/w_2)$, are calculated using EM algorithm. In the E step, we find w_1 in the Hindi sentence, given the occurrence of w_2 in the English sentence.

$$E_{w_1, w_2} = \sum_{w_1 \in h, w_2 \in e} T(w_1/w_2)$$

The M step uses these expectations to re-estimate translation probabilities:

$$T(w_1/w_2) = \frac{E_{w_1, w_2}}{\sum_w E_{w, w_2}}, \text{ where } w \in h$$

This process is repeated to get parameters that assign greater probabilities to sentence pairs that we actually observe.

8.7.2 Example-based Machine Translation

Example-based machine translation (EBMT) systems use past translation examples to generate translations for a given input. Nagao (1984) first emphasized the example as an input sentence in MT. He called the method, translation by analogy. An example-based MT system maintains an example-base consisting of translation examples between source and target languages. When an input sentence is presented to it, the system retrieves a similar source language (SL) sentence from the example-base and its translation. It then adapts the example translation to generate the translation of the input sentence. The EBMT systems rest on the idea that similar sentences will have similar translations. As shown in Figure 8.10, an EBMT system has two main modules—retrieval and adaptation.

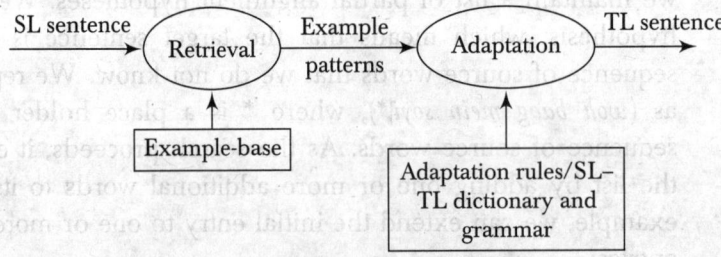

Figure 8.10 Example-based machine translation

Retrieval

The task of this module is to retrieve translation examples from the example-base for a given input. The retrieval strategies attempt to retrieve an example from the base which is similar to the input sentence. This means that a similarity measure has to be defined for sentences. These similarity measures may be based on word similarity or syntactic and semantic similarity.

Adaptation

This module is responsible for carrying out necessary modifications in the retrieved example pair to generate the translation of target sentence. The modification may involve addition, deletion, and replacement of morphological words, constituent words or suffixes. Morphological words are words that are used with the main verb to bring morphological variation, e.g., *chukka, chuki, chuke, raha, rahi,* and *rahe.* These words depend upon the number and gender of the subject (Gupta 2005). Similarly, words like *honga, hongi, honge, hain,* and *ho,* are also morphological words that indicate the notion of singular/plural and person. The various adaptation operations, used in different example-based MT systems, consist of the following basic operations:

Addition: It involves insertion of a new chunk in the retrieved translation example.

Deletion: It involves deletion of some chunk from the retrieved example pair.

Replacement: It involves replacement of some chunk in the retrieved example with some new chunk.

With respect to English and Hindi, Gupta (2005) has differentiated the adaptation operations into two groups, namely word-based (content or morphological words) and suffix-based. Figure 8.11 illustrates how adaptation is performed using these operations.

Input:	Woh	roz	wyayam	karta	hai.
Operations:	WR	copy	copy	SA	MR
Output:	Hum	roz	wyayam	karte	hain.
WR: Word Replacement			SA – Suffix Addition		
MR: Morphological Replacement					

Figure 8.11   Example-based machine translation

We now elaborate on the ideas involved in example-based translation, with the help of few examples. Consider the English–Hindi translations for the following input sentence:

Sheela sings a song.   (8.6a)

Let the example-base contain the following sentences:

Rohit sings a song.   (8.6b)

Sheela is playing.   (8.6c)

Sheela is singing a song.   (8.6d)

Sheela sings a Bengali song.   (8.6e)

Let the system identify the example pair (8.6b) as similar to input:

Rohit sings a song.

Rohit gaana gaata hai

Using the retrieved pair, the system generates the translation of input sentence by replacing Rohit with Sheela and *taa* with *tii* in the target language translation. In this case, the adaptation operations used are word and suffix replacements. Adaptation operations applied on suffixes are usually needed to bring morphological changes. The nature of the adaptation operation depends on the example retrieved by MT system. If the retrieved Example is (8.6e), the adaptation involves deletion of 'Bengali'. If the retrieved sentence is Example (8.6d), then the word *rahi* needs to be deleted and the suffix *tii* needs to be added to the root form of the main verb *gaa*.

Now, consider the following input sentence:

The fan is on.   (8.7)

Suppose the system retrieves the following example in response to the input:

The fan is on the table.   *Pankha mez par hai*

Adapting this translation to generate an output, results in the following incorrect translation for the input sentence.

Pankha par hai

This is an example of translation divergence, which refers to the phenomenon where structurally similar sentences of the source language do not translate into sentences that are similar in structure in the target language (Dorr 1993). The following sentences and their Hindi translations further illustrate this problem (adapted from Gupta 2005).

She is in trouble.   (8.8a)

She is in pain.   *Usse dard ho raha hai*   (8.8b)

She is in shock.   *Woh sadme mein hai*   (8.8c)

Sentences (8.8b) and (8.8c) of the example-base have similar structures but their translations vary significantly. Sentence (8.8c) refers to normal translation, whereas in sentence (8.8b), the prepositional phrase 'in pain' is being translated as '*dard ho raha hai*' (is paining). In such cases, the translation depends on the sentence retrieved to generate the translation of input. If sentence (8.8b) is retrieved by the system when presented with input sentence (8.8a), then the adaptation operation generates the following, syntactically incorrect, translation:

Woh musibata rahi hai.

Language divergence problems in the context of interlingua-based machine translation, involving English, Hindi, and Marathi, have been discussed by Bhattacharyya, who suggested the use of knowledge networks such as WordNet to solve them. A formal treatment of the English–Hindi language divergence is available in Dave et al. (2002).

Gupta (2005) proposed that divergence be handled by partitioning the example-base into two parts: one that contains normal translation patterns and another that contains examples involving divergence. Given an input sentence, examples are retrieved from appropriate part. Gupta proposed algorithms to detect the presence of divergence from a given English–Hindi pair of sentences.

Example-based MT system has certain advantages over rule-based and statistical MT systems. In the rule-based system, knowledge about the syntax and semantics of source and target, need to be represented as rules. Considering that natural language sentences may take a variety of form and involve a number of different types of ambiguities, it seems very difficult to extract all this knowledge in the form of rules. The problem with statistical MT systems is that they need a huge, aligned, parallel corpus. Availability of huge corpora is scarce, except for English and European languages. Therefore, this approach cannot be widely used.

Example-based MT systems require neither a large set of rules, nor a huge parallel corpus. Hence, they provide a useful alternative to the two main MT approaches. Usually in a multilingual society, information is made available to the public in multiple languages, particularly government notices and reports. These documents can be easily used to build an example-base. Consistency is not a problem in example-based systems—we improve a system by adding new examples to the example-base.

8.8 SEMANTIC OR KNOWLEDGE-BASED MT SYSTEMS

Early MT systems are characterized by the use of syntax. Though transfer and interlingual approaches use some semantic information, their central component is syntactic analysis. Semantic features are usually attached to syntactic structures and semantic processing occurs only after syntactic structures have been identified. Synthesis and analysis is restricted to sentences. Semantics-based approaches to language analysis have been introduced by AI researchers. The approaches require a large knowledge-base that includes both ontological and lexical knowledge. The basic AI approaches include: semantic parsing, lexical decomposition into semantic networks and resolution of ambiguities and uncertainties by reference to knowledge-bases. An example of this type of system is the KANT (knowledge-based, accurate natural language translation) system, which is also an interlingual system. KANT uses explicitly coded lexicons, grammars, and semantic rules to perform translation.

8.9   TRANSLATION INVOLVING INDIAN LANGUAGES

Most of the research on translation using Indian languages has focused on the English–Hindi language pair, e.g., ANGLABHARTI (Sinha et al. 2002), SHAKTI (Sangal 2004), and ANUBHARTI (Jain et al. 2001), MaTra (human-aided machine translation) and Mantra. In this section, we discuss these systems. Besides MT system, a number of language accessors have been developed to provide language access to Indian languages.

8.9.1   ANGLABHARTI

ANGLABHARTI (Sinha et al. 1995) is a rule-based MT system. It uses a pseudo-interlingua to facilitate translation between English and a number of Indian languages. Besides using rule-bases, it also uses example-bases and post editing to produce accurate and acceptable translations. ANGLABHARTI is thus a hybrid machine translation system. It takes an English sentence as input and creates an intermediate structure, which is then used to generate sentences in Indian languages. Structural similarity among languages of the same family is exploited to produce translations in a number of Indian languages. Indian language text generation is based on the Paninian framework discussed in Chapter 4. As explained there, free movement of phrases in most of the Indian languages makes CFG unsuitable for them. The Paninian framework employs dependency parsing, which parses based on the relationship between words, and hence takes care of variations of parse.

8.9.2   SHAKTI[1]

SHAKTI (Sangal 2004) is an MT system for translations from English to Indian languages. It follows a rule-based approach and attempts to achieve hybridization by incorporating an example-base and learning in each of the following three main stages:

---

[1]http://gdit.iiit.net/~mt/shakti/index.html

English Sentence Analysis Source language analysis uses a shallow parser and a morphological analyser to get the POS, chunk, and morph information about each word in the sentence; it also includes a number of other steps such as forming a PP, verb+adverb grouping, identifying phrases, word sense disambiguation, etc.

Transfer-grammar for English to Hindi The transfer-grammar component extracts dependency relations of the English sentence and re-orders words to match them with the target language word order.

Hindi Sentence Generation It generates target language sentences using target grammar rules.

SHAKTI implements these stages in terms of the following general steps (Sangal 2007):
1. Part-of-speech tagging
2. Morphological analyser
3. Chunking
4. Parsing
5. Semantic analyser
6. Word sense disambiguation
7. Source to target language dictionary lookup
8. Transfer rules
9. Target language generation

Like ANGLABHARTI, SHAKTI also uses dependency parsing. The phrase structure tree of English sentences is transformed into a dependency tree. Lexical transfer rules are used to select the right word for a given source language word. Transfer grammar is also responsible for selecting tense, aspect, modality and agreement for target language dependency tree.

8.9.3 MaTra^2

MaTra is a human-assisted MT system, which aims at translating English sentences to Indian languages, mainly to Hindi. The core component of the system is a structured editor, which creates a hierarchical frame-like representation of the input sentence. The main verb in a sentence, called pivot, has a template associated with it. This template specifies mandatory and optional slots. The value of a slot may be a sentence fragment or another pivot. This nesting permits the representation of a complex sentence. The system uses rules and heuristics to resolve certain ambiguities. The output produced by the system may serve as a useful aid to translators, editors and content providers.

---

^{2}http://www.ncst.ernet.in/matra/

8.9.4   MANTRA[3] (Machine-Assisted Translation Tool)

MANTRA is an MT system based on tree-adjoining grammar (Joshi and Schabes 1992) formalism. The target language pair includes English–Hindi, Hindi–English, and Hindi–Bengali. Currently, work is focused on a restricted domain (the domain of gazette notifications pertaining to government appointments and parliament proceedings), though the approach is general. The MANTRA technology is being expanded to include languages such as Gujarati, Bengali, and Telugu and other domains such as agriculture, banking, etc.

8.9.5   Anusaarak

The Anusaarak[4] (Bharti et al. 1994, Bharti et al. 2001) project at IIT Kanpur and Tamil–Hindi Anusaarak at Anna University, Chennai, focus on the development of language accessors for Indian languages. Anusaarak developed language accessors for Punjabi, Bengali, Telugu, Kannada, and Marathi into Hindi. It is based on the principles of Paninian grammar, and does not translate an input sentence. It just maps word groups between source and target language. The Tamil-Hindi Anusaarak[5] (Rao 2001) focuses on development of Tamil–Hindi machine-aided translation system.

SUMMARY

- Machine translation is the automatic translation of text from one language to another.
- The different levels of ambiguities create problems in machine translation.
- Direct translation involves a word-by-word translation approach. No intermediate representation is produced.
- Interlingua-based MT systems produce an abstract representation using which, the target language text can be generated.
- Example-based MT systems involve low computational cost and can be extended easily.

REFERENCES

Bharti, Akshar, Rajeev Sangal, Dipti Mishra, Sriram Venkatapathy, and Papi Reddy T., 2004, 'Handling Multi-word expressions without explicit linguistic rules in an MT System,' *Proceedings of 7th International Conference on TEXT, SPEECH, and DIALOGUE–2004,* Brno, Gzech Republic.

---

[3]http://www.cdacindia.com/html/about/success/mantra.asp
[4]http://www.iiit.net/ltrc/Anusaaraka/anu_home.html
[5]http://www.au-kbc.org/

Bharti, Akshar, Vineet Chaitanya, and Rajeev Sangal, 1994, 'Paninian framework and its application to Anusaarak,' *Sadhna*, 19(1), 113–27.

Bharti Akshar, Vineet Chaitanya, Amba P. Kulkarni, and Rajeev Sangal, 2001, 'Overcoming the language barrier in India,' In Rukmini Bhaya (Ed.), *Anuvad: Approaches to Translation*, Sage, New Delhi.

Brown, Peter F., Stephen A. Della Pietra, Vincent J. Della Pietra, and Robert L. Mercer, 1993, 'The mathematics of statistical machine translation: parameter estimation,' *Computational Linguistics*, 19(2), pp. 263–311, MIT Press, Cambridge, MA.

Bhattacharyya, Pushpak, (no date), 'Machine translation, language divergence and lexical resources to the breeming,' *www.ciil.org*.

Chandioux, J., 1976, *METEO: un système operationnel pour la traduction automatique des bulletins météorologiques destinés au grand public*, META, 21, pp.127–33.

Dave, Shachi, Jignashu Parikh, and Pushpak Bhattacharyya, 2002, 'Interlingua-based English Hindi machine translation and language divergence,' *Journal of Machine Translation (JMT)*, 17, Springer, The Netherlands.

Dorr, B. J., 1993, *Machine Translation: A View from the Lexicon*, MIT Press, Cambridge, MA.

Gupta, Deepa, 2005, 'Contributions to English to Hindi machine translation using example-based approach,' *Ph.D. Thesis*, Department of Mathematics, Indian Institute of Technology Delhi, New Delhi.

Hutchins, W. John and Harold L. Somers, 1992, *An Introduction to Machine Translation*, Academic Press, London.

Jain, Renu, R.M.K. Sinha and Ajai Jain, 2001, 'ANUBHARTI: Using Hybrid Example-Based Approach for Machine Translation,' *Proceedings of the Symposium on Translation Support Systems (STRANS2001)*, Kanpur, India.

Joshi, Arvind K. and Yves Schabes, 1992, 'Tree-adjoining grammars and lexicalized grammars,' In Maurice Nivat and Andreas Ladelsk (Eds.), Trte Automata and Languages, Elsevier, North-Holland, pp. 409–32.

Macklovitch, E. and P. Isabelle, 1990, 'Les voies actualles de la traduction atomatique au Canada,' *Tribune des Industries de la language*, OFIL (Observatoire francais des industries de la language), Paris.

Mohanrajah, S. and Veersinghe, 'An English to Tamil translator,' available at www.ucsc.comb.ac.lk/people/rw/papers/EntoTamiTr.htm.

Nagao, M., 1984, 'A Framework of a mechanical translation between Japanese and English by analogy principle,' *Artificial and Human Intelligence*, North-Holland, pp. 173–80.

Nyberg, E. and Mitamura, T., 1992, 'The KANT system: fast, accurate, high-quality translation in practical domains,' *Proceedings of COLING-92*, Association for Computational Linguistics, Morristown, NJ.

Rao, D., 2001, 'Machine translation in India: a brief survey,' *SCALLA 2001 Conference*, Bangalore, www.elda.fr.

Rao, D., P. Bhattacharya, and R. Mamidi, 1998, 'Natural language generation for English to Hindi human-aided machine translation,' *Proceedings of the Knowledge-based Computer System International Conference, KBCS-1998*, NCST, Mumbai.

Sinha, R.M.K. and A. Jain, 2003, 'AnglaHindi: an English to Hindi machine translation system,' *Proceedings of the MT SUMMIT IX, Orleans*, LA, pp. 23–27.

Sangal, R., 2004, 'Shakti: IIIT-Hyderabad machine translation system (experimental),' shakti.iiit.net.

Sinha, R.M.K., R. Jain, and A. Jain, 2002, 'An English to Hindi machine-aided translation system based on ANGLABHARTI technology,' *ANGLAHINDI*, www.anglahindi.iitk.ac.in.

EXERCISES

1. Compare the direct MT system with the transfer system.
2. Choose one of the generation techniques introduced in previous chapter and explain why it would or it would not be useful for machine translation.
3. What makes machine translation hard? Explain.
4. What do you mean by language divergence problem? Explain with the help of appropriate examples.
5. List characteristics that are common among Indian languages.
6. Do you think that an example-base translation model will be more appropriate for translation among Indian languages?

LAB EXERCISES

1. Develop a small Hindi to English dictionary consisting of 50 words. Use it to develop a direct MT system for simple sentences involving words in the dictionary.
2. Write a program to translate simple interrogative sentences in English to Hindi.
3. Write a program that takes an English sentence and reorders it to match word order in Hindi.
4. Use the MT system available on the Web to translate a set of sentences and compare their output. List the problems you noticed.

CHAPTER 9

INFORMATION RETRIEVAL–1

CHAPTER OVERVIEW

The huge amount of information stored in electronic form, has placed heavy demands on information retrieval systems. This has made information retrieval an important research area. This chapter is concerned with the design of information retrieval systems. It discusses design features of systems and introduces various models, such as the classical (boolean, probabilistic, and vector space retrieval models), non-classical (information logic, situation theory, and interaction information retrieval model), and alternative information retrieval (cluster, fuzzy, and LSI) models. A detailed discussion of vector space model is also given. The final topic of discussion in this chapter is information retrieval evaluation models.

9.1   INTRODUCTION

Information retrieval (IR) deals with the organization, storage, retrieval, and evaluation of information relevant to a user's query. A user in need of information formulates a request in the form of a query written in a natural language. The retrieval system responds by retrieving the document that seems relevant to the query. Research in IR is not new. It dates back to the 1960s when text retrieval systems were introduced. Traditionally, however, it is not considered an important application area of NLP. Interest in the field was generated due to the insurgence of the World Wide Web. The interaction between NLP and IR is now strengthening. Many NLP techniques, including the probabilistic model, have found application in IR systems and techniques such as latent semantic indexing, vector space retrieval, etc.

Traditionally, IR systems are not expected to return the actual information, only documents containing that information. As Lancaster (1979) pointed out:

An information retrieval system does not inform (i.e., change the knowledge of) the user on the subject of her inquiry. It merely informs on the existence (or non-existence) and whereabouts of documents relating to her request.

The word *document* is a general term that includes non-textual information such as images and speech.

Our concern in this chapter is only with retrieval of text documents. This excludes question answering systems and data retrieval systems. The three systems differ in the nature of their queries and expected results of the queries. In both question answering systems and data retrieval systems, queries are very specific and precise in nature. On the other hand, queries submitted to IR systems are often vague and imprecise. A question answering system provides users with the answers to specific questions. Data retrieval systems retrieve precise data, usually organized in a well-defined structure. Unlike data retrieval systems, IR systems do not search for specific data. Nor do they search for direct answers to question, as question answering systems do.

9.2 DESIGN FEATURES OF INFORMATION RETRIEVAL SYSTEMS

Figure 9.1 illustrates the basic process of IR. It begins with the user's information need. Based on this need, he/she formulates a query. The IR system returns documents that seem relevant to the query. This is an engineering account of the IR system. The basic question involved is, 'what constitutes the information in the documents and the queries'. This, in turn is related to the problem of representation of documents and queries. The retrieval is performed by matching the query representation with document representation.

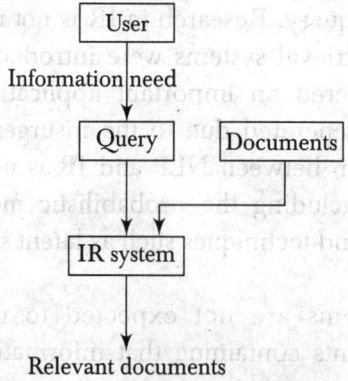

Figure 9.1   Basic information retrieval process

The actual text of the document is not used in the retrieval process. Instead, documents in a collection are frequently represented through a set of index terms or keywords. In this chapter, the word *term* and *keyword* will be used independently. Keywords can be single word or multi-word phrases. They might be extracted automatically or manually (i.e., specified by a human). Such a representation provides a logical view of the document. The process of transforming document text to some representation of it is known as *indexing*. There are different types of index structures. One used data structure, commonly by the IR system, is the inverted index. An inverted index is simply a list of keywords, with each keyword carrying pointers to the documents containing that keyword. The computational cost involved in adopting a full text logical view (i.e., using a full set of words to represent a document) is high. Hence, some text operations are usually performed to reduce the set of representative keywords. The two most commonly used text operations are *stop word elimination* and *stemming*. Stop word elimination removes grammatical or functional words, while stemming reduces words to their common grammatical roots. *Zipf's law* can be applied to further reduce the size of index set. Not all the terms in a document are equally relevant. Some might be more important in conveying a document's content. Attempts have been made to quantify the significance of index terms to a document by assigning them numerical values, called weights. The choice of index terms and weights is a difficult theoretical and practical problem and several technique are used to cope with it. A number of *term-weighting* schemes have been proposed in the literature over the years. We discuss a few of them in this chapter.

9.2.1 Indexing

In a small collection of documents, an IR system can access a document to decide its relevence to a query. However, in a large collection of documents, this technique poses practical problems. Hence, a collection of raw documents is usually transformed into an easily accessible representation. This process is known as indexing. Most indexing techniques involve identifying good document descriptors, such as keywords or terms, which describe the information content of documents. A good descriptor is one that helps describe the content of the document and discriminate it from other documents in the collection. Luhn (1957, 1958) is considered the first person to advance the notion of automatic indexing of documents based on their content. He assumed that the frequency of certain word-occurrences in an article gave meaningful identification of the article's content. He proposed that the discrimination power of index terms is a function of the rank order of the frequency of their occurrence, and that middle frequency terms have the highest discrimination power. This model was proposed for the extraction of salient terms from a document. These extracted terms are then used to represent text. Thus, indexing is simply the representation of text (query and document) as a set of terms whose meaning is equivalent to some content of the original text.

As mentioned earlier, the word *term* can be a single word or multi-word phrases. For example, the sentence, *Design features of information retrieval systems,* can be represented as follows:

Design, features, information, retrieval, systems

It can also be represented by the set of terms:

Design, features, information retrieval, information retrieval systems

These multi-word terms can be obtained by looking at frequently appearing sequences of words, n-grams, part-of-speech tags, or by applying NLP to identify meaningful phrases or handcrafting. POS tagging helps extract meaningful sequences of words; it handles sense ambiguity, as words are assigned POS based on their local (sentential) context. Though statistical approaches to phrase extraction are more efficient, they fail to handle word order changes and structural variations, which are better handled by syntactic approaches. In text retrieval conference (TREC), the method used for phrase extraction is as follows:

1. Any pair of adjacent non-stop words is regarded a potential phrase.
2. The final list of phrases is composed of those pairs of words that occur in, say, 25 or more documents in the document collection.

The NLP is also used in the recognition of proper nouns and the normalization of noun phrases. Ideally, all names in the text need to be recognized and represented as a single entity, e.g. *President Kalam, President of India,* and variants of the same name are recognized as such. Phrase normalization captures structural variations in phrases. For example, the three phrases text categorization, categorization of text, and categorize text, are normalized to give text categorize.

9.2.2 Eliminating Stop Words

The lexical processing of index terms involves elimination of stop words. Stop words are high frequency words which have little semantic weight and are thus, unlikely to help in retrieval. These words play important grammatical roles in language, such as in the formation of phrases, but do not contribute to the semantic content of a document in a keyword-based representation. Such words are commonly used in documents, regardless of topics, and thus, have no topical specificity. Typical example of stop words are articles and prepositions. Eliminating them considerably reduces the number of index terms. The drawback of eliminating stop words is that it can sometimes result in the elimination of useful index terms, for instance the stop word *A* in *Vitamin A*. Some phrases, like *to be or not to be*, consist entirely of stop words. Eliminating stop words in such case, make it impossible to correctly search a document. Table 9.1 lists some of the stop words in English.

Table 9.1 Sample stop words in English

about	above	accordingly	across	after
afterwards	again	against	all	almost
alone	along	already	also	although
am	among	amongst	always	an
and	another	any	anybody	anyhow
anyone	anything	anywhere	apart	are
around	as	aside	at	away
awfully	be	because	been	before
beforehand	behind	being	below	beside
besides	best	better	between	beyond
both	brief	but	by	can
could	did	during	each	even
everybody	everyone	everything	everywhere	else
even	ever	every	for	from
further	had	has	have	her
herself	him	himself	he	furthermore
many	near	shall	she	self
whose	why	will	where	ex
except	far	first	five	former
formerly	over	overall	usually	appropriate

9.2.3 Stemming

Stemming normalizes morphological variants, though in a crude manner, by removing affixes from the words to reduce them to their stem, e.g., the words *compute, computing, computes,* and *computer,* are all be reduced to same word stem, *comput.* Thus, the keywords or terms used to represent text are stems, not the actual words. One of the most widely used stemming algorithms has been developed by Porter (1980). The stemmed representation of the text, *Design features of information retrieval systems*, is

{design, featur, inform, retriev, system}

Note that stop words have been removed in this representation and the remaining terms are in lower case.

One of the problems associated with stemming is that it may throw away useful distinctions. In some cases, it may be useful to help conflate similar terms, resulting in increased *recall*. In others, it may be harmful, resulting in reduced *precision* (e.g., when documents containing the term *computation* are returned in response to the query phrase *personal computer*). *Recall* and *precision* are the two most commonly used measures of the effectiveness of an information retrieval system, and are explained in detail later in this chapter.

9.2.4   Zipf's Law

Zipf made an important observation on the distribution of words in natural languages. This observation has been named Zipf's law. Simply stated, Zipf's law says that the frequency of words multiplied by their ranks in a large corpus is more or less constant. More formally,

Frequent × rank ≈ constant

This means that if we compute the frequencies of the words in a corpus, and arrange them in decreasing order of frequency, then the product of the frequency of a word and its rank is approximately equal to the product of the frequency and rank of another word. This indicates that the frequency of a word is inversely proportional to its rank. This relationship is shown in Figure 9.2.

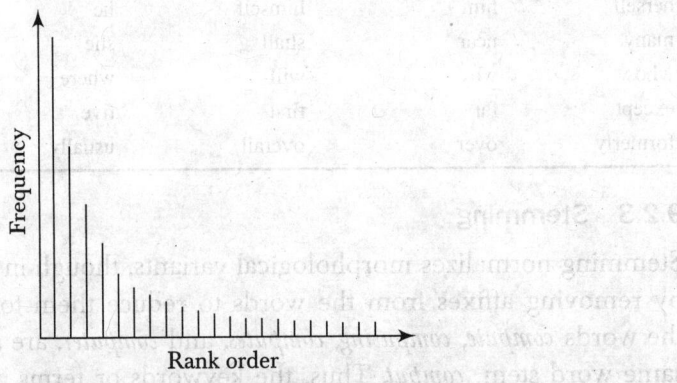

Figure 9.2   Relationship between the frequency of words and their rank order

Empirical investigation of Zipf's law on large corpuses suggest that human languages contain a small number of words that occur with high frequency and a large number of words that occur with low frequency. In between, is a middling number of medium frequency terms. This distribution has important significance in IR. The high frequency words being common, have less discriminating power, and thus, are not useful for indexing. Low frequency words are less likely to be included in the query, and are also not useful for indexing. As there are a large number of rare (low frequency) words, dropping them considerably reduces the size of a list of index terms. The remaining medium frequency words are content-bearing terms and can be used for indexing. This can be implemented by defining thresholds for high and low frequency, and dropping words that have frequencies above or below these thresholds. Stop word elimination can be thought of as an implementation of Zipf's law, where high frequency terms are dropped from a set of index terms.

9.3 INFORMATION RETRIEVAL MODELS

An IR model is a pattern that defines several aspects of the retrieval procedure, for example, how documents and user's queries are represented, how a system retrieves relevant documents according to users' queries, and how retrieved documents are ranked. The IR system consists of a model for documents, a model for queries, and a matching function which compares queries to documents. The central objective of the model is to retrieve all documents relevant to a query. This defines the central task of an IR system.

Several different IR models have been developed. These models differ in the way documents and queries are represented and retrieval is performed. Some of them consider documents as sets of terms and perform retrieval based merely on the presence or absence of one or more query terms in the document. Others represent a document as a vector of term weights and perform retrieval based on the numeric score assigned to each document, representing similarity between the query and the document. These models can be classified as follows:

- Classical models of IR
- Non-classical models of IR
- Alternative models of IR

The three classical IR models—Boolean, vector, and probabilistic—are based on mathematical knowledge that is easily recognized and well understood. These models are simple, efficient, and easy to implement.

Almost all existing commercial systems are based on the mathematical models of IR. That is why they are called classical models of IR.

Non-classical models perform retrieval based on principles other than those used by classical models, i.e., similarity, probability, and Boolean operation. These are best exemplified by models based on special logic technique, situation theory, or the concept of interaction.

The third category of IR models, namely alternative models, are actually enhancements of classical models, making use of specific techniques from other fields. The cluster model, fuzzy model, and latent semantic indexing (LSI) model are examples of alternative models of IR.

9.4 CLASSICAL INFORMATION RETRIEVAL MODELS

9.4.1 Boolean model

Introduced in the 50s, the Boolean model is the oldest of the three classical models. It is based on Boolean logic and classical set theory. In this model, documents are represented as a set of keywords, usually stored in an inverted file. An inverted file is a list of keywords and identifiers of the documents in which they occur. Users are required to express their queries as a Boolean expression consisting of keywords connected with Boolean logical operators (AND, OR, NOT). Retrieval is performed based on whether or not document contains the query terms.

Given a finite set

$$T = \{t_1, t_2, \ldots, t_i, \ldots, t_m\}$$

of index terms, a finite set

$$D = \{d_1, d_2, \ldots, d_j, \ldots, d_n\}$$

of documents and a Boolean expression—in a normal form—represent a query Q as follows:

$$Q = \wedge(\vee \theta_j), \ \theta_i \in \{t_i, \neg t_i\}$$

The retrieval is performed in two steps:

1. The set R_i of documents are obtained that contain or do not contain the term t_i:

   $$R_i = \{d_j \mid \theta_i \in d_j\}, \ \theta_i \in \{t_i, \in t_i\}$$
   where $\neg t_i \in d_j$ means $t_i \notin d_j$

2. Set operations are used to retrieve documents in response to Q:

   $$\cap R_i$$

Example 9.1 Let the set of original documents be

$$D = \{D_1, D_2, D_3\}$$

where

D_1 = Information retrieval is concerned with the organization, storage, retrieval, and evaluation of information relevant to user's query.

D_2 = A user having an information needs to formulate a request in the form of query written in natural language.

D_3 = The retrieval system responds by retrieving the document that seems relevant to the query.

Let the set of terms used to represent these documents be

$$T = \{\text{information, retrieval, query}\}$$

Then, the set D of document will be represented as follows:

$$D = \{d_1, d_2, d_3\}$$

where

d_1 = {information, retrieval, query}
d_2 = {information, query}
d_3 = {retrieval, query}

Let the query Q be

$$Q = \text{information} \wedge \text{retrieval}$$

First, the sets R_1 and R_2 of documents are retrieved in response to Q, where

$$R_1 = \{d_j \mid \text{information} \in d_j\} = \{d_1, d_2\}$$
$$R_2 = \{d_j \mid \text{retrieval} \in d_j\} = \{d_1, d_3\}$$

Then, the following documents are retrieved in response to query Q

$$\{d_j \mid d_j \in R_1 \cap R_2\} = \{d_1\}$$

This results in the retrieval of the original document D_1 that has the representation d_1. If more than one document have the same representation, every such document is retrieved. Boolean information retrieval does not differentiate between these documents. With an inverted index, this simply means taking an intersection of the list of the documents associated with the keywords *information* and *retrieval*.

Boolean retrieval models have been used in IR systems for a long time. They are simple, efficient, and easy to implement and perform well in terms of recall and precision if the query is well formulated. However, the model suffers from certain drawbacks.

First, the model is not able to retrieve documents that are only partly relevant to user query; all information is 'to be or not to be'.

Second, a Boolean system is not able to rank the returned list of documents. It distinguishes between presence and absence of keywords but fails to assign relevance and importance to keywords in a document.

Third, users seldom formulate their query in the pure Boolean expression that this model requires.

Numerous extensions of the Boolean model have been suggested to overcome these weaknesses. The best of these are the P-norm model developed by Salton et al. (1983) and a fuzzy-set model suggested by Paice (1984).

9.4.2 Probabilistic Model

The probabilistic model applies a probabilistic framework to IR. It ranks documents based on the probability of their relevance to a given query (Robertson and Jones 1976). Retrieval depends on whether probability of relevance (relative to a query) of a document is higher than that of non-relevance, i.e. whether it exceeds a threshold value. Given a set of documents D, a query q, and a cut-off value α, this model first calculates the probability of relevance and irrelevance of a document to the query. It then ranks documents having probabilities of relevance at least that of irrelevance in decreasing order of their relevance. Documents are retrieved if the probability of relevance in the ranked list exceeds the cut off value.

More formally, if $P(R/d)$ is the probability of relevance of a document d, for query q, and $P(I/d)$ is the probability of irrelevance, then the set of documents retrieved in response to the query q is as follows.

$$S = \{d_j \mid P(R/d_j) \geq P(I/d_j)\}\ P(R/d_j) \geq \alpha$$

Development of the probabilistic model was carried out largely by Maron and Kuhns (1960), Robertson and Sparck-Jones (1976), Robertson et al. (1982). Different mathematical methods for calculating the probabilities of relevance and irrelevance, as well as properties and applications, are discussed by Van Rijsbergen (1992), Callan et al. (1992), and Fur (1992).

Most of the systems assume that terms are independent when estimating probabilities for the probabilistic model. This assumption allows for accurate estimation of parameter values and helps reduce computational complexity of the model. However, this assumption seems to be inaccurate, as terms in a given domain usually tend to co-occur. For example, it is more likely that 'match point' will co-occur with 'tennis' rather than 'cricket'. Different forms of assumption are discussed in Robertson (1977). A comparison of the Boolean and probabilistic models can be found in Losee (1997). A comprehensive review of the probabilistic model is presented by Crestani et al. (1998). A baysian network version of the probabilistic model forms the basis of the INQUERY system (Callan et al. 1992).

The probabilistic model, like the vector model, can produce results that partly match the user query. Nevertheless, this model has drawbacks, one of which is the determination of a threshold value for the initially retrieved set; the number of relevant documents by a query is usually too small for the probability to be estimated accurately.

9.4.3   Vector Space Model

The vector space model is one of the most well-studied retrieval models. Important contribution to its development was made by Luhn (1959), Salton (1968), Salton and McGill (1983), and van Rijsbergen (1977). The vector space model represents documents and queries as vectors of features representing terms that occur within them. Each document is characterized by a Boolean or numerical vector. These vectors are represented in a multi-dimensional space, in which each dimension corresponds to a distinct term in the corpus of documents. In its simplest form, each feature takes a value of either zero or one, indicating the absence or presence of that term in a document or query. More generally, features are assigned numerical values that are usually a function of the frequency of terms. Ranking algorithms compute the similarity between document and query vectors, to yield a retrieval score to each document. This score is used to produce a ranked list of retrieved documents. Given a finite set of n documents

$$D = \{d_1, d_2, ..., d_j, ..., d_n\}$$

and a finite set of m terms

$$T = \{t_1, t_2, ..., t_i, ..., t_m\}$$

each document is represented by a column vector of weights as follows:

$$(w_{1j}, w_{2j}, w_{3j}, ..., w_{ij}, ..., w_{mj})^t$$

where w_{ij} is the weight of the term t_i in document d_j. The document collection as a whole is represented by an $m \times n$ term-document matrix as

$$\begin{pmatrix} w_{11} & w_{12} & \cdots & w_{1j} & \cdots & w_{1n} \\ w_{21} & w_{22} & \cdots & w_{2j} & \cdots & w_{2n} \\ w_{i1} & w_{i2} & \cdots & w_{ij} & \cdots & w_{in} \\ w_{m1} & w_{m2} & \cdots & w_{mj} & \cdots & w_{mn} \end{pmatrix}$$

Different term-weighting functions have been introduced (Salton and Buckley 1988). We discuss some of them in this section.

Example 9.2   Consider the documents and terms in Example 9.1. Let the weights be assigned based on the frequency of the term within the document. Then, the associated vectors will be

(2, 2, 1)
(1, 0, 1)
(0, 1, 1)

The vectors can be represented as a point in Euclidean space, where the coordinates of point P_j are the components of d_j. The term-document matrix is

$$\begin{pmatrix} 2 & 1 & 0 \\ 2 & 0 & 1 \\ 1 & 1 & 1 \end{pmatrix}$$

This raw term frequency approach gives too much importance to the absolute values of various coordinates of each document. For example, a document with weights 4, 4, and 2, would be quite similar to document 1, except for the differences in magnitude of term weights.

To reduce the importance of the length of document vectors, we normalize document vectors. Normalization changes all vectors to a standard length. We convert document vectors to unit length by dividing each dimension by the overall length of the vector. Normalizing the term-document matrix shown in this example, we get the following matrix:

$$\begin{pmatrix} 0.67 & 0.71 & 0 \\ 0.67 & 0 & 0.71 \\ 0.33 & 0.71 & 0.71 \end{pmatrix}$$

Elements of each column are divided by the length of the column vector given by $\sqrt{\sum_i w_{ij}^2}$. The values shown in this matrix have been rounded to two decimal digits.

9.4.4   Term Weighting

Each term that is selected as an indexing feature for a document, acts as a discriminator between that document and all other documents in the corpus. Luhn (1958) attempted to quantify the discriminating power of the terms by associating the frequency of their occurrence (term frequency) within the document. He postulated that the most discriminating (content bearing) terms are mid frequency terms. This postulate can be refined by noting the following facts:

1. The more a document contains a given word, the more that document is about a concept represented by that word.
2. The less a term occurs in particular document in a collection, the more discriminating that term is.

The first factor simply means that terms that occur more frequently represent the document's meaning more strongly than those occurring less frequently, and hence should be given high weights. In the simplest form, this weight is the raw frequency of the term in the document, as discussed earlier. The second factor actually considers term distribution across the document collection. Terms occurring in a few documents are useful for distinguishing those documents from the rest of the collection. Similarly, terms that occur more frequently across the entire collection are less helpful while discriminating among documents. This requires a measure that favours terms appearing in fewer documents. The fraction n/n_i—where n is the total number of the document in the collection and n_i the number of the document in which the term i occurs—exactly gives this measure. This measure assigns the lowest weight 1 to a term that appears in all documents and the highest weight of n to a term that occurs in only one document. As the number of documents in any collection is usually large, the log of this measure is usually taken, resulting in the following form of inverse document frequency (idf) term weight:

$$\text{idfi} = \log\left(\frac{n}{n_i}\right)$$

Inverse document frequency (idf) attaches more importance to more specific terms. If a term occurs in all documents in a collection, its idf is 0. Researchers have attempted to include term distribution in the weighting function (Sparck-Jones 1972, Salton 1971) to give a more accurate quantification of term importance. Sparck-Jones showed experimentally that a weight of $\log(n/n_i)+1$, termed as inverse document frequency, leads to more effective retrieval. Later researchers attempted to combine term frequency (tf) and idf weights, resulting in a family of tf $\times$ idf weight schemes having the following general form:

$$w_{ij} = \text{tf}_{ij} \times \log\left(\frac{n}{n_i}\right)$$

According to this scheme, the weight of a term t_i in document d_j is equal to the product of document frequency of the term and log of its inverse document frequency within the collection. The tf $\times$ idf weighting scheme combines both the 'local' and 'global' statistics to assign term weight. The tf component is a document specific statistic that measures the importance of a term within the document, whereas the idf is a global statistic that attempts to include distribution of the term across the document collection. These weighting schemes are well suited to the ad-hoc retrieval environment but pose problems in routing environment, where no fixed document collection exists. Usually, a training set of documents is used to compute statistics in such an environment and it is assumed that subsequent documents arriving at the system have the same statistical properties as the training set. We now give an example to explain how term weights can be calculated using the tf–idf weighting scheme.

Example 9.3 Consider a document represented by the three terms {tornado, swirl, wind} with the raw tf 4, 1, and 1 respectively. In a collection of 100 documents, 15 documents contain the term *tornado*, 20 contain *swirl*, and 40 contain *wind*. The idf of the term *tornado* can be computed as

$$\log\left(\frac{n}{n_i}\right) = \log\left(\frac{100}{15}\right) = 0.82$$

The idf of other terms are computed in the same way. Table 9.2 shows the weights assigned to the three terms using this approach.

Table 9.2   Computing tf-idf weight

Term	Frequency (tf)	Document frequency (n_i)	idf [$\log(n/n_i)$]	Weight (tf × idf)
Tornado	4	15	0.824	0.296
Swirl	1	20	0.699	0.699
Wind	1	40	0.398	0.389

Many variations of tf × idf measure have been reported. Some of these attempt to normalize tf and idf factors in different ways to allow for variations in document length (Salton and Buckley 1988). One way to normalize tf is to divide it by the frequency of the most frequent term in the document. This kind of normalization, often termed as maximum normalization, yields a value between 0 and 1. Normalization is needed because using absolute (raw) term frequency to weight terms favours longer documents over shorter ones. After frequency normalization, the weight of a term in a given document depends on the frequency of its occurrence in relation to the other terms in the same document, instead of its absolute frequency. Similarly, idf can be normalized by dividing it by the logrithm of the collection size (n).

$$w_{ij} = \frac{tf_{ij}}{\max(tf_{ij})} \times \log\left(\frac{n}{n_i}\right)/\log n$$

A third factor that may affect weighting function is the document length. A term appearing the same number of times in a short document and in a long document, will be more valuable to the former. Most weighting schemes can thus be characterized by the following three factors:

- Within-document frequency or term frequency (tf)
- Collection frequency or inverse document frequency (idf)
- Document length

Any term weighting scheme can be represented by a triple ABC. The letter A in this triple represents the way the tf component is handled, *B* indicates the way the idf component is incorporated, and *C* represents the length normalization component. Possible options for each of the three dimensions of the triple are shown in Table 9.3. Different combinations of options can be used to represent document and query vectors. The retrieval model themselves can be represented by a pair of triples like nnn.nnn (doc = 'nnn', query = 'nnn'), where the first triple corresponds to the weighting strategy used for the documents and the second triple to the weighting strategy used for the query term.

Table 9.3 Calculating weight with different options for the three weighting factors

Term frequency within document		
n	$\text{tf} = \text{tf}_{ij}$	Raw term frequency
b	$\text{tf} = 0$ or 1 (binary weight)	
A		
a	$\text{tf} = 0.5 + 0.5\left(\dfrac{\text{tf}_{ij}}{\max \text{tf in } D_j}\right)$	Augmented term frequency
l	$\text{tf} = \ln(\text{tf}_{ij}) + 1.0$	Logarithmic term frequency
L	$\text{tf} = \dfrac{\ln(\text{tf}_{ij} + 1.0)}{1.0 + \ln[\text{mean}(\text{tf in } D_j)]}$	Average term frequency-based normalization
Inverse document frequency		
n	$\text{wt} = \text{tf}$	No conversion
B		
t	$\text{wt} = \text{tf} \cdot \ln\left(\dfrac{n}{n_i}\right)$	Multiply tf with idf
Document length		
n	$w_{ij} = \text{wt}$	(no conversion)
C		
c	w_{ij} is obtained by dividing each wt by sqrt [sum of (wts squared)]	

There are many ways to compute each component. The simplest is to use either binary weight or raw term frequency. As shown in Table 9.3, the first occurrence of a term is more important than successive repeating occurrences. Thus, tf can be computed as $0.5 + 0.5\ (\text{tf}_{ij}/\text{max tf in } d_j)$ in which normalization is achieved by dividing tf by maximum tf value for any term in the document, or as $\ln(\text{tf}_{ij}) + 1.0$, which is known as logarithmic term frequency. The former computation is called augmented normalized term frequency. It causes tf to vary between 0.5 and 1. The problem with maximum normalization and augmented normalization of the tf component is that a single term in a document, with an unusually high frequency, may degrade the weights of the other terms significantly. However, this effect is not too pronounced with the augmented tf, because the highest frequency term cannot degrade the frequency of other terms below 0.5. The logarithmic term frequency reduces the effect of unusually frequent terms within a document, and also the importance of raw term frequency in a collection of documents with significant variations in length. It actually decreases the effect of all sorts of variations in tf, because for any two term frequencies tf_1 and $\text{tf}_2 > 0$, such that $\text{tf}_1 > \text{tf}_2$, the ratio of the logarithmic term frequencies will be always less than the ratio of the raw term frequencies, i.e.,

$$\frac{\log(\text{tf}_1) + 1}{\log(\text{tf}_2) + 1} < \frac{\text{tf}_1}{\text{tf}_2}$$

Different choices for *A*, *B*, and *C* for query and document vectors yield different retrieval modes, for example, ntc-ntc, lnc-ltc, etc. The choices for tf are *n* (use the raw term frequency), *b* (binary, i.e., neglect term frequency, term frequency will be 1 if term is present in the document, otherwise 0), *a* (augmented normalized frequency), *l* (logarithmic term frequency), and *L* (logarithmic frequency normalized by average term frequency). The options for idf are *n* (use 1.0, ignore idf factor) and *t* (use idf). The possible options listed in Table 9.3 for document length normalization are *n* (no normalization) and *c* (cosine normalization). To achieve cosine normalization, every element of the term weight vector is divided by the Euclidean length of the vector. This is called cosine normalization because the length of the normalized vector is 1 and its projection on any axis in document space gives the cosine of the angle between the vector and the axis under consideration.

The widely known weighting scheme, ntc-ntc, normalizes both the document and query term weight in the range 0–1 and may prove beneficial. The weighting scheme lnc-ltc means that document term weights are computed as the product of the logarithmic tf (l) of the given term, 1.0 (n) and cosine normalization (c) of the document vector. The query term weights are computed in the same way, except that each query term weight is also multiplied by the idf (t) of the given term in the document collection.

More recent weighting schemes integrate document length within the weighting formula yielding more complex retrieval models, for example, Okapi probabilistic search model (Robertson et al. 1995) and doc= "Lnu" model (Buckley et al. 1996). In TREC-3, the three systems with the best base performance were Okapi, INQUERY, and Cornell's Smart. The best performance was reported by Okapi system. Okapi uses the BM25 weighting algorithm introduced by developers of the probabilistic model during TREC-2 (Robertson et al. 1994) and TREC-3 (Robertson et al. 1995). Robertson and Walker (1994) developed the best match (BM) algorithms using the probabilistic model and some simple approximations to two-poisson model.

Considerable research efforts have been devoted to refining term weighting methods. As a result, a large number of term weighting schemes have been proposed in IR literature. (Salton and McGill 1983, Rijsbergen 1979). Some recent weighting schemes also consider the structure of the document.

A simple automatic method for obtaining indexed representation of the documents is as follows.

Step 1 Tokenization This extracts individual terms form a document, converts all the letters to lower case, and removes punctuation marks. The output of the first stage is a representation of the document as a stream of terms.

Step 2 Stop word elimination This removes words that appear more frequently in the document collection.

Step 3 Stemming This reduces the remaining terms to their linguistic root, to obtain the index terms.

Step 4 Term weighting This assigns weights to terms according to their importance in the document, in the collection, or some combination of both.

Table 9.4 shows the document vectors obtained after the application of these steps on sample documents shown in Figure 9.3.

Document 1:	Vector space model
Document 2:	Probabilistic retrieval model
Document 3:	Intelligent techniques in information retrieval

Figure 9.3   Sample documents

Table 9.4   Vector representation of sample documents after stemming

Stemmed terms	Document 1	Document 2	Document 3
inform	0	0	1
intellig	0	0	1
model	1	1	0
probabilist	0	1	0
retriev	0	1	1
space	1	0	0
technique	0	0	1
vector	1	0	0

9.4.5   Similarity Measures

Vector space model represents documents and queries as vectors in a multi-dimensional space. Retrieval is performed by measuring the 'closeness' of the query vector to document vector. Documents can then be ranked according to the numeric similarity between the query and the document. In the vector space model, the documents selected are those that are geometrically closest to the query according to some measure. The model relies on the intuitive notion that similar vectors define semantically related documents. Figure 9.4 gives an example of document and query representation in two-dimensional vector space. These dimensions correspond to the two index terms t_i and t_j. Document d_1 has two occurrences of t_i, document d_2 has one occurrence of t_i, and document d_3 has one occurrence of t_i and t_j each.

Documents d_1, d_2, and d_3 are represented in this space using term weights—raw term frequency being used here—as coordinates. The angles between the documents and query are represented as θ_1, θ_2, and θ_3 respectively.

The simplest way of comparing document and query is by counting the number of terms they have in common. One frequently used similarity measure

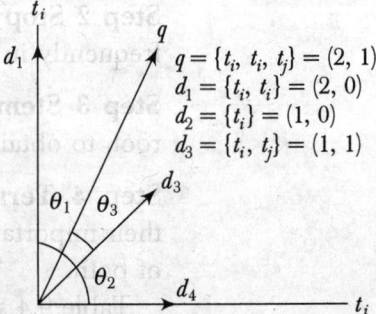

$q = \{t_i, t_i, t_j\} = (2, 1)$
$d_1 = \{t_i, t_i\} = (2, 0)$
$d_2 = \{t_i\} = (1, 0)$
$d_3 = \{t_i, t_j\} = (1, 1)$

Figure 9.4   Representation in two-dimensional vector space

is to take the 'inner product' between the query and the document vector. The inner product is given by

$$\text{sim}\,(d_j,\,q_k) = (d_j,\,q_k) = \sum_{i=1}^{m} w_{ij} \times w_{ik}$$

where m is the number of terms used to represent documents in the collection.

Other measures, e.g., dice coefficient, Jaccard's coefficient, and cosine coefficient, attempt to normalize the similarity by the length of document and query. The dice coefficient defines the similarity between query k and document j as

$$\text{sim}\,(d_j,\,q_k) = \frac{2 \times \left(\sum\limits_{i=1}^{m} w_{ij} \times w_{ik} \right)}{\sum\limits_{i=1}^{m} w_{ij}^2 + \sum\limits_{i=1}^{m} w_{ik}^2}$$

Jaccard's coefficient is defined as

$$\text{sim}\,(d_j,\,q_k) = \frac{\sum\limits_{i=1}^{m} w_{ij} \times w_{ik}}{\sum\limits_{i=1}^{m} w_{ij}^2 + \sum\limits_{i=1}^{m} w_{ik}^2 - \sum\limits_{i=1}^{m} w_{ij} \times w_{ik}}$$

The cosine measure is commonly used for measuring similarity in IR. It computes cosine of the angle between the document and query vector, to give a similarity value between 0 and 1. A minimum value of 0 (angle 90°) indicates that the vectors are unrelated (i.e., they have no terms in common), and a value of 1 means that the vectors share common terms. If d_j and q_k are the document and query vector respectively, then the cosine similarity is computed as

$$\text{sim}\,(d_j,\,q_k) = \frac{(d_j,\,q_k)}{\|d_j\| \|q_k\|} = \frac{\sum\limits_{i=1}^{m} w_{ij} \times w_{ik}}{\sqrt{\sum\limits_{i=1}^{m} w_{ik}^2} \times \sqrt{\sum\limits_{i=1}^{m} w_{ij}^2}}$$

If both the document and the query vectors have been cosine-normalized, then the inner product yields the cosine similarity.

The cosine measure divides the numerator by the product of the length of vectors. This tends to give low similarities to long vectors, i.e. vectors with many terms. The overlap coefficient compensates for this by dividing by the vector having smaller sum of weights. More formally, this measure is defined as

$$\text{sim}\,(d_j,\,q_k) = \frac{\sum\limits_{i=1}^{m} w_{ij} \times w_{ik}}{\min\left(\sum\limits_{i=1}^{m} w_{ij},\, \sum\limits_{i=1}^{m} w_{ik}\right)}$$

9.5   NON-CLASSICAL MODELS OF IR

Non-classical IR models are based on principles other than similarity, probability, Boolean operations, etc., on which classical retrieval models are based. Examples include information logic model, situation theory model, and interaction model.

The *information logic model* is based on a special logic technique called logical imaging. Retrieval is performed by making inferences from document to query. This is unlike classical models, where a search process is used. Unlike usual implication, which is true in all cases except that when antecedent is true and consequent is false, this inference is uncertain. Hence, a measure of uncertainty is associated with this inference. The principle put forward by van Rijsbergen is used to measure this uncertainty. This principle says:

Given any two sentences x and y, a measure of the uncertainty of y → x relative to a given data set is determined by the minimal extent to which one has to add information to the data set in order to establish the truth of y → x.

In fact, this model was developed in response to van Rijsbergen's realization that classical models were unable to enhance effectiveness and that new meaning-based models were required to do so.

The *situation theory* model is also based on van Rijsbergen's principle. Retrieval is considered as a flow of information from document to query. A structure called *infon,* denoted by ι, is used to describe the situation and to model information flow. An infon represents an *n*-ary relation and its polarity. The polarity of an infon can be either 1 or 0, indicating that the infon carries either positive or negative information.

For example, the information in the sentence, *Adil is serving a dish,* is conveyed by the infon

$\iota = \langle\langle serving\ Adil,\ dish;\ 1\rangle\rangle$

The polarity of an infon depends on the support. The support of an infon is represented as $s \models \iota$ and means that the situation s makes the infon ι true. For example, the infon, $\iota = \langle\langle serving\ Adil,\ dish;\ 1\rangle\rangle$ is made true by a situation $s1 = $ "I see Adil serving a dish."

A document d is relevant to a query q, if $d \models q$

If a document does not support the query q, it does not necessarily mean that the document is not relevant to the query. Additional information, such as synonymys, hypernyms/hyponyms, meronyms, etc., can be used to transform the document d into d' such that $d' \models q$. Semantic relationships in a thesaurus, like WordNet, are useful sources for this information. The transformation from d to d' is regarded as flow of information between situations.

The *interaction IR* model was first introduced in Dominich (1992, 1993) and Rijsbergen (1996). In this model, the documents are not isolated; instead, they are interconnected. The query interacts with the interconnected documents. Retrieval is conceived as a result of this interaction. This view of interaction is taken from the concept of interaction as realized in the Copenhagen interpretation of quantum mechanics. Artificial neural networks can be used to implement this model. Each document is modelled as a neuron, the document set as a whole forms a neural network. The query is also modelled as a neuron and integrated into the network. Because of this integration, new connections are built between the query and the documents, and existing connections are changed. This restructuring corresponds to the concept of interaction. A measure of this interaction is obtained and used for retrieval. Detailed mathematical treatments of the model have been discussed by Dominich (1992, 1993, 2001) and van Rijsbergen (1996).

9.6 ALTERNATIVE MODELS OF IR

9.6.1 Cluster Model

The cluster model is an attempt to reduce the number of matches during retrieval. The need for clustering was first pointed out by Salton. Before we discuss the cluster-based IR model, we would like to state the cluster hypothesis that explains why clustering could prove efficient in IR.

Closely associated documents tend to be relevant to the same clusters.

This hypothesis suggests that closely associated documents are likely to be retrieved together. This means that by forming groups (classes or clusters) of related documents, the search time reduced considerably. Instead of matching the query with every document in the collection, it is matched with representatives of the class, and only documents from a class whose representative is close to query, are considered for individual match.

Clustering can be applied on terms instead of documents. Thus, terms can be grouped to form classes of co-occurrence terms. Co-occurrence terms can be used in dimensionality reduction or thesaurus construction. A number of methods are used to group documents. We discuss here, a cluster generation method based on similarity matrix. This method works as follows:

Let $D = \{d_1, d_2, ..., d_j, ..., d_m\}$ be a finite set of documents, and let $E = (e_{ij})_{n,n}$ be the similarity matrix. The element E_{ij} in this matrix, denotes a similarity between document d_i and d_j. Let T be the threshold value. Any pair of documents d_i and d_j $(i \neq j)$ whose similarity measure exceeds the threshold $(e_{ij} \geq T)$ is grouped to form a cluster. The remaining documents form a single cluster. The set of clusters thus obtained is

$$C = \{C_1, C_2, ..., C_k, ..., C_p\}$$

A representative vector of each class (cluster) is constructed by computing the centroid of the document vectors belonging to that class. Representation vector for a cluster C_k is

$$r_k = \{a_{1k}, a_{2k}, ..., a_{ik}, ..., a_{mk}\}$$

An element a_{ik} in this vector is computed as

$$a_{ik} = \frac{\sum\limits_{d_j \in C_k} a_{ij}}{|C_k|}$$

where a_{ij} is weight of the term t_i, of the document d_j, in cluster C_k. During retrieval, the query is compared with the cluster vectors

$$(r_1, r_2, ..., r_k, ..., r_p)$$

This comparison is carried out by computing the similarity between the query vector q and the representative vector r_k as

$$s_{ik} = \sum\limits_{i=1}^{m} a_{ik} q_i, \quad k = 1, 2, ..., p$$

A cluster C_k whose similarity s_k exceeds a threshold is returned and the search proceeds in that cluster.

Example 9.4
Let

$$A = \begin{pmatrix} 1 & 1 & 0 \\ 0 & 1 & 0 \\ 1 & 1 & 1 \\ 0 & 0 & 1 \\ 1 & 1 & 0 \end{pmatrix}$$

be the term-by-document matrix. The similarity matrix corresponding to these documents is

1.0

0.9 1.0

0.4 0.4 1.0

Using a threshold of 0.7, we get the following two clusters:

$C_1 = \{d_1, d_2\}$

$C_2 = \{d_3\}$

The cluster vectors (representatives) for C_1 and C_2 are

$r_1 = (1\ 0.5\ 1\ 0\ 1)$

$r_2 = (0\ 0\ \ 1\ 1\ 0)$

Retrieval is performed by matching the query vector with r_1 and r_2.

9.6.2 Fuzzy Model

In the fuzzy model, the document is represented as a fuzzy set of terms, i.e., a set of pairs $[t_i, \mu(t_i)]$, where μ is the membership function. The membership function assigns to each term of the document a numeric membership degree. The membership degree expresses the significance of term to the information contained in the document. Usually, the significance values (weights) are assigned based on the number of occurrences of the term in the document and in the entire document collection, as discussed earlier. Each document in the collection

$$D = \{d_1, d_2, ..., d_j, ..., d_n\}$$

can thus be represented as a vector of term weights, as in the following vector space model

$$(w_{1j}, w_{2j}, w_{3j}, ..., w_{ij}, ..., w_{mj})^t$$

where w_{ij} is the degree to which term t_i belongs to document d_j.

Each term in the document is considered a representative of a subject area and w_{ij} is the membership function of document d_j to the subject area represented by term t_i. Each term t_i is itself represented by a fuzzy set f_i in the domain of documents given by

$$f_i = \{(d_j, w_{ij})\} \mid i = 1,. ..., m; j = 1, ..., n$$

This weighted representation makes it possible to rank the retrieved documents in decreasing order of their relevance to the user's query.

Typically, queries are Boolean queries. For each term that appears in the query, a set of documents is retrieved. Fuzzy set operators are then applied to obtain the desired result.

For a single-term query $q = t_q$, those documents from the fuzzy set $f_q = \{(d_j, w_{iq})\}$, are retrieved for which w_{iq} exceeds a given threshold. The threshold may also be zero.

Consider the case of an AND query $q = t_{q1} \wedge t_{q2}$

First, the fuzzy sets f_{q1} and f_{q2} are obtained and then, their intersection is obtained, using the fuzzy intersection operator $f_{q1} \vee f_{q2} = \min \{(d_j, w_{iq1}), (d_j, w_{iq})\}$

The documents in this set are returned.

Similarly, for an OR query $q = t_{q1} \wedge t_{q2}$, the union of fuzzy sets f_{q1} and f_{q2} is computed to retrieve documents as follows:

$$f_{q1} \vee f_{q2} = \max \{(d_j, w_{iq1}), (d_j, w_{iq})\}$$

Example 9.5   Consider the following three documents:

$d_1 = \{\text{information, retrieval, query}\}$
$d_2 = \{\text{retrieval, query, model}\}$
$d_3 = \{\text{information, retrieval}\}$

where the set of terms used to represent documents is

$$T = \{\text{information, model, query, retrieval}\}$$

The fuzzy sets induced by these terms are

$f_1 = \{(d_1, 1/3), (d_2, 0) \ (d_3, 1/2)\}$
$f_2 = \{(d_1, 0), (d_2, 1/3) \ (d_3, 0)\}$
$f_3 = \{(d_1, 1/3), (d_2, 1/3) \ (d_3, 0)\}$
$f_4 = \{(d_1, 1/3), (d_2, 1/3) \ (d_3, 1/2)\}$

If the query is $q = t_2 \wedge t_4$, then document d_2 will be returned.

9.6.3   Latent Semantic Indexing Model

Latent semantic indexing model is the application of single value decomposition to IR. The use of latent semantic indexing (LSI) is based on the assumption that there is some underlying 'hidden' semantic structure in the pattern of word-usage across documents, rather than just surface level word choice. LSI attempts to identify this hidden semantic structure through statistical techniques and use it to represent and retrieve information. This is done by modelling the association between terms and documents based on the manner in which terms co-occur across documents. LSI transforms the term-document vector space into a more compact latent semantic space. Each dimension in the reduced space corresponds to an 'artificial concept'. These concepts loosely correspond to a set of terms. It is believed that in the vector space of reduced dimensionality, the words referring to related concepts, i.e., words that co-occur, are collapsed into the same dimension. Latent semantic space is thus able to capture similarities that go beyond term similarity. In the latent semantic space, a query and a document can have high similarity even if the document does not contain a query term, provided the terms are semantically related.

Now we discuss how the LSI technique is actually employed in IR. The document collection is first processed to get a $m \times n$ term-by-document matrix, W, where m is the number of index terms and n is the total number of documents in the collection. Columns in this matrix represent document vectors, whereas the rows denote term vectors. The matrix element W_{ij} represents the weight of the term i in document j. The weight may be assigned based on term frequency or some combination of local and global weighting, as in the case of vector space model. Singular value decomposition (SVD) of the term-by-document matrix is then computed. Using SVD, the matrix is represented as a product of three matrices

$$W = TSD^T$$

where T corresponds to term vectors and has m rows and r columns and $r = \min(m, n)$. S corresponds to singular values. D^T is the transpose of D and has r rows and n columns. D corresponds to the document vector.

T and D are orthogonal matrices containing the left and right singular vectors of W. S is a diagonal matrix, containing singular values stored in decreasing order. We eliminate small singular values and approximate the original term-by-document matrix using truncated SVD. For example, by considering only the first k number of the largest singular values, along with their corresponding columns in T and D, we get the following approximation of the original term-by-document matrix in a space of k orthogonal dimensions, where k is sufficiently less than n:

$$W_k = T_k S_k D_k^T$$

where T_k is the first k columns of T, D_k^T is the first k columns of D^T, and S_k is the k largest singular values.

The matrix W_k is used for retrieval. The idea is that the elimination of small singular values throws out the 'noise' resulting from term usage variation, and captures the underlying 'hidden' semantic structure (i.e., concepts). Each dimension in the reduced space corresponds to artificial or derived concepts. Each such concept loosely represents a set of terms in the original term-document matrix. Documents with varying word usage patterns are collapsed to the same vector in k-space.

The queries are also represented in k-dimensional space. Let $q = (q_1, q_2, \ldots, q_m)$ be the original query vector, where each element q_i is the frequency of term i in the query q. The query q is represented in the k-dimensional space as

$$q_k = q^T T_k S_k^{-1}$$

where q^T is the transpose of the query vector, and T_k and S_k are the weights. $q^T T_k$ denotes the sum of k-dimensional term vectors and S_k^{-1}, the weights of each dimension. Thus, the query is represented as the weighted sum of its constituent term vectors.

Retrieval is performed by computing the similarity between query vector and document vector. For example, we can use the cosine similarity measure to rank documents to perform retrieval. In a keyword-based retrieval, relevant documents that do not share any term with the query are not retrieved. The LSI-based approach is capable of retrieving such documents, as similarity is computed based on the overall pattern of term usage across the document collection rather than on term overlap.

We now give an example to explain how a document in high-dimensional space is represented in a low, reduced, latent semantic space.

Example 9.6 Consider the matrix shown in Figure 9.5. This matrix defines five-dimensional space in which six documents, $d_1, d_2, d_3, ..., d_6$, have been represented. The five dimensions correspond to five index terms *tornado, storm, tree, forest,* and *farming.* For simplicity, tf has been used to weight index terms. Figure 9.6 shows the documents in a two-dimensional space. The vectors in the figure correspond to document vectors in the matrix R, which is the representation of X in reduced two-dimensional space. The two dimensions correspond to derived concepts obtained through the application of truncated SVD.

$$
X = \begin{pmatrix}
 & d_1 & d_2 & d_3 & d_4 & d_5 & d_6 \\
\text{tornado} & 1 & 1 & 0 & 0 & 0 & 0 \\
\text{storm} & 1 & 0 & 1 & 0 & 1 & 0 \\
\text{tree} & 1 & 0 & 1 & 0 & 0 & 0 \\
\text{forest} & 0 & 0 & 1 & 1 & 0 & 0 \\
\text{farming} & 0 & 0 & 0 & 1 & 0 & 1
\end{pmatrix}
$$

Figure 9.5 A term-document matrix representing six documents in five-dimensional space

We now explain how to arrive at the reduced dimensionality representation of X. First, the SVD of X is computed to get the three matrices T, S, and D.

$$X_{5 \times 6} = T_{5 \times 5} \, S_{5 \times 5} \, (D_{6 \times 5})^T$$

These matrices are shown in Figures 9.7, 9.8, and 9.9 respectively. Consider the first two largest singular values of S, and rescale $D^T_{2 \times 6}$ with singular values to get matrix $R_{2\times6} = S_{2\times2}D^{T}_{2\times6}$, as shown in Figure 9.10, where $S_{2\times2}$ is S restricted to two dimensions and $D^{T}_{2\times6}$ is D^{T} restricted to two columns. R is a reduced dimensionality representation of the original term-by-document matrix X and is used to plot the vectors in Figure 9.6.

To find out the changes introduced by the reduction, we compute document similarities in the new space and compare them with the similarities between documents in the original space. The document–document correlation matrix for the original n-dimensional space is given by the matrix $Y = X^{T}X$. Here, Y is a square, symmetric $n \times n$ matrix. An element Y_{ij} in this matrix gives the similarity between documents i and j. The correlation matrix for the original document vectors is shown in Figure 9.12. This matrix is computed using X, after normalizing the lengths of its columns. The document–document correlation matrix for the new space is computed analogously using the reduced representation R. Let N be the matrix R with length-normalized columns. Then, $M = N^{T}N$ gives the matrix of document correlations in the reduced space. The correlation matrix M is given in Figure 9.11. The similarity between document d_1, $d_4(-0.0304)$, and $d_6(-0.2322)$ is quite low in the new space because document d_1 is not topically similar to documents d_4 and d_6. In the original space, the similarity between documents d_2 and d_3 and between documents d_2 and d_5 is 0. In the new space, they have high similarity values (0.5557 and 0.8518 respectively) although documents d_3 and d_5 share no term with the document d_2. This topical similarity is recognized due to the co-occurrence of patterns in the documents.

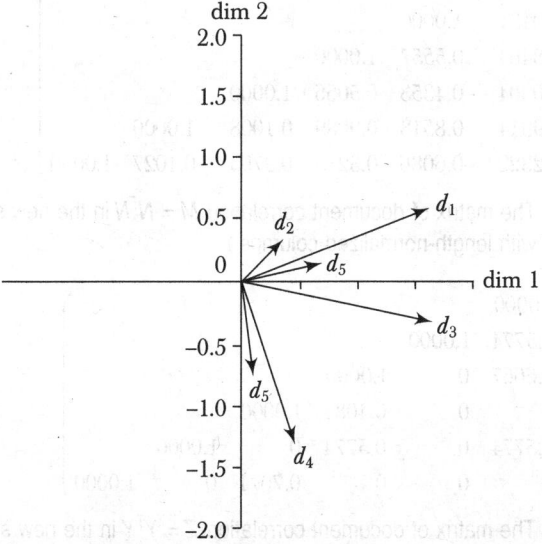

Figure 9.6 Documents in reduced two-dimensional space

$$T = \begin{pmatrix} 0.3318 & 0.3338 & 0.8064 & -0.2426 & -0.2634 \\ 0.6693 & 0.1616 & -0.2737 & 0.5853 & -0.3293 \\ 0.5514 & 0.1038 & -0.0961 & -0.2667 & 0.7777 \\ 0.3583 & -0.5745 & -0.2148 & -0.5778 & -0.4021 \\ 0.0974 & -0.7223 & 0.4684 & 0.4400 & 0.2362 \end{pmatrix}$$

Figure 9.7 Matrix *T* for the SVD of the term-document matrix *X* shown in Figure 9.5

$$S = \begin{pmatrix} 2.3830 & 0 & 0 & 0 & 0 \\ 0 & 1.6719 & 0 & 0 & 0 \\ 0 & 0 & 1.2415 & 0 & 0 \\ 0 & 0 & 0 & 0.8288 & 0 \\ 0 & 0 & 0 & 0 & 0.5454 \end{pmatrix}$$

Figure 9.8 The matrix *S* for singular values of the SVD of the term-document matrix *X*

$$D^T = \begin{pmatrix} 0.6515 & 0.1392 & 0.6626 & 0.1912 & 0.2809 & 0.0409 \\ 0.3584 & 0.1996 & -0.1848 & -0.7756 & 0.0967 & -0.4320 \\ 0.3516 & 0.6495 & -0.4710 & 0.2042 & -0.2205 & 0.3773 \\ 0.0916 & -0.2927 & -0.3127 & -0.1662 & 0.7062 & 0.5309 \\ 0.3392 & -0.4829 & 0.0849 & -0.3042 & -0.6037 & 0.4330 \end{pmatrix}$$

Figure 9.9 The matrix D^T for singular values of the SVD of the term-document matrix

$$R = \begin{pmatrix} 1.5526 & 0.3318 & 1.5790 & 0.4557 & 0.6693 & 0.0974 \\ 0.5992 & 0.3338 & -0.3090 & -1.2967 & 0.1616 & -0.7223 \end{pmatrix}$$

Figure 9.10 The matrix $R_{2\times6} = S_{2\times2}D^T_{2\times6}$ representing documents in two-dimensional space

$$M = \begin{pmatrix} 1.0000 \\ 0.9131 & 1.0000 \\ 0.8464 & 0.5557 & 1.0000 \\ -0.0304 & -0.4353 & 0.5066 & 1.0000 \\ 0.9914 & 0.8518 & 0.9089 & 0.1008 & 1.0000 \\ -0.2322 & -0.6086 & 0.3215 & 0.9793 & -0.1027 & 1.0000 \end{pmatrix}$$

Figure 9.11 The matrix of document correlation $M = N^T N$ in the new space (*N* is matrix *R* with length-normalized columns.)

$$Z = \begin{pmatrix} 1.0000 \\ 0.5774 & 1.0000 \\ 0.6667 & 0 & 1.0000 \\ 0 & 0 & 0.4082 & 1.0000 \\ 0.5774 & 0 & 0.5774 & 0 & 1.0000 \\ 0 & 0 & 0 & 0.7071 & 0 & 1.0000 \end{pmatrix}$$

Figure 9.12 The matrix of document correlation $Z = Y^T Y$ in the new space (*Y* is matrix *X* with length-normalized columns.)

The LSI performs IR based on concept. It is completely automatic and has been applied successfully (Deerwester et al. 1990, Foltz 1990) in many IR systems. However, it is costly in terms of computation.

9.7 EVALUATION OF THE IR SYSTEM

The evaluation of IR systems is the process of assessing how well a system meets the information needs of its users (Voorhees 2001). Evaluating an IR system is a difficult task involving a number of areas including cognition, statistics, and man-machine interactions. IR evaluation models can be broadly classified as system driven models and user-centered models. System driven models (Cleverdon et al. 1966) measure how well a system ranks documents; user-centered models measure user satisfaction. Cleverdon listed the following six criteria that can be used for evaluation:

1. *Coverage of the collection*: The extent to which the system
2. *Time lag*: The time that elapses between submission of a query and getting back the response
3. *Presentation format*
4. *User effort*: The effort made by the user to obtain relevant information
5. *Precision*: The proportion of retrieved documents that are relevant
6. *Recall*: The proportion of relevant documents that are retrieved

Of these criteria, recall and precision have most frequently been applied in measuring IR. Both are related to effectiveness, i.e., the ability of a system to retrieve relevant documents in response to user query. A number of effectiveness measures have been formulated (van Rijsbergen 1979). We discuss them in the following section. To better understand the relationship between aspects of retrieval process and different measures, see (Voorhees and Harman 1999), where correlations between pairs of measures are estimated.

The major goal of IR is to search for documents that are relevant to a user's query. It is necessary to understand what constitutes relevance, as the evaluation of IR systems relies on the notion of relevance.

9.7.1 Relevance

Relevance is subjective in nature (Saracevic 1991), i.e., it depends on the individual judgements of users. Given a query, the same document may be judged as relevant by one user and non-relevant by another. It is not possible to measure this 'true relevance' because no human can read all documents in a collection and provide a relevance assessment. Most evaluations of IR systems have so far been done on test document collections with known relevance judgments. These test document collections contain documents from a particular discipline; for a set of questions representing information needs, relevance assessments are obtained from experts of that discipline. This provides an experimental setup for evaluating the performance of a retrieval strategy. If a retrieval strategy performs well under these situations, it is expected to perform well in an operational environment where relevance is not known.

Another issue with relevance is the degree of relevance. Traditionally, relevance has been visualized as a binary concept, i.e., a document is either relevant or not relevant; whereas relevance is actually a continuous function (a document may be exactly what the user wants or it may be closely related). This is an attractive but difficult proposition and current evaluation techniques do not support it.

A number of relevance frameworks have been proposed by Saracevic (1996). This includes system, communication, psychological, and situational frameworks. The most inclusive of these is the situational framework, which is based on a cognitive view of the information seeking process and considers the importance of situation, context, multi-dimensionality, and time. A survey of relevance studies has been discussed by Mizzarro, (1996).

9.7.2 Effectiveness Measures

Effectiveness is purely a measure of the ability of a system to satisfy the user in terms of the relevance of documents retrieved (Rijsbergen 1979). Aspects of effectiveness include whether the documents returned are relevant to the user, whether they are presented in order of relevance, whether a significant number of relevant documents in the collection are returned to the user, etc. A number of measures have been proposed to quantify effectiveness. As stated earlier, the most commonly used measures of effectiveness are precision and recall. These measures are based on relevance judgments.

Precision and Recall

Precision is defined as the proportion of relevant documents in a retrieved set. This can be seen as the probability that a relevant document is retrieved. *Recall* is the proportion of relevant documents in a collection that have actually been retrieved. Precision measures the accuracy of a system while recall measures its exhaustiveness. Precision and recall can be computed as follows:

$$\text{Precision} = \frac{\text{Number of relevant document retrieved } (NR_{\text{ret}})}{\text{Total number of documents retrieved } (N_{\text{ret}})}$$

$$\text{Recall} = \frac{\text{Number of relevant documents retrieved } (NR_r)}{\substack{\text{Total number of relevant documents in the} \\ \text{collection } (NR_{\text{rel}})}}$$

These definitions of precision and recall are based on binary relevance judgment, which means that every retrievable item is recognizably 'relevant', or recognizably 'not relevant'. Hence, for every search result, all retrievable documents will be either (i) relevant or non-relevant and (ii) retrieved or not retrieved. Thus, each document will fall into one, and only one, of four cells of the matrix, as sown in Figure 9.13. This matrix is used to derive a number of measures.

	Relevant	Non-relevant	
Retrieved	$A \cap B$	$\bar{A} \cap B$	B
Not-retrieved	$A \cap \bar{B}$	$\bar{A} \cap \bar{B}$	$\bar{B}$
	A	$\bar{A}$	

Figure 9.13   Relevant matrix

Referring to Figure 9.13, precision and recall will be given as follows:

$$\text{Precision} = \frac{|A \cap B|}{|B|} = \frac{NR_{\text{ret}}}{N_{\text{ret}}}$$

$$\text{Recall} = \frac{|A \cap B|}{|A|} = \frac{NR_{\text{ret}}}{NR_{\text{rel}}}$$

where $A =$ Set of relevant documents
$\quad |A| =$ No. of relevant documents in the collection (NR_{rel})
$\quad B =$ Set of retrieved documents
$\quad |B| =$ No. of retrieved documents (NR_{ret})

It is clear from the preceding definitions that the total number of relevant documents in a collection must be known in order for recall to be calculated. The amount of effort and time required from the user makes this almost impossible in most operating environment. To provide a framework of evaluation for IR systems, a number of test collections have been developed (Cranfield and TREC). These collections are accompanied by a set of queries and relevance judgements. These test collections make it possible for IR researchers to efficiently evaluate their experimental approaches and compare the effectiveness of their system with that of others. In Table 9.5, basic statistics for a number of test collections are presented.

Table 9.5   IR test collections

Collection	Number of documents	Number of queries
Cranfield	1400	225
CACM	3204	64
CISI	1460	112
LISA	6004	35
TIME	423	83
ADI	82	35
MEDLINE	1033	30
TREC-1	742,611	100

There exists a trade-off between precision and recall, though a high value of both at the same time is desirable. The trade-off is shown in Figure 9.14. Precision is high at low recall values. As recall increases, precision decreases. The ideal case of perfect retrieval requires that all relevant documents be retrieved before the first non-relevant document is retrieved. This is shown in the figure by the line parallel to *x*-axis having a precision of 1.0 at all recall points. Recall is an additive process. Once the highest recall (1.0) is achieved, it remains 1.0 for any subsequent document retrieved. We can always achieve 100% recall by retrieving all documents in the collection, but this defeats the intent of an IR system.

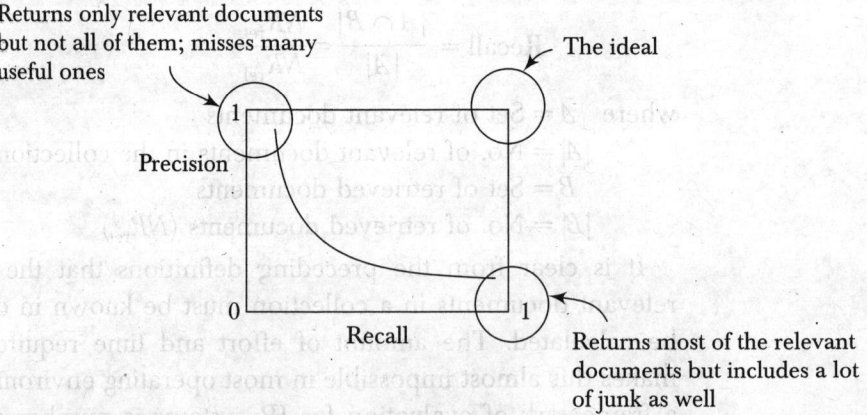

Figure 9.14   The trade-off between recall and precision

A number of researchers have discussed the relationship between recall and precision (Cleverdon 1972, Robertson 1975, Gordon and Kochen 1989, Buckland and Gey 1994). Some of them modelled precision and recall as continuous functions (Robertson 1975, Gordon and Kochen 1989), while others (Bookstein 1974) described recall and precision in terms of two-Poisson discrete model. Buckland and Gey (1994) studied the relationship between precision and recall, and suggested that a two-stage, or more generally, a multi-stage retrieval procedure is likely to achieve the goal of improving both precision and recall simultaneously, even though the trade-off between them cannot be avoided.

In order to evaluate the performance of an IR system, recall and precision are almost always used together. One measure is to calculate precision at a particular cut-off. Typical cut-offs are 5 documents, 10 documents, 15 documents, etc.

Yet another measure is *non-interpolated average precision*, which is average of the precision at observed recall points. Observed recall points correspond to points where a relevant document is retrieved. We first compute precision at each point where a relevant document is found and then compute average of these precision numbers to get a single number. Precision at relevant documents not in the returned set is assumed to be zero. We now give an example to illustrate how precision is calculated.

Table 9.6 An example of retrieval

Rank	Document #	Relevance
1	10	x
2	8	x
3	5	
4	3	
5	1	x
6	2	
7	4	
8	7	
9	9	x
10	6	x

Example 9.7 Table 9.6 shows the ranking of 10 documents for a particular retrieval. The crossed documents are those that are relevant. Let the total number of relevant document be five.

Precision values at 5 and 10 documents are given as follows:

Precision at 5   3/10 = 0.3
Precision at 10   5/10 = 0.5

Non-interpolated Average Precision   The observed recall points are 0.2, 0.4, 0.6, 0.8, and 1.0. These recall values correspond to the documents marked relevant in the table. We have one of five relevant documents retrieved after retrieving only one document. This correspond to a recall value of $1/5 = 0.2$. After two documents, the recall is $2/5 = 0.4$. As the third document retrieved is not relevant, recall value does not change. The next relevant document is found after five documents have been retrieved, resulting in a recall value of $3/5 = 0.6$. Similarly, the other two recall points are calculated. The precision values at these points are as follows:

Precision at recall point 0.2: $1/1 = 1.0$
Precision at recall point 0.4: $2/2 = 1.0$
Precision at recall point 0.6: $3/5 = 0.6$
Precision at recall point 0.8: $4/9 = 0.4$
Precision at recall point 1.0: $5/10 = 0.5$
Non-interpolated average precision $= 0.7$

Considering the fact that a system may not always retrieve all the relevant documents, and that the number of relevant documents is not the same for all queries, precision values are interpolated for a set of recall points. The most widely used recall levels are 0.0, 0.1, 0.2, 0.3... 1.0. The precision values are calculated at each of these 11 recall levels and then averaged to get a single value. This is known as 11-point interpolated average precision. This has become almost a standard in evaluating the performance of an IR system. The interpolation used at TREC states that, precision at a given recall level is the greatest known precision at any recall level greater than or equal to this given level. For example, if the observed recall points are 0.25, 0.4, 0.55, 0.8, and 1.0, then precision at recall level 0.3 will be the maximum of the precision at recall levels 0.4, 0.55, 0.8, and 1.0, and not precision at recall point 0.4 where the 30% recall (i.e., recall level 0.3) is first reached. The interpolated precision at standard recall points for the documents shown in the Table 9.5 is computed in Example 9.8.

Example 9.8   Consider the following precision values at observed recall points:

0.25	1.0
0.4	0.67
0.55	0.8
0.8	0.6
1.0	0.5

The interpolated precision at the standard 11 recall levels will be

0.0	1.0
0.1	1.0
0.2	1.0
0.3	0.8
0.4	0.8
0.5	0.8
0.6	0.6
0.7	0.6
0.8	0.6
0.9	0.5
1.0	0.5

Interpolated average precision = 0.745

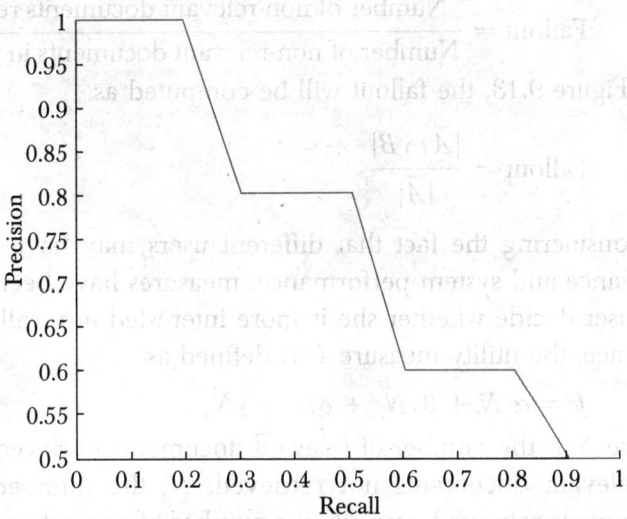

Figure 9.15 Recall–precision curve for interpolated precision

Often, precision values are calculated at different recall levels and a recall–precision graph, like the one shown in Figure 9.15, is plotted. As a retrieval system is evaluated over several queries, such a graph is usually plotted using precision figures averaged over all queries (Salton and McGill 1983, van Rijsbergen 1979). The most standard method for deriving a recall–precision graph is to plot the average over all queries in the interpolated precision values for a set of 11 standard recall points, namely 0.0, 0.1, 0.2, 0.3, ..., 1.0.

Instead of the recall–precision graph, the mean average precision is sometimes used to evaluate an IR system. The non-interpolated average precision is averaged over all queries to get the mean average precision (MAP). Geometrically, MAP is the area below the non-interpolated recall–precision curve (Voorhees and Harman 1999). The 11-point interpolated average precisions (11 avgP) can also be used for calculating MAP, though the non-interpolated measure has the advantage that it rewards systems that quickly retrieve (give high ranks to) relevant documents.

The R-precision is the precision after a total number of R documents relevant to the query have been retrieved.

Recall is not defined if there is no relevant document in a collection. An alternative measure is fallout, which may be seen as the inverse of recall. It is not defined only if all the documents in the collection are relevant (Salton 1983). Fallout is the ratio of non-relevant documents retrieved to non-relevant documents in the collection.

$$\text{Fallout} = \frac{\text{Number of non-relevant documents retrieved } (N_n)}{\text{Number of non-relevant documents in the collection}}$$

For Figure 9.13, the fallout will be computed as

$$\text{Fallout} = \frac{|\bar{A} \cap B|}{|\bar{A}|}$$

Considering the fact that different users may have different ideas of relevance and system performance, measures have been developed to let the user decide whether she is more interested in recall or precision. For instance, the utility measure U is defined as

$$U = \alpha \cdot N_r + \beta \cdot \bar{N}_r + \delta N_n + \gamma \bar{N}_n$$

where N_r is the number of relevant documents retrieved, $\bar{N}_r$ the number of relevant documents not retrieved, N_n the number of non-relevant documents retrieved, and $\bar{N}_n$ the number of non-relevant documents not retrieved (α, β, δ, and γ are positive weights specified by the user). This measure was later simplified by considering only retrieved documents.

The F-measure takes into account both precision and recall. It is defined as the harmonic mean of recall and precision.

$$F = \frac{2PR}{P + R}$$

Compared to the arithmetic mean, both recall and precision need to be high for harmonic mean to be high.

The E-measure is a variant of the F-measure. It allows weighting to emphasize precision rather than recall. It is defined as

$$E = \frac{(1 + \beta^2) PR}{\beta^2 P + R} = \frac{(1 + \beta^2)}{\dfrac{\beta^2}{R} + \dfrac{1}{P}}$$

where P is precision, R is recall, and β is the relative importance of P compared to R. The value of β controls the trade-off between precision and recall. Setting β to 1 gives equal weight to precision and recall, resulting in a harmonic mean of recall and precision ($E = F$). $\beta > 1$ gives more weight to precision, and $\beta < 1$ gives more weight to recall.

Swets (1969) developed a model that received attention in the literature (Heine 1974; Bookstein 1977). None of these alternative measures, however, received as widespread acceptance as did the recall–precision model.

Normalized recall measures how close the set of retrieved documents is to an ideal retrieval, in which the most relevant NR_{rel} document appears in the first NR_{rel} position. Relevant documents are ranked 1, 2, 3, ..., NR_{rel}, where NR_{rel} is the number of relevant documents. The ideal rank is given by

$$IdR = \frac{\sum\limits_{r=1}^{NR_{rel}} r}{NR_{rel}}$$

Let the average rank (AvR) over the set of relevant documents retrieved by a system be

$$AvR = \frac{\sum\limits_{r=1}^{NR_{rel}} Rank_r}{NR_{rel}}$$

where $Rank_r$ represents the rank of the rth relevant document. The difference between AvR and IdR given by AvR – IdR, represents a measure of the effectiveness of the system. This difference can range from 0, for the perfect retrieval (AvR – IdR), to $(N - NR_{rel})$, for the worst case (N is the total number of documents in the collection). The worst case is when all the N documents are retrieved and the relevant documents, NR_{rel}, are the last retrieved. The expression AR–IR can be normalized by dividing it by $(N - NR_{rel})$ and then subtracting the result from 1. The normalized recall (NR) is given by

$$NR = 1 - \frac{AvR - IdR}{N - NR_{rel}}$$

This measure ranges from 1 for the best case, to 0 for the worst case. If the value of NR is close to 1, the ranks of relevant documents in average case deviate very little from the ideal case. A high value of NR indicates that the ranks of the relevant documents in the average case deviate considerably from the ideal case.

9.7.3 User-centred Evaluation

The system-driven model is still the dominant approach followed in IR research for evaluation of IR systems. The evaluation here, is made on a test collection having known relevance judgments. These relevance judgements were usually provided by problem domain experts, and are binary, objective, topical, and static in nature, and lack a user's viewpoint. There is also major disagreement among experts in providing relevance judgements (Haynes et al. 1990, Hersh and Hickam 1994). Further, these judgements of relevance are affected not only by the expertise of the judge, but also by the order of the documents (Eisenberg and Barry 1988; Schamber et al. 1990). It has been argued that relevance is not fixed, that it varies over time (Meadow 1992). The meaning and the relevance of a document can thus be different for different users and can be inferred only in the context of the user's situation. Relevance, therefore, is subjective, dynamic, and multi-dimensional in nature (Saracevic 1975, Mizzaro 1998, Harter 1992).

Another drawback of the system-driven approach is that it removes the end users from the retrieval process, substituting them with the queries and judgements provided with the test collection. This allows fast experimentation but makes it difficult to evaluate the effect of interactive IR techniques and is suitable only for non-interactive environment (Draper and Dunlop). In an interactive setting, a user normally starts with a query, which goes through many refinements to eventually get the desired documents. The test-collection approach poses a problem in such an environment. As performance of the IR system will eventually be measured in terms of it ability to retrieve documents relevant to a user's query, it seems realistic to follow a user-centered approach to evaluation. Such an approach will result in a much more direct measure of the overall goal. A number of measures have been proposed for interactive IR including relative relevance (RR), ranked half life (RHL), and cumulated gain (CG). Details of these measures can be found in Hersh et al. (1995), Borlund and Ingwerson (1998), and Järvelin and Kekäläinen (2000).

The subjective nature of interactive IR has been highlighted (Borlund and Ingwerson 1997) and attempts have been made to integrate cognitive theory into IR evaluation. However, efforts in this direction have been limited. A task oriented, user-centred, non-interactive evaluation methodology has been proposed by Reid (2000), in which the basic unit of evaluation is task rather than query. More recently, an interactive IR evaluation model has been proposed by Borlund (2000, 2003) to evaluate interactive IR systems. The key elements of an interactive IR model are the use of realistic scenarios (simulated work tasks) and alternative performance measures such as RR and RHL. However, user-centred evaluation methods are expensive both in terms of time and resources. A properly designed user-centred evaluation, with a few exceptions, requires a sufficiently large, representative sample of actual users of retrieval systems. The systems to be compared must be equally well-developed and equipped with the appropriate user interface. The subject must be trained with these systems. Further, it is difficult to develop a standard interactive evaluation methodology that will allow for comparison across different systems and users (Reid 2000). Because of these considerations, recall and precision remain the most popular and standard measure for evaluating IR system performance.

SUMMARY

- Information retrieval (IR) deals with the organization, storage, retrieval, and evaluation of information relevant to a user's query.
- An IR system does not return the actual information but returns the documents containing that information.
- The actual text of the document is not used in the retrieval process. Instead, documents in a collection are frequently represented through a set of index terms or keywords.
- The process of transforming document text to some representation of it is known as *indexing*.
- A common lexical processing of index terms involves elimination of stop words.
- An IR model is a pattern that defines several aspects of the retrieval procedure, for example, how the documents and users' queries are represented, how the system retrieves relevant documents according to users' queries, and how retrieved documents are ranked.
- Classical IR models, such as Boolean, vector space, and probabilistic, are based on mathematical knowledge that is easily recognized and well understood.

- Non-classical IR models are based on principles other than similarity, probability, Boolean operations, etc., on which classical IR models are based. Examples include information logic model, situation theory model, and interaction model.
- Latent semantic indexing (LSI) attempts to identify hidden semantic structures using statistical techniques and then uses this structure to represent and retrieve information.
- The evaluation of an IR system is the process of assessing how well the system meets the information needs of its users. Recall and precision are the two most widely used evaluation measure.

REFERENCES

Bookstein, 1974, 'The anonymous behavior of precision in the Swets model, and its resolution,' *Journal of Documentation*, 21, pp. 374–80.

Borlund, P., 2000, 'Experimental components for the evaluation of interactive information retrieval systems,' *Journal of Documentation*, 56(1).

_____2003, 'The IIR evaluation model: a framework for evaluation of interactive information retrieval systems,' *Information Research*, 8(3).

Borlund, P. and Ingwersen, 1997, 'The development of a method for the evaluation of interactive information retrieval systems,' *Journal of Documentation*, 53, pp. 225–50.

Buckland, M. and F. Gey, 1994, 'The relationship between recall and precision,' *Journal of the American Society for Information Science*, 45, pp. 12–19.

Buckley, C., A. Singhal, M. Mitra, and G. Salton, 1996, 'New retrieval approaches using SMART,' *Proceedings of the 4th Text Retrieval Conference (TREC-4)*, NIST Special Publication (500-236), pp. 25–48.

Callan, J.P., W.B. Croft, and S.M. Harding, 1992, 'The INQUERY retrieval system,' *Proceedings of the 3rd International Conference on Database and Expert System's Applications*, Valencia, Spain, pp. 78–83.

Crestani, F., M. Lemas, C.J. van Rijsbergen, and I. Campbell, 1968, 'Is the document relevant? ... probably: a survey of probabilistic models in information retrieval,' *ACM Computing Surveys*, 30(4), pp. 528–52.

Deerwester, S., T. Dumais, George W. Furnas, and Thomas K. Landauer, 1990, 'Indexing by latent semantic analysis,' *Journal of the American Society of Information Science*.

Dominich, S., 1992, 'The Copenhagen interpretation to handle relevancy and meaning in information retrieval,' *Symposium on Informatics*,

Technical University Clausthal–Zellerfeld, Institute for informatics, Germany.

_____1993, 'The formulation of the interaction information retrieval model as a new and complementary framework for information retrieval,' *PhD Thesis*, The Hungarian Academy of Sciences, Budapest, Hungary.

_____2001, *Mathematical Foundation of Information Retrieval*, Kluwer Academic, the Netherlands.

Eisenberg, C. Barry, 1988, 'Order effects: a study of the possible influence of presentation order on user judgements of document relevance,' *Journal of the American Society for Information Science*, 39, pp. 293–300.

Foltz, Peter W., 1990, 'Using latent semantic indexing for information filtering,' *The ACM Conference on Office Information System (COSIS'90)*.

Fuhr, 1992, 'The probabilistic models in information retrieval,' *The Computer Journal*, 35(3), pp. 243–55.

Gordon M. and M. Kochen, 1989, 'Recall-precision trade-off: a derivation,' *Journal of the American Society for Information Science*, 40, pp. 145–51.

Harter, S.P., 1992, 'Psychological relevance and information science,' *Journal of the American Society for Information Science*, 43, pp. 602–15.

Haynes, R.B., K.A. McKibbon, and C.J. Walker, 1990, 'Online access to MEDLINE in clinical settings,' *Annals of Internal Medicine*, 112(1), pp. 78–84.

Hersh, R. and D.H. Hickam, 1994, 'A performance and failure analysis of SAPHIRE with a MEDLINE test collection,' *Journal of the American Medical Informatics Association*, 1, Elsevier, New York, pp. 51–60.

Järvelin K. and J. Kekäläinen, 2000, 'IR evaluation methods for retrieving highly relevant document,' *Proceedings of the 23rd ACM SIGIR Conference on Research and Development of Information Retrieval*, Athens, Greece, ACM Press, New York, pp. 41–48.

Lancaster, F.W., 1979, *Information Retrieval Systems: Characteristics, Testing and Evaluation*, Wiley, New York.

Losee, Robert M., 1997, 'Comparing Boolean and probabilistic information retrieval systems across queries and disciplines,' *Journal of the American Society for Information Science*, 48(2), pp.143–56.

Luhn, H.P., 1957, 'A statistical approach to mechanized encoding and searching of literary information,' *IBM Journal of Research and Development*, 1(4), pp. 309–17.

_____1958, 'The automatic creation of literature abstracts,' *IBM Journal of Research and Development*, 2(2).

Maron, E. and J.L. Kuhns, 1960, 'On relevance, probabilistic indexing and information retrieval,' *Association for Computing Machinery*, 7(3), pp. 216–44.

Meadow, C.T., 1992, 'Text information retrieval systems,' *Academic Press*, San Diego.

Mizzaro, S., 1998, 'How many relevance's in information retrieval?,' *Interacting with Computers*, 10(3), pp. 305–22.

Paice, C.P., 1984, 'Soft evaluation of Boolean search queries in information retrieval systems,' *Information Technology: Research and Development*, 3(1), pp. 33–42.

Porter, M.F., 1980, 'An algorithm for suffix stripping,' *Program*, 14(3), pp. 130–37.

Reid, Jane, 2000, 'A task-oriented non-interactive evaluation methodology for information retrieval systems,' *Information Retrieval*, 2, pp. 115–29.

Robertson, S.E. and Sparck Jones, 1998, 'Relevance weighting of search terms,' *Journal of American Society for Information Science*, 27, pp. 129–46.

Robertson, S.E., M.E. Maron, and W. S. Cooper,1982, 'Probability of relevance: a unification of two competing models for document retrieval,' *Information Technology: Research and Development*, 1(1–21).

Robertson, S.E., S. Walker, S. Jones, M. Hancock-Beaulieu, and M. Gatford, 1995, 'Okapi at TREC-3,' *The 3rd Text Retrieval Conference (TREC-3)*, NIST Special Publication 500-225, Gaithensbang, MD.

Robertson, S.E. and S. Walker, 1994, 'Some simple effective approximations to the 2-poisson model for probabilistic weighted retrieval,' *Proceedings of the 17th Annual International ACM SIGIR Conference on Research and Development in Information Retrieval*, Springer-Verlag, New York, pp. 232–41.

Salton, G., 1968, *Automatic Information Organization and Retrieval*, McGraw-Hill, New York.

_____1971, *The SMART Retrieval System: Experiments in Automatic Document Processing*, Prentice Hall, NJ.

Salton, G. and C. Buckley, 1968, 'Term weighting approaches in automatic text retrieval,' *Information Processing and Management*, 24(5), pp. 513–23.

Salton, G., E.A. Fox, and H. Wu, 1983, 'Extended Boolean information retrieval,' *Communications of the ACM*, 26(11), pp.1022–36.

Salton, G. and M.J. McGill, 1983, *Introduction to Modern Information Retrieval*, McGraw-Hill, New York.

Saracevic, 1995, 'Evaluation of evaluation in information retrieval,' *Proceedings of the 18th ACM SIGIR Conference on Research and Development of Information Retrieval,* Seattle, ACM Press, NY, pp. 138–46.

Schamber, L., M.B. Eisenberg, and M.S. Nilan, 1990, 'A re-examination of relevance: toward a dynamic, situational definition,' *Information Processing and Management.*

Sparck-Jones, K., 1972, 'A statistical interpretation of term specificity and its application in retrieval,' *Journal of Documentation,* 28, pp. 111–21.

Swets, 1969, 'Effectiveness of information retrieval methods,' *American Documentation,* 10, pp. 72–89.

Van Rijsbergen, C.J., 1977, 'A theoretical basis for the use of co-occurrence data in tin formation retrieval,' *Journal of Documentation,* 33, pp. 106–119.

_____1979, '*Information Retrieval,*' 2nd ed., Butterworths, London.

_____1992, 'Probabilistic retrieval revisited,' *The Computer Journal,* 35(3), pp. 291–98.

_____1996, 'Quantum logic and information retrieval,' *Proceedings of the 12th Annual International ACMSIGIR Conference on Research and Development on Information Retrieval,* University of Glasgow, Scotland.

Voorhees, E.M., 1994, 'Query expansion using lexical-semantic relations,' *Proceedings of the 17th Annual International ACM-SIGIR Conference on Research and Development in Information Retrieval,* Dublin, Ireland, Springer Verlag, London, pp. 61 69.

EXERCISES

1. What is the difference between data retrieval and information retrieval?
2. How does stemming affect the performance of an IR system?
3. What are the benefits of eliminating stop words. Give examples in which stop word elimination may be harmful.
4. Given the following document:

   The oldest Chinese language we know about is on oracle bones. Priests scratched questions on animal bones and then held the bones in a fire so that they cracked. The places where the cracks crossed the pictograms were thought to give the answers from the god.

   Assume that raw term frequency is used and the stop words are 'the', 'we', 'is', 'on', 'and', 'then', 'in', 'a', 'so', 'that', 'they', 'were', 'to', 'where', 'but', 'only', 'out'.

   Find the vector representation of the above document. Use porter stemmer for stemming.

5. In a collection of 10,000 documents, the following words occurs in the following number of documents:

   'oasis' occurs in 400 documents
   'place' occurs in 3,500 documents
   'desert' occurs in 800 documents
   'water' occurs in 800 documents
   'comes' occurs in 800 documents
   'beneath' occurs in 800 documents
   'ground' occurs in 800 documents

   Calculate tf-idf term vector for the following document:

   'An oasis is a place in a desert where water comes out from beneath the ground'.

   Perform stop word removal using the stop word list in exercise 4, and order tokens in the vector alphabetically.

6. Use the stop words given in Exercise 4 to construct vectors (assume simple term frequency weight) for the following documents:

   'An oasis is a place in a desert where water comes out from beneath the ground'.

   'Most people think of desert as a vast sandy region, but only 20% of the world's deserts are sandy'.

   Find cosine similarity between the two vectors.

7. Using Zipf's law, estimate the following in terms of constant K.
   (i) Number of distinct terms that have a frequency equal to f.
   (ii) Number of distinct terms in the collection.
   (iii) Number of distinct terms that appear only once in the corpus.

8. How well does LSI work? What will happen if k is too big or too small?

9. Define recall and precision. What will happen if there is no relevant document in a collection for a given query? What will the recall–precision curve look like if all the documents in a collection were relevant?

10. A user submitted a query to an IR system. Out of the first 15 documents returned by the system, those ranked 1, 2, 5, 8, and 12 were relevant. Compute non-interpolated average precision for this retrieval. Assume that the total number of relevant documents is six.

11. Interpolate precision for the retrieval situation described in Exercise 10 at the 11- recall points, viz., {0.0, 0.1, 0.2, 0.3, 0.4, 0.5, 0.6, 0.7, 0.8, 0.9, 1.0} and draw recall–precision curve.

LAB EXERCISES

1. Write a program that extracts tokens from a document, removes stop words, and lists the remaining tokens and their frequencies.

2. Prepare a small collection, say of 50–100 documents, and extract all the unique tokens and their frequencies from the collection. Rank the tokens based on their frequency and create a table showing the token, its frequency, its rank, and the product of rank and frequency.

3. Write a program to count the number of stop words in each document of the collection developed in Question 2. Find the percentage of stop words in each document.

4. Write a program to find document frequency of each token identified in Question 2. Use this to compute their inverse document frequencies and output a text file containing token, frequency, and inverse document frequency.

5. Using the output of Question 5, prepare a vector representation of documents in the collection.

CHAPTER 10

INFORMATION RETRIEVAL—2

This chapter continues the discussion started in Chapter 9. In particular, it talks about semantic approaches to IR. The chapter begins with an introduction to the problems associated with keyword-based retrieval models and various levels of NLP involved in IR. Relation matching and knowledge-based approaches are discussed next. A simplified conceptual graph-based IR model is introduced. Finally, cross-lingual IR approaches are discussed.

10.1 INTRODUCTION

Most of the information available online is textual in nature. As discussed earlier, in order to be retrieved, the information first needs to be represented in a way amenable to processing. The choice of representation put constraints on the retrieval process. Most of the current retrieval models are based on keyword representation. This representation creates problems during retrieval due to polysemy, homonymy, and synonymy. Polysemy occurs when a lexeme has multiple meanings. Keyword matching may not always include word sense matching. Homonymy is an ambiguity in which words that appear the same have unrelated meanings. Such ambiguity makes it difficult for a computer to automatically determine the conceptual content of documents. Studies indicate that human beings use different expressions to convey the same meaning (Blair and Maron 1990). Synonymy creates problems when a document is indexed under one term, the query uses a different term, and the two terms share a common meaning. Statistical approaches to handling semantics have their own limitation. The recent work in developing extensive lexicons is an attempt to improve the situation.

Another problem associated with traditional retrieval models is that they ignore semantic and contextual information in the retrieval process (Dick 1992, Ounis and Huibers 1997). This information is lost in the extraction of keywords from the text and cannot be recovered by the retrieval algorithms. Improving IR, demands an improved representation of text. One of the ways to improve representation of documents and queries is to capture semantic aspects of the documents. A number of knowledge representation formalisms have been proposed in AI literature, e.g., semantic networks, frames, and conceptual graphs. We use them to represent semantic aspects of documents, so that the retrieval system can understand and use semantics. In addition, the early IR models deal with monolingual IR, i.e., both the query and documents are of the same language. Cross-lingual IR, in which a user framing a query in one language is provided with documents in other languages, is becoming more commonplace now.

In Chapter 9, we discussed the basic concepts involved in the development of an IR system and its evaluation as well as various IR models and NLP techniques employed in retrieval models. We are now in a position to discuss semantic approaches to IR. We will also talk about cross-lingual approaches to IR. Before we discuss these approaches, let us take a quick look at the various levels of NLP analysis that has been used in IR.

10.2 NATURAL LANGUAGE PROCESSING IN IR

Processing textual information available in electronic form and retrieving information intelligently in response to users' queries has emerged as one of the great challenges in IR. NLP plays a vital role in both storage and retrieval of documents. There are seven interdependent levels of analysis for NLP. These levels are

- Phonological
- Morphological
- Lexical
- Syntactic
- Semantic
- Discourse
- Pragmatic

The phonological level is concerned with analysis of speech and sounds, e.g., phonemes, and is of little interest in textual IR. The morphological level deals with the meaning of units, i.e., morphemes. It is concerned with the analysis of different forms of a given word in terms of its prefixes, roots, and suffixes. This level of NLP has traditionally been incorporated into IR systems. Stemming techniques that reduce words to some root forms (stems) for query-document similarity are examples of this level of processing.

The next higher level is the lexical level, which deals with word level processing involving the analysis of structure and meaning of words and part-of-speech-tagging. Lexical operation in IR includes elimination of stop words, and generation and use of a thesaurus for expanding queries and handling abbreviations and acronyms. Part-of-speech tagging is another lexical level processing that is now being used in IR.

Next is the syntactic level, which deals with the grammar and structure of sentences. There are many types of sentence structures. Identification of the correct structure from various alternatives requires higher-level knowledge. Attempts to use syntactic analysis to understand the meaning of natural languages have been based on the assumption that the meaning is inherent in the syntactic structure. The limitation of the meaning, which can be drawn from the syntactic analysis, is discussed by Salton et al. (1990). Syntactic level processing is rarely used in traditional IR. Identification of phrase units is an example of this level of processing. Although sophisticated parsers have been developed, statistical methods, such as co-occurrence and proximity methods, have been preferred over NLP, for phrase identification in IR.

The semantic level is concerned with the meaning of units larger than words, such as clauses and sentences. This involves the use of contextual knowledge to represent meaning. Word sense disambiguation is a task that requires semantic level processing. This is because a word can be disambiguated only in the context of larger textual units. Due to sophisticated levels of processing, and the need for real world and domain specific knowledge, most IR systems prefer statistical keyword matching to semantic level processing.

Discourse level processing attempts to interpret the structure and meaning of even larger units, e.g., paragraphs and documents, in terms of words, phrases, clusters, and sentences. The highest level is the pragmatic level, which deals with outside world knowledge (i.e., knowledge external to the document and/or query). This level analysis has not been used in IR so far. Even in AI, research at the pragmatic level is only experimental in nature.

Although it is difficult, the potential benefits of syntactic and semantic processing have led researchers to investigate their use in IR. There are contending views regarding the use of lexical methods in IR. Supporters consider IR as an early stage of question-answering and believe that this will improve retrieval. Opponents believe that IR and other areas of language processing are two different processes. They are apprehensive about the usefulness of any attempt of meaning representation in IR and believe that one can get quite good result with statistical, probabilistic or vector space techniques. Salton et al. (1990) remarked that to analyse the meaning of a text automatically, syntactic analysis need to be supplemented by a knowledge base consisting of world knowledge and semantic knowledge, and that it is impossible to put semantic and world knowledge into a computer.

After examining the potential contribution of knowledge-based techniques, NLP and expert systems, in particular, Sparck Jones (1991) pointed out: "Although AI techniques can contribute to specialized systems, one should not overestimate the power of these techniques for IR."

She remarked that *for really hard tasks, it will not be possible to replace humans by machines,* and argued that *many information processing tasks are rather shallow linguistics tasks, which do not involve elaborate reasoning or complex knowledge.*

As direct application of NLP results in lack of robustness and efficiency (Strazalkowski 1995), it is almost always used with existing systems for indexing, query expansion and modification, and document categorization. Table 10.1 summarizes NLP techniques that have found useful applications in IR.

Table 10.1 NLP techniques in IR

Level of Analysis	Description	Examples of application in IR
Phonological level	Concerned with analysis of speech sounds, e.g., phonemes	Is of little interest in textual IR
Morphological level	Deals with meaning units, i.e., morphemes; concerned with analysis of the different forms of a given word in terms of its prefixes, roots, and suffixes.	Stemming
Lexical level	Deals with word level processing, involving analysis of structure and meaning of words and part-of speech tagging, etc.	Stop words, generation and use of thesaurus for expanding queries and handling abbreviations and acronyms; part-of-speech tagging
Syntactic level	Deals with the grammar and structure of sentences	Identification of phrase units
Semantic level	Concerned with meaning of units larger than words, such as clauses and sentences	Word sense disambiguation

(Contd.)

(Contd.)

| Discourse level | Interprets the structure and meaning of even larger units, e.g., paragraphs and documents, in terms of words, phrases, clusters, and sentences | Anaphor resolution; not much used in traditional IR |
| Pragmatic level | Deals with outside world knowledge, i.e., knowledge external to the document and/or query | No evidence of pragmatic level analysis in IR |

10.3 RELATION MATCHING

Keyword-based systems perform retrieval merely on the presence or absence of certain terms (keywords) in the document. Considering terms in isolation creates a problem, as meaning in a document does not come merely by putting words together. Out of context, a word does not provide useful information about its importance to the text. It is the arrangement of words in a sentence that gives it its meaning. Further, correct semantic and pragmatic interpretation is possible only by accounting for the relationship between terms. It makes sense, therefore, that by considering relationships in the matching process, retrieval effectiveness can be improved. This brings semantics into the matching process and makes intelligent retrieval possible. There can be sentences comprising the same set of words involved in different relationships. As an example, consider the following two sentence fragments:

- The farmer exploits worker…
- Exploitation of farmers and workers…

These two sentences have entirely different meanings, though they contain the same set of words. It is the relationship existing between the words that gives them their different meanings. Thus, consideration of the local context of the words being used, helps capture the underlying semantics. In relation matching, both the terms and the relation between them (as expressed in the query) are matched with terms and relations in the documents.

One of the ways of using relations is to expand the query, using different types of relations, e.g., synonym or hypernym/hyponym relation, to add terms. The term 'driving' can be expanded using

Synonyms: travelling, steering, etc.
Hyponyms: motoring
Object: car, truck, vehicle, etc.

Other approaches include the use of proximity search and multi-word phrases. The proximity search approach is based on the assumption that words that are related tend to appear close together. But words appearing together may express different relationships. One of the often-quoted examples is 'Library School' and 'School Library'. Multi-word phrase matching is simpler and existing methods for single term matching can be applied to multi-words. However, this fails to capture variations in syntactic structure unless the phrases have been normalized. For example, 'extraction of roots' might be transformed into 'root extraction'. Further, the use of multi-word phrases has not yielded significant improvement (Lewis and Jones 1996, Mittendorf, Mateev, and Schäuble 2000). A number of researchers have used syntactic relation matching. Syntactic relations may fail to identify similarities existing between semantic relations expressed using different syntactic structures. This suggests that the use of semantic relations can provide better results. Identifying semantic relations requires domain knowledge. Identifying and coding this knowledge is a labour intensive and time consuming task, and it is difficult to port this knowledge to a different domain. If any significant improvement in retrieval is to be achieved, then non-specific procedures have to be developed for identifying semantic relations.

10.4 KNOWLEDGE-BASED APPROACHES

Knowledge bases have also been used in IR systems to enhance performance, e.g., by providing methods for query expansion. A knowledge base is a representation and collection of knowledge, usually specific to an application area (Dominich 2001). Knowledge bases are represented using semantic networks, rules, frames, etc. WordNet (Miller 1990, 1995) is one of the most widely used knowledge base; originally designed as semantic networks in IR research. It was developed at Princeton University, inspired by psycholinguistic theories. Information is organized as a set of synonymous words called synsets. Some 54,000 word forms are organized into 49,000 synonym sets, each representing one base concept. These synsets are linked with each other by means of lexical and semantic relations. These relations include synonymy, hypernymy/hyponymy, antonymy, meronymy/holonymy, etc. An example of the WordNet hyponym hierarchy is illustrated for the word 'ocean', as follows:

Ocean—(a large body of water constituting a principal part of the hydrosphere)
⇒ Antarctic Ocean
⇒ Arctic Ocean

⇒ Atlantic, Atlantic Ocean
⇒ Bermuda Triangle
⇒ North Atlantic
⇒ South Atlantic
⇒ deep
⇒ Indian Ocean—
⇒ Pacific, Pacific Ocean
⇒ North Pacific
⇒ South Pacific

WordNet has been widely used in NLP research. In IR, WordNet has been used to disambiguate query terms and to expand queries using its semantic relations. A query is expanded by adding terms using different types of relation, e.g., synonyms or hypernym/hyponym relation. For example, the term 'driving' can be expanded using

Synonyms: travelling, steering, etc.

Hyponyms: motoring

Longman Dictionary of Contemporary English (LDOCE) and *Roget's Thesaurus* are some of the other lexical resources used in IR research.

More recently, ontologies have received increasing interest, particularly with the development of the semantic web. Ontology refers to the exhaustive and rigorous conceptual schema within a given domain, in which concepts are organized hierarchically by semantic relations and sub-sumption relations. Ontologies attempts to provide knowledge reuse and sharing by explicitly encoding shared understanding of a domain.

Application of ontologies in specific domains has been discussed by many researchers (Shum et al. 2000, Slade and Bokma 2001). However, their usefulness in a general domain is an issue of debate, and how general-purpose ontology can be utilized for improving information retrieval is still an open question. There are both opponents and proponents for the use of general-purpose ontology in IR and other NLP related tasks. The opponents argue that people believe in different things, speak in different ways, and, therefore, cannot have the same ontology. The proponents say that people may believe in different things, speak in different languages, and use different words and phrases to refer to the same concepts, even in the same language. They believe that most of these differences can be traced to mixing of ontology, language, and knowledge and can be dealt with by separating naming and definitions of concepts from language, knowledge, and belief. The studies conducted on concept acquisition processes and social/linguistic interaction behaviours of human beings also suggest that general-purpose ontology is not much helpful in the learning process or for achieving intelligence. With the advent of the semantic web, the use of ontology in IR is gaining interest and effort is being made to develop ontologies. Some of the freely available lexicons that have found useful applications in IR research are discussed in Chapter 12. OpenCyc and suggested upper merged ontology (SUMO) are two public-domain ontologies. Cyc is a well-known and quite comprehensive ontology. It includes foundation ontology and many domain specific ontologies, called micro theories. A subset of Cyc kown as OpenCyc has been released for public access.

The SUMO is another comprehensive ontology developed by the IEEE working group. It is freely available at *http://suo.ieee.org*. An upper or foundation ontology is one that is not tied to a particular problem domain but attempts to describe general entities. It extends to many domain ontologies and a complete set of links to WordNet.

Existing lexical resources and knowledge bases are not sufficient for in-depth processing of natural languages. Attempts have been made to develop knowledge bases (lexicons, taxonomies, ontologies, etc.) automatically, and several methods have been proposed. These methods, however, are still premature. Developing knowledge bases manually is a labour-intensive task and requires a lot of knowledge. These constraints typically restrict the knowledge bases to specific domains.

The application of NLP to a large amount of unrestricted text can be reduced to a two-step process: a coarse ranked retrieval of candidate documents using statistical and shallow NLP technique, followed by a more sophisticated NLP applied only to the much smaller list of high-ranking documents retrieved during the first stage. By combining both the approaches, the benefits of each can be achieved in a single system.

10.5 CONCEPTUAL GRAPHS IN IR

Conceptual graphs (CGs) are closely related to natural languages and can therefore be used for representing text. Such a representation holds the promise of extracting more information from documents by explicitly capturing relationship between terms; unlike word-statistical approaches that merely count nouns and noun phrases. Early use of conceptual graphs in IR occurred in CoDHIR (Marega and Pazienza 1994), DR-LINK (Liddy and Myaeng 1994), RELIEF (Ounis and Pasca 1998), and ITELS (Dicheva and Dimitrova 1998) systems. CGKAT (Martin 1997) and WebKB (Martin and Eklud 2000) use conceptual graphs to index document elements (chapters, paragraphs, etc.). Rama and Srinivasan (1993) underlined the utility value of conceptual roles in information retrieval. Their work provides strong support for the use of conceptual roles in information retrieval. Marega and Pazienza (1994) also emphasized the use of the contextual role of words in CoDHIR (content-driven hypertext information retrieval) and concluded that this results in an improvement in retrieval precision. They used conceptual graphs in CoDHIR to represent semantic information extracted from the text. They work on identifying contextual roles of words and extending vector models to consider compound descriptors (contextual role-word). The DR-LINK (document retrieval using linguistic knowledge) system (Liddy and Myaeng 1994) is one of the attempts to use conceptual graphs to represent documents. It uses conceptual graphs to extract and use semantic relations for IR, which involves the processing and representation of text at the lexical, syntactic, semantic, and discourse levels. The image retrieval system, RELIEF (relational logical approaches based on inverted files), uses conceptual graphs as the indexing language. Martin and Eklund (2000) argued in favour of general knowledge representation languages for indexing Web documents and suggested the use of concise and easily comprehensible CGs. They argued that CG representations have the advantage over metadata language, which are based on extensible markup language (XML). Intelligent terminology learning system (ITELS) is an intelligent tutorial system that helps Bulgarians learn English terminology. It uses conceptual and linguistic knowledge as well as conceptual graphs, to represent relations between concepts. Both CGKAT and WebKB are knowledge annotation systems. They allow indexing and retrieval of documents using a knowledge markup language either embedded in documents or stored in WWW-accessible databases. The document indices are built using annotations in a knowledge representation language (e.g., conceptual graphs). Conceptual graphs permit semantic content, and relationships to other documents, to be described precisely. WebKB is an extension of CGKAT. It is freely accessible at *http://www.int.gu.edu.au/ kvo/software.*

The major source of difficulty in the use of CGs is the development of an automated system to extract CG representation of text. Siddiqui and Tiwary (2005) attempt to overcome this difficulty by proposing a simplified CG model for IR, which enhances the capabilities of the existing statistical information retrieval models, instead of replacing them.

Conceptual graphs (CGs) are the basic building blocks of conceptual structures. CGs have evolved out of conceptual structures theory, as set down by John F. Sowa (1984, 2000). Sowa's conceptual structures are notations for knowledge representation in text analysis. It is highly expressible, mathematically well founded, and computationally tractable. CGs are based on the logic of Charles Sanders Peirce, Tesńierés dependency graphs, and the semantic networks of AI. Peirce developed algebraic notations for logic but was never satisfied with them. He believed that graph notations were more flexible and readable. Peirce experimented with graph notations for logic and proposed existential graphs (EGs). He stated rules of inferences and defined an interpreter for it. An EG is a graphical representation of logic which can represent both modality and quantification. However, it does not capture the details of the linguistic structure correctly. Existential graphs (EGs) have a single canonical form, rather than the multiple synonymous sentences in languages with built-in operators, such as English and predicate calculus. This makes it difficult to use them to translate natural language sentences. Peirce called the graphic notation 'the luckiest finding of my carier' and believed that it is the 'logic of the future' (Sowa 1993). Peirce proved to be right, and several developers and theorists formed a group called 'Peirce Project' to implement and extend Peirce's logic. Conceptual graphs (CGs) are synthesis of Peirce's logic with Tesńierés dependency graphs and semantic networks. They are a graphic notation for logic based on EGs but with extended features that supports more direct translations to natural languages. The advantage with graph notation is that diverse graph-manipulation operations are available, and it is easy to determine the global status.

A CG can be used to represent natural language text. The choice of CG formalism has a number of advantages. This formalism provides a way to extract and represent the meaning of natural language text. CG theory provides a framework in which all components of an IR system can be represented adequately. It can be easily extended to accommodate specific knowledge and needs, without a revision of its semantics. More importantly, it leads to a scalable representation.

However, the development of a system to capture truly conceptual representation is beyond the scope of this book. It is still very difficult to do an indexing at a conceptual level. It is a challenge because of the need to transform text into meaning, syntax into semantics.

10.5.1 Definition and Notation

Definition 1 Conceptual graph "A conceptual graph is a finite, bipartite graph. It has two types of nodes: concept node and relation node" (Sowa 1984). In the graph, concept nodes represent entities, attributes, states, and events, and relation nodes show how concepts are interconnected.

Example 10.1   Consider the sentence, *A cup is on the table.* Figure 10.1 shows a conceptual graph of the sentence.

Figure 10.1   Conceptual graph of sentence

The conceptual graph in Figure 10.1 consists of two concept nodes connected with a conceptual relation. A single concept node may constitute a conceptual graph. However, a single relation node cannot constitute a conceptual graph. In the graph, the boxes represent concept nodes and the circles, conceptual relations. In text, linear notation is usually preferred over graphic notation, in which boxes are replaced with square brackets and circles with parentheses. The linear form of the graph shown in Figure 10.1 will be

[cup] → (on) → [table]

Concept nodes have two types of field—type field and referent field. The two fields are separated using a colon. In the box, type field is shown on the left and referent field, on the right, separated by a colon as in Figure 10.2. Concepts that do not identify a specific individual are called generic concepts. The referent part of these concepts is omitted. The existential quantifier (∃) is assumed to apply on concepts with blank referent field, as in [cup]. For individual concepts, the referent field is a specific entity, such as a name.

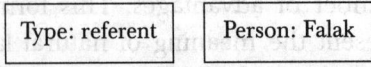

Figure 10.2   A concept with referent

Table 10.2 lists the referents and their descriptions. The concepts are organized in a hierarchy called type hierarchy. The type hierarchy constitutes a partial ordering and becomes a type lattice when all the intermediate types are introduced. The type lattice has the undefined type ⊤ at the top and the absurd type ⊥ at the bottom. Although the type hierarchy represents a model of the real world, or at least of the domain, its significance is more implementational than ontological (Dick 1992). The choice of types depends on the nature of the domain and on the specific requirement.

Table 10.2 Different types of referents

Kind of referent	Example	Description
Universal	[Student: ∀]	Every student
Generic or existential	[Student] or [Student :*]	A student or some student
Definite referent	[Student: #]	The student
Named individual	[Student: Zeeshan]	Student Zeeshan
Singular	[Student:@1]	Exactly one student
Generic set	[Students: {*}]	Students or some students
Counted set	[Student:{*}@3]	Three students
Set of individuals	[Student: {Sana, Aman}]	Sana and Aman
Question	[Student: ?]	Which student
Measure	[distance: 6 km]	Distance 6 km

There exists a defined mapping between a conceptual graph and a corresponding first-order predicate logic (FOPL) formula. A conceptual graph can be easily translated into its predicate logic equivalent using the formula operator Φ. This operator maps the boxes to quantified variables with monadic predicates to specify type. Circles are mapped to predicates with each arc as one of its arguments. The arrow pointing towards the circle becomes the first argument and the arrow pointing away represents the second argument. When applied to Figure 10.1, Φ assigns a variable x to represent the concept [cup] and a variable y to represent the concept [table]. The type labels CUP and TABLE are represented as monadic predicates cup (x) and table (y). The relation 'on' is represented by a dyadic predicate on (x, y). The equivalent predicate calculus representation is $(\exists x)\,(\exists y)\,(cup(x) \wedge table(y) \wedge on(x, y))$.

Now, consider the CG representation of the sentence, *Danish is playing football.*

g: [Person: Danish] ← (agnt) ← [Play] → (obj) → [Football]

When the translator operator Φ is applied to *g*, it is translated into a first-order predicate calculus, as follows:

$$\varphi g = \exists x(\exists y(\exists z(Person(x) \wedge play(y) \wedge football(z) \wedge agnt(y, x)$$
$$\wedge obj(y, z) \wedge name(x, \text{'Danish'}))))$$

10.5.2 Conceptual Graph Representation

CGs are formally defined in an abstract syntax that is independent of any notation but can be represented in three different forms. These are linear form (LF), display form (DF), and conceptual graph interchange form (CGIF). The DF has the advantage of being more intuitive for average user to comprehend. Linear form (LF) is a compact readable form, which is normally followed in text. In LF, concepts are represented by square brackets instead of boxes, and the conceptual relations are represented by parentheses instead of circles. In the LF notation, a concept that is central to the proposition being represented is selected as the head concept. Most of the other concepts in the assertion are related to the head concept. In order to get a CG representation of a sentence, the main verb of the sentence is usually selected as the head concept. Alternatively, the head concept may be the concept with the maximum number of related nodes. A '–' separate the head node from other conceptual relations and concepts occurring in the graph. Other CGs may be embedded within the main graph in a similar fashion.

A formal context-free grammar description of linear form, as given by Sowa (1984), is shown in Figure 10.3. The first production rule of the grammar says, 'The non-terminal CGGRAPH represents a conceptual graph consisting of either a CONCEPT followed by an optional relational link (RLINK) or a RELATION followed by a required concept link (CONLINK). Recall that a single concept node constitutes a conceptual graph but single relation node does not.

Both DF and LF are designed for communication with humans or between humans and machines. CGIF was developed for communication between machines that use CG as their internal representation. CGIF has a concrete syntax, which makes it usable for implementation. In CGIF, co-reference labels are used to represent the arcs.

```
CGGRAPH → {CONCEPT [RLINK] | RELATION COLINK} {"." | ";"}
RLINK → ARC RELATION [CONLINK] | "-" RLIST ","
COLINK → ARC CONCEPT [RLINK] | "-" CONLIST ","
RLIST → NEWLINE RELATION [CONCEPT]} ...
CONLIST → {NEWLIST ARC CONCEPT [RLINK]} ...
CONCEPT → "["TYPEFIELD [":" REFFIELD] "]"
RELATION → "("TYPELABEL")"
ARC → [NUMBER] {"←" | "→"}
NUMBER → DIGIT ...
DIGIT → "0" | "1" ... | "9"
```

Figure 10.3 CFG description of linear form

In CGIF, the sentence given in Example 10.1 can be represented as
[Cup: *x] [Table: *y] (On ?x ?y)

The symbols *x and *y are called defining labels. The matching symbols ?x and ?y are the bound labels that indicate references to the same instance of a cup x or a table y. CGIF also permits concepts to be nested inside the relation nodes. Nesting of concepts helps reduce the number of co-reference labels.

(On [Cup] [Table])

For communication with systems that use other internal representations, CGIF can be translated into different logical languages, such as knowledge interchange format (KIF). Hence, it is better to use CGIF for storage and retrieval of CGs. The KIF representation of the preceding example is:

(exists ((x Cup) (y Table)) (On $?x$ $?y$))

Although DF, LF, CGIF, and KIF seem to be quite different, their semantics is defined by the same logical foundations. They can all be translated to a statement of the following form, in typed predicate calculus.

($\exists x$: cup) ($\exists y$: Table)on(x, y)

Any statement, expressed in any one of these notations, can be automatically translated to a logically equivalent statement in any of the other notations.

Example 10.2 Consider the following sentence:

Sana goes to school by bus.

The displayed form of its CG representation is shown in Figure 10.4.

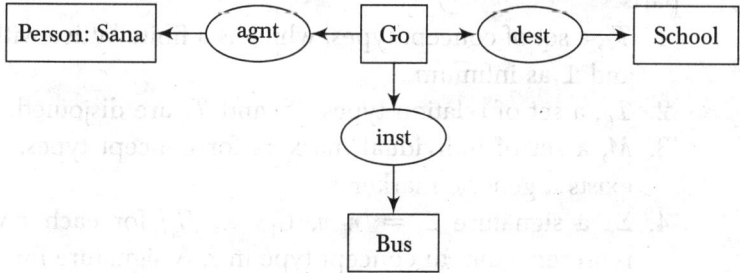

Figure 10.4 Conceptual graph of sentence in Example 10.2

It has three conceptual relations: (agnt) relates [Go] to the agent Sana, (dest) relates [Go] to the destination School, and (inst) relates [Go] to the instrument bus.

In DF, concepts are represented by boxes. Conceptual relations are represented by circles or ovals.

The linear form for CGs is intended for more compact notation than DF, but with good human readability. The LF for Figure 10.4 is as follows:

[Go]-
    (agnt)->[Person: Sana]
    (dest)->[School]
    (inst)->[Bus].

In this form, the concepts are represented by square brackets and the conceptual relations are represented using parentheses. A hyphen at the end of a line indicates that the relations attached to the concept are continued on subsequent lines. The CGIF of the sentence in Example 10.2 is given by

[Go: *x] [Person: Sana *y] [School *z] [Bus: *w] (agnt ?x ?y) (dest ?x ?z) (inst ?x ?w)

or

[Go: *x] (Agnt ?x [Person: Sana]) (Dest ?x [School]) (Inst ?x [Bus])

10.5.3 Conceptual Graph Operations

Before discussing the operations that can be performed on CGs we first introduce the notion of support, and give a more formal definition of a conceptual graph. Each conceptual graph is defined in relation to a support, which defines syntactic constraints, and provides background knowledge on a specific application domain. A support consists of the following parts:

1. T_C, a set of concept types, which is a finite lattice with $\top$ as supremum and $\bot$ as infimum.
2. T_R, a set of relation types, T_C and T_R are disjoined.
3. M, a set of individual markers for concept types. In addition, there exists a generic marker *.
4. Σ_r, a signature $\Sigma_r = (r, n, C_1, ..., C_n)$ for each r with arity n. $\Sigma_i(r)$ represents the ith concept type in r. A signature for each relation type thus fixes the arity of a relation type and shows the greatest concept types this relation type can link.

More conceptually, CG can be defined as follows.

Definition 2 Conceptual graph   A CG $g = (R, C, E, \text{ord}, \text{label})$ is a bipartite (recall that they are not necessarily connected) and finite graph with $C \neq \Phi$. R and C denote its relation and concept nodes respectively. E is the set of edges, and the edges adjacent to each relation node r are totally ordered by the function ord. The ith neighbour node r in g is denoted by $g_i(r)$. Every concept node in the conceptual graph has a label defined by the mapping label. A label of a concept node type $c \in C$ is a pair $\text{label}(c) = [c, m(c)]$ with $c \in T_C$ and $m(c) \in M$, where M is a finite set of individual markers.

There are four basic operations that can be performed on conceptual graphs (Sowa, 1993).

Copy This creates a conceptual graph v as an exact copy of another conceptual graph u.

Restrict Let c be a concept of v with a constant or existential quantifier as a referent. Then, a conceptual graph w, can be derived by restricting c using the type or referent: restriction using type replaces the type label of c with a sub-type; and restriction using referent replaces an existential quantifier with a constant.

Join Let c_1 be a concept of u, and c_2 be a concept of v, where neither c_1 nor c_2 are nested inside a context, and both c_1 and c_2 have identical type and referent fields. Then a join of u and v is obtained by deleting c_2 and linking all arcs of conceptual relations that had been previously linked to c_2 to c_1.

Simplify A conceptual graph can be simplified by deleting duplicate conceptual relations. Two conceptual relations r_1 and r_2 are said to be duplicate if they are of exactly the same type, and each arc of one conceptual relation is attached to the same concept as the arc of another conceptual relation.

A new conceptual graph v, can be derived from other conceptual graphs by applying copy, join, and restrict operations. The conceptual graph v thus derived is said to be a specialization of every graph u from which it has been derived. More formally, this is represented as $v \leq u$. Every specialization sequence from a graph u to v, is associated with a projection from u to v. A derivation sequence for the graph v representing, 'A person x loves another person y,' from the graph u_1 and u_2, is shown in Figure 10.5. In this case, v is the specialization of graph u_1 and u_2, i.e., $v \leq u_1$ and $v \leq u_2$.

Two alternative notions are introduced in conceptual graph theory: graph derivation and projection. The basic rule for graph derivation is the restriction rule, in which a concept C is restricted to a sub-concept C' ($C' \leq C$), together with suitable substitutes for the existential quantifiers of the individuals. As the main operation for deriving graphs is given by restriction, Sowa's idea was to obtain the graph derivation $h \leq| g$ (h derives from g) by projecting all the concept nodes of the starting graph g on to nodes of the final graph h. Thus, both graph derivation and projection proceed in the same direction (Amati and Ounis 2000).

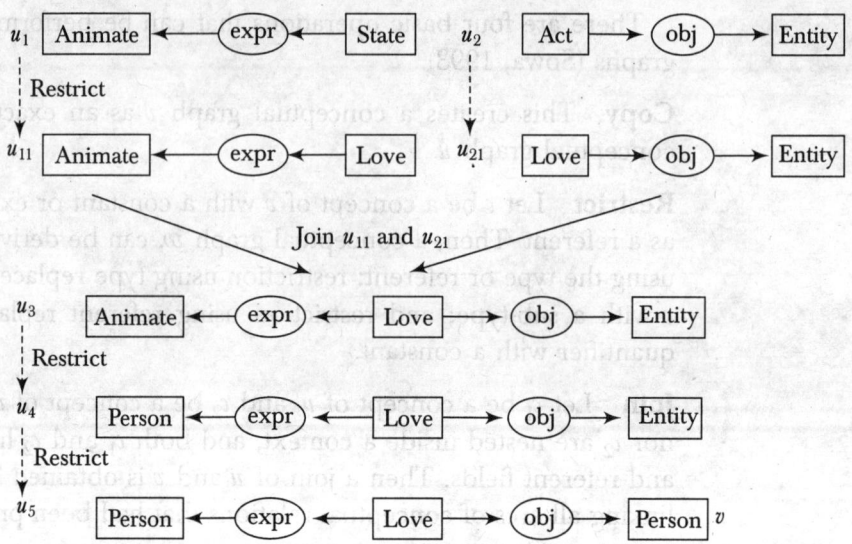

Figure 10.5   Derivation graph of conceptual graph *v*

If a graph *v* is derivable from another graph *u* (i.e., $v \leq u$), then there exists a projection mapping $\pi : u \rightarrow v$ with the following properties:

- For each concept *c* in *u*, πc is a concept in *v*, such that type $(\pi c) \leq$ type (*c*). If *c* is an individual concept, then referent (πc) = referent (*c*).
- For each relation node *r* in *u*, its image πc is a relation node in *v*, such that type (πr) = type (*r*). If the *i*th arc of *r* is linked to a concept *c* in *u*, then the *i*th arc of πr must be linked to πc in *v*.

πr is a sub-graph of *v*, called a projection of *u* in *v*.

However, projection is not sufficient for IR because it retrieves only the exact answer. IR requires a search mechanism that retrieves and ranks approximate answers (Gilnest and Chein 2004). The computational complexity involve in graph derivations also makes them impractical for IR. IR researchers have proposed a number of CG similarity measures for retrieval; we discuss few of them in this chapter.

10.5.4   Representing Documents as a Set of Conceptual Graphs

The NLP techniques are used to create CG representations of textual document. Figure 10.6 depicts the general architecture of CG construction. Documents are first tagged. The tagged representation is then parsed to generate structured representation. The CG is constructed for each parsed sentence. For constructing a CG, the syntactic patterns in a sentence are identified. A sentence itself is thus considered as consisting of many fragments (syntactic patterns). The CG is constructed for each such pattern. Subordinate clauses are considered as independent sentences and their

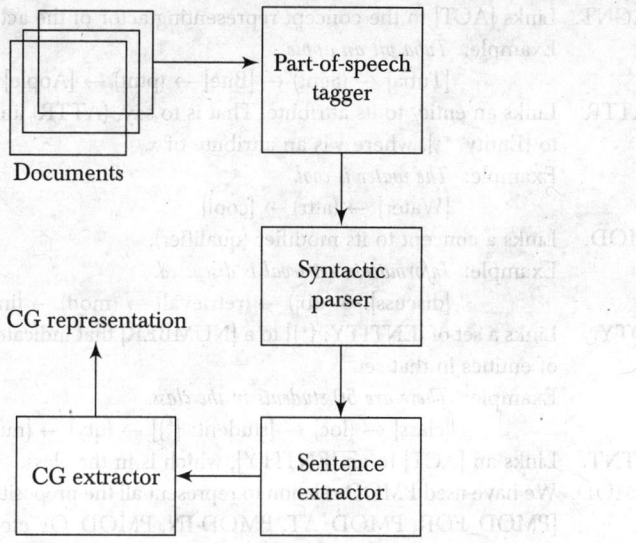

Figure 10.6 General architecture of conceptual graph construction

conceptual graphs are similarly constructed. These various segments are related to each other through the relationships identified by verbs, prepositions, etc.

The CG formalism can be used to perform search by content. For example, the specialization relation between CGs allows them, and the documents they index, to be retrieved by specifying parts of their content with any super-type of the types they use. Additionally, if the calculation of the specialization relation takes into account type definitions, CGs may be retrieved even if their structure differs from the query structure.

10.5.5 Identifying Relations

Sowa (1984) provided a set of fundamental conceptual relations (conrels). Appendix C contains a subset of that catalogue with appropriate descriptions and examples wherever needed. The basic types of relations in which verbs are involved are 'agnt' (agent) and 'ptnt' (patient), or 'obj' (object). However, they can exhibit many different semantic roles. We have used syntactic patterns to identify the constituents of a sentence. Some of the syntactic patterns used for noun phrase are shown in Table 10.3. The conceptual graphs of these noun phrases depend on their syntactic structure. For Pattern 6, we create concepts representing the noun and the adjective, and link them by an 'attr' relation. For example, consider the following pattern:

An (DT) efficient (JJ) structure (NN)

318 Natural Language Processing and Information Retrieval

AGNT.	Links [ACT] to the concept representing actor of the act.
	Example: *Tuba bit an apple.*
	[Tuba] ← (agnt) ← [Bite] → (ptnt) → [Apple]
ATTR.	Links an entity to its attribute. That is to say, (ATTR) links (Entity. *x] to [Entity. *y], where y is an attribute of x.
	Example: *The water is cool.*
	[Water] → (attr) → [cool]
MOD.	Links a concept to its modifier (qualifier).
	Example: *Information retrieval is discussed.*
	[discuss] → (obj) → (retrieval] → (mod) → [information]
QTY.	Links a set of [ENTITY: {*}] to a [NUMBER] that indicates the number of entities in that set.
	Example: *There are 50 students in the class.*
	[class] ← (loc) ← [student: {*}] → (qty) → [number: 50]
PTNT.	Links an [ACT] to an [ENTITY], which is in the class.
PMOD.	We have used PMOD relation to represent all the prepositional relations [PMOD_FOR, PMOD_AT, PMOD-IN, PMOD_OF etc.].
MNR.	Links an [ACT] to its attribute.
	Example: *Shereen drives slowly.*
	[Shereen] ← (agnt) ← [Drive] → (mnr) → (slowly)

Figure 10.7 A subset of conceptual relations

Its CG will be

[structure] → (attr) → [efficient]

The determiners provide information on the type of referents. However, we have not considered referents while creating CGs. Complex nominals can be represented as a single concept as in Smeaton and Rijsbergen (1999). Following this notation, the representation of Pattern 9 will be a single concept node, where the concept corresponds to the noun sequence separated by a dash (–). The representation of 'The {DT} information {NN} retrieval {NN}' will, therefore, be

[information–retrieval]

Table 10.3 Syntactic patterns for noun phrase

1.	NP → NN
2.	NP → NNP
3.	NP → PRP
4.	NP → DT NN
5.	NP → NP PP
6.	NP → DT JJ NN
7.	NP → NNP NNP
8.	NP → CD NNS
9.	NP → DT NN NN
10.	NP → NP SBAR

However, this representation creates a problem because it fails to match 'retrieval of information' or 'retrieve information'. In order to capture these structural variations, we represent both nouns as separate concepts and combine them by the generic relation 'mod'. Therefore, its CG representation will be

[retrieval] → (mod) → [information]

To handle structural variations, we create additional CGs corresponding to different variants. For documents, we create just a single CG for noun sequences, depicted by the 'mod' relation. For queries, we additionally include alternative CGs corresponding to possible structural variants.

In case of an adjective qualifying a compound noun sequence (a pattern such as JJ NN NN), the last noun in the sequence is used in 'attr' relation. For example, the conceptual graph generated for 'flexible {JJ} information {NN} retrieval {NN}' is as follows.

[flexible] ← (attr) ← [retrieval] → (mod) → [information]

This gives an exact match with 'flexible retrieval'.

Pattern 10 is transformed to a CG by creating concept nodes corresponding to the words tagged as NN and CD, and linking them with the generic relation 'MEAS'. In Sowa's original notation, this is represented by referents. Many different semantic roles are represented by 'MEAS'. It may represent quantities as in '10 passengers', distances as in '10 kilometres', and weights as in '5 kilograms'.

We express all prepositional relations using the generic relation 'PMOD'. The actual semantic relation represented by these generic relations can be different. For instance, in the conceptual graph, the prepositional relation specified using 'at' can be assigned the semantic role 'location' in the phrase 'arrived at station', but not in 'available for sale at a reasonable cost'. Similarly the semantic relation 'source' and 'destination' can be used to replace the prepositional relations (PMOD) expressed by 'from' and 'to' in the phrase 'flight from Delhi to Chennai'. Some constraints need to be checked while assigning semantic roles to relations. For example, a common requirement for constituents of the role of 'agent' is that they be of type animate or organization. Similarly, for a constituent to fill the role of 'instrument', it has to be inanimate. It is difficult to identify semantic relations automatically. For example, in each of the following sentences, the preposition 'for' is involved in different semantic relations between constituent concepts.

(i) Danish bought a book for Daniya.
(ii) Adiba bought a barbie doll for 500 rupees.

(iii) The train left for Agra.

(iv) The customer asked for quick service.

(v) There is no justification for attacking the findings of the judge.

The basic patterns for identifying propositional phrase are as follows:

$$PP \rightarrow IN\ NP$$
$$PP \rightarrow TO\ NP$$

A CG is constructed for the noun phrase, and the preposition is used to mark the relation between this phrase and the other relevant phrases in the sentence.

Compound sentences involving Wh-determiners (*which* and *that*) are broken into two independent sentences, and CGs of the constituents of these sentences are created. Whenever possible, the Wh-determiners are replaced by the concepts they refer to. These concepts provide a means to relate the constituent sentences.

The level of semantic detail to be captured within the conceptual graph depends on the nature of application and the domain in which it is to be used. The generic relations used to represent relations between prepositions, verbs and noun concepts, can be replaced by semantic relations with the help of relation ontology. The assignment of semantic role is a domain dependent process and is very difficult to perform automatically. To some extent, the traditional parsing approaches suggest semantic relations corresponding to syntactic ones, but they require that all possible semantic roles be recognized and manually encoded in grammars and lexicons. Developing such grammars is time consuming and tedious, and such systems well work only within a limited domain. Resorting to a manual solution for assignment of semantic roles may be a viable solution for specific domains but when considering IR in general, it is almost impossible. This is particularly so because of the size of the document collection with which an IR system has to deal.

Further, detailed automatic analysis of the semantics of a text requires full text analysis, a preliminary defined description of every word expected in the input text, and at least an upper level ontology. Full text analysis is problematic for large document collections. Ontologies are still in the stage of development. Some existing lexical resources, such as WordNet, do provide both the description of words and a number of semantic relations between them. Still, they are not sufficient. For example, it is not possible to identify the 'is_a' relationship that exists between 'concurrent program' and 'parallel program', nor is it possible to identify that 'genetic algorithm' and 'neural networks' are AI techniques involving WordNet. This clearly shows the limitation of existing lexical resources.

We argue that deep semantic analysis of documents is neither possible nor desired for general IR problems. The amount of text to be analyzed is so huge that it seems impractical to carry out detailed semantic interpretation. Existing lexical resources cannot be utilized effectively in different domains. However, in a specific domain, it may be feasible to manually build domain ontology. This is because of the following characteristics of domain specific sentences (Zhang and Yu 2001):

(i) Vocabulary is limited.
(ii) Word usage has pattern.
(iii) Semantic ambiguities are rare.
(iv) Terms and jargon of the domain appear frequently.

These characteristics imply that there are limited relations and that sentences with similar meaning have similar patterns. This makes it relatively easy to construct an ontology for a specific domain. However, in an unrestricted domain, the same process becomes a herculean task. We therefore focus on techniques that use information that can be easily extracted and that can be found within the document itself, without requiring much domain knowledge. We treat a document as a set of words involved in different types of relationships with each other. For the purpose of IR, we avoid detailed semantic interpretation of text and rely on a CG model that can be easily extracted from the text without requiring much domain knowledge.

10.5.6 Algorithm

The step-by-step description to derive a simplified CG representation of a document is given below.

1. First, the document is tagged. The tagged representation of the document is processed to make necessary substitutions and modifications. These include the removal of determiners, pre-determiners, modal words, wh-determiners, wh-pronouns and few patterns such as 'such that', 'so that', and 'as well as' (replaced with 'and').

2. The tagged and processed document is used as the input to a sentence extractor. Each extracted sentence of the tagged text is passed through four modules. Each of these modules is devoted to identifying certain types of relationships between concepts. These modules correspond to the following:

   - *Preposition and adverb handlers* to extract prepositional and adverbial relations

- *Noun handlers* to extract relations between noun sequences and cardinals.
- *Adjective handlers* to extract adjectival relations.
- *Verb handlers* to extract relations between verbs and their subject and object.

Similar steps are followed to obtain the CG representation of a query.

Example 10.3 We now explain these steps by considering the conceptual graph representation for the sample document shown in Figure 10.8.

> document 1
> Title
> analysis of the role of the computer in the reproduction and distribution of scientific papers.
> Abstract
> The American chemical society has begun an analysis of the role of the computer in related aspects of the reproduction, distribution, and retrieval of scientific information.

Figure 10.8 Sample document text

The tagged representation of the document is as follows:

%%document	1		
%% title			
analysis	NN	computer	NN
of	IN	in	IN
the	DT	the	DT
role	NN	reproduction	NN
of	IN	and	CC
the	DT	distribution	NN
of	IN		
scientific	JJ		
papers	NNS		
%% abstract			
the	DT		
american	JJ		
chemical	JJ		
society	NN		
has	VBZ		
begun	VBN		
an	DT		
analysis	NN		
of	IN		

the	DT
role	NN
of	IN
the	DT
computer	NN
in	IN
related	JJ
aspects	NNS
of	IN
the	DT
reproduction	NN
,	,
distribution	NN
,	,
and	CC
retrieval	NN
of	IN
scientific	JJ
information	NN
.	.

where %%% represents comments.

Each sentence of the tagged text is then passed through the four modules. The outputs produced by these modules are shown here on the sentences of the sample documents.

The first sentence

Analysis {NN} of {IN} the {DT} role {NN} of {IN} the {DT} computer {NN} in {IN} the {DT} reproduction {NN} and {CC } computer {NN} in {IN} the {DT} distribution {NN} of {IN} scientific {JJ} papers {NNS}.

The LF of the CG produced for the above sentence by module 1 is

[Analysis] → (PMOD_OF) → [role: #] → (PMOD_OF) → [computer: #] –

→ (PMOD_IN) → [reproduction] → (PMOD_OF) → [papers: #]
→ (PMOD_IN) → [distribution] → (PMOD_OF) → [papers: #]

No CG will be produced by the second and fourth module in this case. The CG produced by the adjective handler is

[paper] → (attr) → [scientific]

The second sentence

The {DT} American {JJ} chemical {JJ} society {NN} has {VBZ} begun {VBN} an {DT} analysis {NN} of {IN} the {DT} role {NN} of {IN} the

{DT} computer {NN} in {IN} related {JJ} aspects {NN} of {IN} the {DT} reproduction {NN}, distribution {NN}, and {CC}retrieval {NN} of {IN} scientific {JJ} information {NN}.

Module 1:

[Analysis] → (PMOD_OF) → [role: #] → (PMOD_OF) → [computer: #] → (PMOD_IN) → [aspects] –

(PMOD_OF) → [reproduction] → (PMOD_OF) → [information: #]
→ (PMOD_OF) → [distribution] → (PMOD_OF) → [information: #]
→ (PMOD_OF) → [retrieval] → (PMOD_OF) → [information: #]

Module 3:

[society : #]–
→ (attr) → [American]
→ (attr) → [chemical]
[aspect] → (attr) → [related]
[information] → (attr) → [scientific]

Module 4:

[begin] → (agnt) → [society]
→ (ptnt) → [analysis]

The displayed form of the resulting conceptual graph is shown in Figure 10.9.

10.5.7 Conceptual Graph Matching Algorithms

Most of the research on CG focuses on graph theory and graph algorithms (Amati and Ounis 2000, Sowa 1984), which have been considered the essence of CG theory (Chein and Mugneir 1995). Researchers have proposed sound and complete graph derivation with respect to first order logic semantics and graph theory (Ghosh and Wuwongse 1995, Wermelinger 1995). But these algorithms are non-deterministic and it seems unreasonable to use non-deterministic graph derivation algorithms for IR applications. Efficient graph algorithms make use of CG normal form (Amati and Ounis 2000, Mugnier 1995). However, there is a widespread opinion that any reasonably complete method of structural comparison is not computationally affordable (Mugnier 1995).

A number of researchers have attempted to use CGs in IR and have proposed computationally tractable CG matching algorithms (Liddy and Myaeng 1994). While the general sub-graph isomorphism problem is known to computationally intractable, matching CGs containing conceptual information seems to be practical.

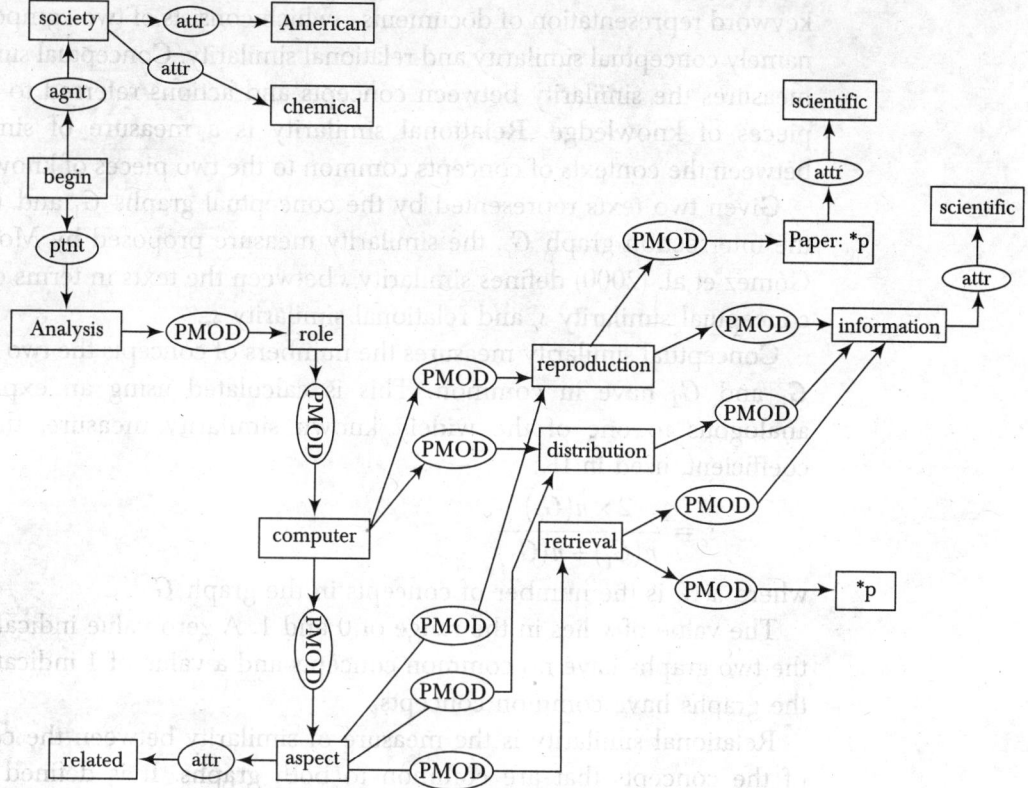

Figure 10.9 Conceptual graph for the sample document

Rau (1988) proposed a CG matching algorithm, which uses semantic relations between concepts to retrieve relevant information in the absence of an exact match. Her algorithm employs the semantic information in the inheritance hierarchy for partial matching. However, no syntactic information is taken into account to retrieve the relevant information in the absence of exact information.

Levinson and Ellis (1992) developed algorithms that could search a lattice of graphs in logarithmic time. They use a partially ordered hierarchy to reduce the search space. The hierarchy however, does not cover the full generalization or full specialization if we consider a complete graph instead of a sub-graph.

Myaeng and López-López (1992) presented a flexible algorithm for CG partial matching. Later, a flexible criterion was proposed by Mntes-y-Gómez et al. (2000) to quantify the approximate matching expressed in sub-graphs. The similarity measure proposed by them is analogous to dice coefficient—a well-known similarity measure used for weighted keyword representation of documents—which consists of two components, namely conceptual similarity and relational similarity. Conceptual similarity measures the similarity between concepts and actions referred to in two pieces of knowledge. Relational similarity is a measure of similarity between the contexts of concepts common to the two pieces of knowledge.

Given two texts represented by the conceptual graphs G_1 and G_2 and the intersection graph G_c, the similarity measure proposed by Montes-y-Gómez et al. (2000) defines similarity s between the texts in terms of their conceptual similarity s_c and relational similarity s_r.

Conceptual similarity measures the numbers of concepts the two graphs G_1 and G_2 have in common. This is calculated using an expression analogous to one of the widely known similarity measure, the dice coefficient, used in IR:

$$s_c = \frac{2 \times n(G_c)}{n(G_1) + n(G_2)}$$

where $n(G)$ is the number of concepts in the graph G.

The value of s_c lies in the range of 0 and 1. A zero value indicates that the two graphs have no common concepts and a value of 1 indicates that the graphs have common concepts.

Relational similarity is the measure of similarity between the contexts of the concepts that are common to both graphs. It is defined as the proportion of the degree of connection of concept nodes in G_c, and the degree of the connection of the same node in initial graphs G_1 and G_2.

$$s_r = \frac{2 \times n(G_c)}{mG_c(G_1) + mG_c(G_2)}$$

Here, $m(G_c)$ is the number of arcs in the graph G_c, and $mG_c(G)$ is the number of arcs in the immediate neighbourhood of graph G_c in the graph G.

The two similarity values are combined to give the similarity s between graphs G_1 and G_2:

$$s = s_c \times (a + b \times s_r)$$

The value of the coefficients a and b depends on the structure of the document and is computed as follows:

$$a = \frac{2 \times n(G_c)}{2 \times n(G_c) + mG_c(G_1) + mG_c(G_2)}$$

and $b = 1 - a$

Example 10.3 Consider the conceptual graphs G_1 and G_2 and their overlap graph G_c shown in Figure 10.10.

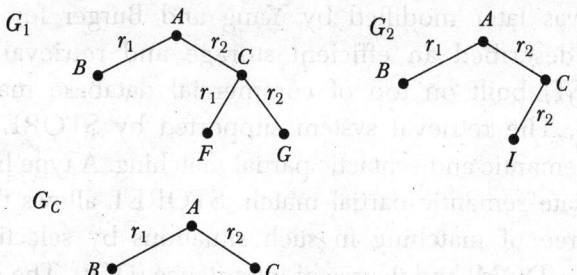

Figure 10.10 Example conceptual graphs and their intersection

The number of concepts in the intersection graph G_c is 3, and in the conceptual graph G_1 and G_2 is 5 and 4 respectively. Therefore, the conceptual similarity between G_1 and G_2 is

$$\frac{3}{(5+4)} = \frac{3}{9} = 0.33$$

The number of arcs in the intersection graph is 2, and the number of arcs adjacent to Gc in G_1 and G_2 are 4 and 3 respectively. Therefore, the relational similarity between G_1 and G_2 is

$$\frac{2}{(4+3)} = \frac{2}{7} = 0.29$$

The values of the coefficients a and b for the conceptual graphs in Figure 10.10 is computed as follows.

$$a = \frac{6}{(6+4+3)} = \frac{6}{13} = 0.46$$

$$b = 0.54$$

This similarity measure was later modified by Montes-y-Gómez (2001) to take into account synonymy and subtype/supertype relationships between concepts and relations used in CGs, and to allow for different weights to different types of nodes (entity, action, and attribute).

Liddy and Myaeng (1994) also used CG matching algorithm proposed by Myaeng and Lopez-Lopez (1992) and attempted to normalize the score by taking 'connectivity' into account. Their experimental results show that $(1.05) \wedge (1 - x)$ is the best normalization factor, where x is the number of test units that contain one or more matching CG fragments.

Liu (1997) investigated partial relation matching. Instead of trying to match the concept-relation-concept group, he attempted to match individual concepts with their semantic role. Yang et al. (1993) presented a CG matching algorithm that allowed both partial and exact matching, which was later modified by Yang and Burger for CGIF. Yang and Burger described an efficient storage and retrieval system for CGs (STORET) built on top of commercial database management system (DBMS). The retrieval system supported by STORET allows exact as well as semantic and syntactic partial matching. A type hierarchy is utilized to facilitate semantic partial match. STORET allows the user to control the degree of matching in such situations by selecting the degree of matching (DOM) and degree of inheritance (DOI). The DOM is calculated as the ratio of the number of matched relations to the number of relations in the query graph. DOI indicates inheritance length in the type hierarchy. Both these measures specify the closeness of the query graph with the matched graph, from different points of view.

A number of conceptual graph matching algorithms have been proposed and utilized for information retrieval. However, little has been done to integrate these measures with existing statistical similarity measures. Siddiqui and Tiwary (2005) proposed a simplified CG-based information retrieval model that performs retrieval based on a cumulative measure.

10.6   CROSS-LINGUAL INFORMATION RETRIEVAL

The purpose of cross-lingual information retrieval (CLIR) is to support queries in one language with a collection in other languages. The rapid growth and availability of online information in many languages have made the task of finding relevant information from large, multilingual text collections a field of active research in the IR and NLP communities. In many respects, CLIR shares the same goals as machine retrieval. Well-known IR techniques, such as vector space retrieval, latent semantic indexing, similarity measures for matching documents and query processing procedures, are equally useful in CLIR.

However, CLIR differs from IR in several significant ways. Most importantly, IR requires no translation phase. Since the queries and documents are in different languages, CLIR requires a machine translation phase along with the usual monolingual retrieval phase. For this purpose, CLIR adopts various techniques explored in NLP research. It seems that CLIR is simply a matter of coupling IR and MT. However, it is not so. The special nature of CLIR places constraints on the input to MT, which makes a straightforward coupling unfeasible. The central challenge in CLIR is to find the most effective way to bridge the language barrier between queries and documents.

Methods of CLIR are typically divided into approaches based on the use of bilingual dictionaries or MT; on corpuses and a range of IR-specific statistical measures; and concept-driven approaches.

10.6.1 Issues

Query processing

A query is a statement of information need from the user. There are three main problems associated with the processing of query terms for CLIR.

- How can a query term in L1 be expressed in L2?
- What mechanisms determine which of the possible translations of text from L1 to L2 should be retained?
- In cases where more than one translation is retained, how are the different translation alternatives to be weighted?

Indexing

Choosing good terms and weighting them is hard enough in one language, but when the problem is extended to documents in multiple languages, it becomes considerably more difficult. Standard IR methods for indexing involve a small amount of language-specific processing. Multilingual document preparation and pre-processing techniques have been the focus of much recent research. This includes tokenizing (separating the continuous character stream into individual words), part-of-speech tagging, stemming and demorphing (for example, converting inflected words into their root forms plus associated information, a task that can be quite complex in highly inflected languages such as Arabic). In addition, new techniques for extracting collocational and phrasal information, both monolingually and multilingually, are being developed.

Matching and ranking

The matching problem is primarily difficult in CLIR due to imperfect correspondence between languages. Similarity metrics based on keyword matching may create problems due to conceptual mismatches.

10.6.2 Approaches to CLIR

Figure 10.11 shows the general approach to CLIR. The user submits a query in one language and the retrieval system responds by returning documents in a language different from the query language. As the query language is different from the language of the documents in the collection, it has to be translated before the actual retrieval. The various approaches that can be used for CLIR are as follows.

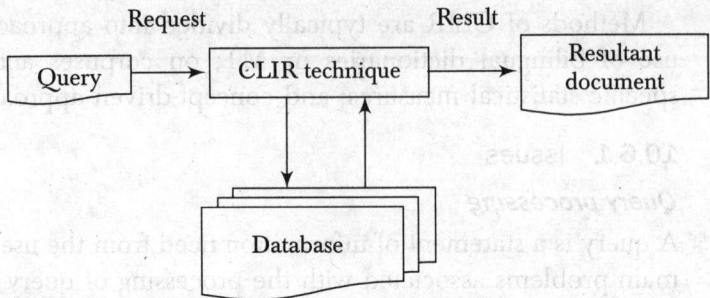

Figure 10.11   Cross-lingual information retrieval

Dictionary-based Query Translation

Query keywords are translated to the target language using machine readable dictionaries (MRD). The main problems associated with this approach are phrase identification and translation, source ambiguity, translation ambiguity, the coverage of the dictionary and processing of inflected words (Pirkola et al. 2004, Helhund et al. 2001).

Machine Translation

In CLIR, MT can be implemented in two ways. The first is to use an MT system to translate foreign language documents into the language of the user's query. The second way is to translate the user's query into the target language (the language of the documents in the stored collection). The target language query is then used to retrieve target language documents using classical IR techniques.

Both methods require an MT stage that is separate from the retrieval stage. An ambiguity problem exists in the MT component, since the translated query does not necessarily represent the sense of the original query. MT systems normally attempt to determine the correct word sense for translation by using context analysis (Braschler 2004). However, a typical search engine query is small and lacks context.

Corpus-based

A corpus is a repository for a collection of natural language material, such as text, paragraphs, and sentences from one or many languages. Two types of corpuses have been used in query translation: parallel and comparable. Parallel corpuses consist of the same text in more than one language. When retrieving text from a parallel corpus, the source language query need not be translated. The source language query is used to match against the source language component of the corpus, and the target language component aligned to it is retrieved. A number of alignment methods have been proposed in literature (Carpuat et al. 2006). Comparable corpuses contain text in more than one language. The texts in each language are not translations of each other, but cover the same topic area, and hence contain equivalent vocabulary.

Concept-driven

Concept-driven approaches exploit multilingual ontologies or thesauri to bridge the gap between surface linguistic forms and meanings. The terms and phrases from multiple languages that refer to the same concept are mapped into a language-independent scheme. In this approach, both documents and queries are mapped into the conceptual interlingua, which, permits matching and retrieval based on any combination of languages involved, rather than relying on pair-wise translations. This seems particularly appropriate for domains (and languages) for which extensive multilingual ontologies are available, such as UMLS (unified medical language system) in the medical domain.

SUMMARY

- Keyword-based IR systems perform retrieval merely on the presence or absence of terms (keywords) in the document. Synonymy and polysemy create a problem in such retrievels.
- Semantic approaches can be useful in extending the capabilities of keyword-based retrieval systems.
- WordNet can be used to exploit semantic relationships between words.
- Keyword-based representation does not consider the relationship between terms. Conceptual graph-based representations can be used to capture the relationship between concepts (terms) appearing in the text.
- A cross-lingual information retrieval (CLIR) system supports queries in one language with a collection from other languages.
- Concept-driven approaches to multilingual information retrieval exploit multilingual ontologies or thesauri to bridge the gap between surface linguistic forms and meanings.

REFERENCES

Amati, G. and I. Ounis, 2000, 'Conceptual graphs and first order logic,' *The Computer Journal*, 43(1), Oxford University Press.

Bemrules, 1993, 'Relating diagram to logic,' *Proceedings of First International Conference on Conceptual Structures, ICCS'93*, Quebec, Canada.

_____2000, *Knowledge Representation*, Brooks/Cole Publishing, CA, USA.

Blair, D. and M. Maron, 1990, 'Full text information retrieval: further analysis and classification,' *Information Processing and Management*, 26(3), pp. 437–77.

Braschler, Martin, 2004, 'Combination approaches for multilingual text retrieval,' *Information Retrieval*, 4(1–2), Springer, the Netherlands, pp. 183–204.

Carpuat, Marine, Pascale Fung, and Grace Ngai, 2006, 'Aligning word senses using bilingual corpora,' *ACM Transactions on Asian Language Information Processing (TALIP)*, 5(2), ACM, New York, NY, pp. 89–120.

Dicheva, Darina and Vania Dimitrova, 1998, 'An approach to representation and extraction of terminological knowledge in ICALL,' *Journal of Computing and Information Technology CIT*, 6(1), University Computing Centrer Zagreb, Crotia, pp. 39–52.

Dick, Judith P., 1992, 'A conceptual case relation representation of text for intelligent retrieval,' *Technical Report*, CSRI-265.

Dominich, S., 2001, 'Mathematical foundation of information retrieval,' *Mathematical Modeling: Theory and Applications*, Kluwer Academic.

Genest, David and Michael Chein, 2004, 'A content-search information retrieval process based on conceptual graphs,' *Knowledge and Information Systems*.

Ghosh, B.C. and V. Wuwongse, 1995, 'A direct proof procedure for definite conceptual graphs programs,' *Proceedings of 3rd International Conference on Conceptual Structures, ICCS'95*, Lecture notes in Artificial Intelligence, 954, Springer-Verlag, Santa Cruz, CA, pp.158–72.

Jones, K. Sparck, 1991, 'The role of artificial intelligence in information retrieval,' *Journal of the American Society for Information Science*, 42(8), pp. 558–65.

Levinson, R. and G. Ellis, 1992, 'Multi-level hierarchical retrieval,' *Knowledge-based Systems*, 5(3), Elsevier, pp. 233–44.

Liddy, E.D. and S.H. Myaeng, 1994, 'DR-LINK: a system update for TREC-2,' *Second Text Retrieval Conference, NIST-SP 500-215*, NIST, Washington DC, USA, pp. 85–99.

Liu, G.Z., 1997, 'Semantic vector space model: implementation and evaluation,' *Journal of the American Society of Information Science*, 48(5), pp. 395–417.

Marega, R. and M. T. Pazienza, 1994, 'CoDHIR: an information retrieval system based on semantic document representation,' *Journal of Information Science*, 20(6), UK, pp. 399–412.

Martin, P., 1997, 'CGKAT: A knowledge acquisition and retrieval tool using structured documents and ontologies,' *Proceedings of ICCS'97*, Lecture notes in artificial intelligence 1257, Springer-verlag, Heidelberg, New York, pp. 581–84.

Martin, P. and P. Eklund, 2000, 'Embedding knowledge in Web documents,' Griffith University, Australia, http://decweb.ethz.ch/WWW8/data/2145/html/bindex.htm.

Miller, G., 1990, 'WordNet: an online lexical database,' *International Journal of Lexicography*, 24, pp. 513–23.

————1995, 'WordNet: a lexical database for English,' *Communication of the ACM*, 38(11).

Mittendorf, E., B. Mateev, and P. Schäuble, 2000, 'Using the co-occurrence of words for retrieval weighting,' *Information Retrieval*, 3, pp. 243–51.

Montes-y-Gómez, M., A. Gelbukh, A. López-López, and Baeza-Yates, 2001, 'Flexible comparison of conceptual graphs,' *Lecture Notes in Computer Science 2113*, Springer-Verlag.

Montes-y-Gómez, M., A. López-López, and A. Gelbukh, 2000, 'Comparison of conceptual graphs,' *MICAI 2000: Advances in Artificial Intelligence*, Lecture notes in Springer-Verlag, pp. 548–56.

Mugnier, M.L., 1995, 'On generalization/specialization for conceptual graphs,' *Journal of Experimental and Theoretical Artificial Intelligence*, 7, Taylor and Francis, pp. 325–44.

Mugnier, M.L. and Marie-Laure, 1995, 'Conceptual graphs are also graph,' *Research Report*, 95-004, LIRMM.

Myaeng, Sung H. and Aurelio Lopez-Lopez, 1992, 'Conceptual graph matching: a flexible algorithm and experiments,' *Journal of Experimental and Theoretical Artificial Intelligence*, 4, Taylor and Francis.

Ounis, I. and T.W.C. Huibers, 1997, 'A logical relational approach for information retrieval indexing,' *19th Annual BCS-IRSG Colloquium on IR Research*, EWIC, Springer-Verlag, Aberdeen, Scotland.

Ounis, I. and Marius Pasca, 1998, 'RELIEF: Combining expressiveness and rapidity into a single system,' *SIGIR'98*, Melbourne, Australia.

Pirkola, Ari, Turid Hedlund, Heikki Keskustalo, and Kalervo Järvelin, 2004, 'Dictionary-based cross-language information retrieval: problems, methods, and research findings,' 4(3–4), September, Springer, the Netherlands

Rama, D.V. and P. Srinivasan, 1993, 'An investigation of content representation using text grammars,' *ACM Transactions on Information Systems*, 11(1), ACM, New York, NY, pp. 51–75.

Rau, L., 1988, 'Exploring the semantics of conceptual graphs for efficient graph matching,' *Proceedings of the 3rd Annual Workshop on Conceptual Graphs*, Boston.

Salton, G. and C. Buckley, 1990, 'Improving retrieval performance by relevance feedback,' *Journal of the American Society for Information Science*, 51(4), pp. 288–97.

Salton, G., C. Buckley, and M. Smith, 1990, 'On the application of syntactic methodologies in automatic text analysis,' *Information Processing and Management*, 26, pp. 73–92.

Shum, S.B., E. Motta, and John Domingue, 2000, 'ScholOnto: an ontology-based digital library server for research documents and discourse,' *Int. Journal of Digital Library*, 3, Springer-Verlag, Heidelberg, pp. 237–48.

Siddiqui, Tanveer and U.S. Tiwary, 2005, 'Integrating relation and keyword matching in information retrieval,' *Proceedings of 9th International Conference on Knowledge-based Intelligent Information and Engineering Systems: Data mining and soft computing applications-II*, Lecture notes in Computer Science, 3684, Springer-Verlag, Melbourne, Australia, pp. 64–71.

Slade, Andrew J. and Albert F. Bokma, 2001, 'Conceptual approaches for personal and corporate information and knowledge management,' *Proceedings of the 34th Hawaii International Conference on System Sciences*, IEEE Computer Society Press.

Smeaton, A.F., 1999, 'Using natural language processing or natural language processing resources for information retrieval,' In T. Strzalkowsi, *Natural Language Information Retrieval, Kluwer*, pp. 99–112.

Sowa, J.F., 1984, *Conceptual Structures: information processing in mind and machine*, Addison-Wesley.

———1993, 'Relating diagram to logic,' *Proceedings of First International Conference on Conceptual Structures*, ICCS'93, Quebec, Canada.

———2000, *Knowledge Representation*, Brooks/Cole Publishing, CA, USA.

Strzalkowski, T., 1995, 'Natural language information retrieval,' *Information Processing and Management*, 31(3), pp. 397–417.

Turid Hedlund, Heikki Keskustalo, Ari Pirkola, Mikko Sepponen, and Kalervo Järvelin, 2001, 'Bilingual tests with Swedish, Finnish, and German queries: dealing with morphology, compound words, and query structure,' *LNCS*, 2069, Springer, Berlin/Heidelberg

Wermelinger, M., 1995, 'Conceptual graphs and first order logic,' *Proceedings of 3rd International Conference on Conceptual Structures, ICCS'95, Lecture notes in Artificial Intelligence*, 954, Springer-Verlag, Santa Cruz, CA, pp. 323–37.

Yang, Gi-Chul, Y. Choi, and J. Oh, 1993, 'CGMA: a novel conceptual graph matching algorithm,' *Conceptual Structures: Theory and Implementation*, LNCS, Springer, Berlin/Heidelberg.

Zhang Lei, and Y. Yu, 2001, 'Learning to generate CGs from domain specific sentences,' *Proceedings of 9th International Conference on Conceptual Structures*, LNAI, 2120, Springer-Verlag.

EXERCISES

1. Discuss the problems associated with the keyword-based IR systems.
2. What are the advantages of using semantic relations over syntactic ones?
3. Give evidence where the use of WordNet relations for query expansion might have an adverse effect on the performance of an IR system.
4. Given here is the tagged representation of the sentence:
   A methodology for calculating system performance:
   A *DT* methodology *NNP* for *IN* calculating *VBG* system *NNP* performance *NNP*.
   Construct its conceptual graph using the basic relations discussed in this chapter.
5. Construct the conceptual graph for the following tagged sentence.
   A *DT* methodology *NNP* for *IN* calculating *VBG* square *JJ* root *NNP*.
6. Compute the conceptual and relational similarity between the conceptual graphs constructed in Questions 4 and 5.
7. Distinguish between concept–based and corpus-based approach to cross lingual information retrieval.

LAB EXERCISES

1. Write a program to create the conceptual graph for an input sentence.
2. Write a program to compute relational similarity between two conceptual graphs G1 and G2.
3. Download Opencyc and SUMO in your lab. Find out their possible applications.
4. Write a program to implement the join operation on conceptual graph.

CHAPTER 11

OTHER APPLICATIONS

CHAPTER OVERVIEW

The Web has made it possible to access a large amount of information quickly and automatically. This has led to the development (or more appropriately, the resurgence) of a number of information processing applications. We discuss three such applications in this chapter, namely information extraction, text summarization, and question-answering. An *information extraction system* identifies a subset of information within a document that fits a pre-defined template. Automatic *text summarization systems* attempt to produce a summary of the document. A *question-answering system* attempts to find precise answers or at least, the precise portion of a text in which the answers appear (unlike an IR system, in which the entire document that contains relevant information is returned). The chapter provides sufficient details on these applications to allow readers to implement them.

11.1 INTRODUCTION

Information retrieval systems and search engines were developed to help users find relevant information from large collections, such as those found on the Web. However, rather than return precise answers, such systems retrieve ranked lists of documents containing all or some of the terms from user-queries. This has led to the development of a number of other information processing applications. Among these are information extraction, text summarization, and question-answering systems.

These applications are related to IR but different from it. Although the term *information* generally includes all sorts of information—text, images, audio, video, etc.—we restrict our discussion to text processing applications. These applications have a number of processing steps in common with IR (their evaluation measures also bear a strong resemblance to IR evaluation measures), but there are significant differences, such as the fact that these applications involve more NLP than IR.

11.2 INFORMATION EXTRACTION

An information extraction system captures and outputs factual information contained within a document. Like an IR system, it responds to a user's information need. However, unlike an IR system, the information need is not expressed as a keyword query; it is specified as a pre-defined template. An IR system identifies a subset of documents in a large repository of text database, whereas an information extraction system identifies a subset of information within a document that fits the pre-defined template. This subset of information is not necessarily a summary of the contents of the document, as it would be in document summarization. Rather, it corresponds to specific instances of pre-defined templates representing information need, found in the text.

Suppose an analyst wants to identify and maintain a database of all the products launched by a company, in some country, in the format, *A launched B in X for amount y*. It would be difficult to express such an information need as a keyword query. Keyword-based search engines respond to queries that seek documents containing information about an entity expressed as a noun or noun phrase. They do not respond to queries for specific information, matching a template of relationships between entities whose names are to be extracted from documents. An information extraction system on the other hand, can extract and output all of the occurrences of such products and their companies, along with their target countries and price.

An information extraction system does not try to understand the text, as a question-answering system would. Its output is less ambitious. It tries to identify assertions about very specific kinds of facts. Instead of generating answers, the information is distilled into a structured form in which individual facts are accessible. Let us emphasize the difference between and information extraction system and a question-answering system with the help of an example.

The following text excerpt from *The Hindu*, May 26, 2007, is used as an input:

> *Tata Tea Ltd. (TTL) will sell its 30 percent stake in Energy Brands Inc. (EBI), US, to the Coca-Cola Company for a consideration of around $1.2 billion. The transaction is in response to an offer received by EBI from the Coca-Cola Company for acquiring 100 per cent of the capital of EBI. Tata Tea Investment Ltd. (UK) and Tata Tea Ltd. (UK), two Tata subsidiary companies, will sell their shares to the Coca-Cola Company by the end of the year.*

A question-answering system is expected to answer questions such as 'In which year did Coca-Cola Company buy EBI?' and 'Why is Tata Tea Ltd. selling its stake in EBI?'.

In contrast, an information extraction system would extract the fact that there is a sell event, that the seller is Tata Tea Ltd, the fraction of stake entity in EBI being sold is 25 per cent, and the sale amount is $1.2 billion.

To summarize, the two important characteristics of an information extraction task are as follows:

- The desired information is expressed as a relatively simple and fixed format, usually a template or frame.
- Only a small fraction of the text qualifies for filling slots in the template or frame; the rest are discarded.

The input to an information extraction system can be expressed as a database schema or template, specifying the output format. Here is an example.

Example 11.1 Consider the following text taken from MUC-7:

A relevant article refers to a vehicle launch that is scheduled, in progress or has actually occurred and must minimally identify the payload, the date of the launch, whether the launch is civilian or military, the function of the mission and its status.

Figure 11.1 shows the information sought by the text in the form of a template.

```
task:    Launch Event
Vehicle:
Payload:
Mission_Date:
Mission_Site:
Mission_Type (Military, Civilian)
Mission_Function (Test, Deploy, Retrieve)
Mission_Status (Succeeded, Failed, In Progress, Scheduled)
```

Figure 11.1   Information to be extracted expressed as a template

The output of an information extraction system can be a single template with a certain number of slots filled from input text. MUC tasks output a more complex, hierarchically related set of templates. Figure 11.2 shows sample text and the information extracted using the template in Figure 11.1.

The second developmental test flight of India's Geosynchronous Satellite Launch Vehicle, GSLV, was successfully carried out this evening May 8, 2003 from Satish Dhawan Space Centre, Sriharikota, about 100 km north of Chennai, marking a major milestone in the Indian space programme. The payload is 573 pounds (260 kilograms) heavier than the one launched on the first GSLV flight in April 2001. With this launch, India has moved further in establishing its capability to launch geo-synchronous communication satellites.

Task:	Launch Event
Vehicle:	Geosynchronous Satellite Launch Vehicle
Payload:	573 pounds
Mission_Date:	May 8, 2003
Mission_Site:	Satish Dhawan Space Centre, Sriharikota: Chennai
Mission_Type (Military, Civilian):	Civilian
Mission_Function (Test, Deploy, Retrieve):	Test
Mission_Status (Succeeded, Failed, In Progress, Scheduled):	Succeeded

Figure 11.2 Sample text and extracted template

11.2.1 Design of an Information Extraction System

An information extraction system can be designed using the knowledge-engineering approach (rule-based approach) or the trainable approach. The knowledge engineering approach is suitable when human experts and linguistic resources are available, whereas the training approach is suitable when training data is available or can be acquired easily.

In the knowledge-engineering approach, a knowledge engineer writes rules for extracting the desired information. The knowledge engineer is supposed to be familiar with the information extraction system and the formalism for expressing rules. The knowledge engineer examines a moderate-sized corpus of domain-relevant texts, consults domain experts and uses his own intuition to write the rules. The skill of the knowledge engineer plays an important role in the performance of the overall system.

Building a high performance system is an iterative process. The knowledge engineer writes an initial set of rules. Using these, the system is run over a training corpus of texts and information is extracted. The output is examined to see where the rules can be modified. The knowledge engineer then makes the appropriate modifications, and iterates the process. The knowledge-engineering approach, therefore, has the advantage that it results in a high performance system. Most of the best-performing information extraction systems to date have been handcrafted.

However, this approach is fairly time consuming and labour intensive, and requires human expertise to write rules.

The automatic training approach eliminates the need for human expertise. However, it requires an annotated corpus for training. For example, to train a name recognizer, a corpus of texts annotated with the domain-relevant proper names is needed. Similarly, a co-reference component requires a corpus indicating the co-reference equivalence classes for each text for training. Such a corpus can be built manually, i.e., by a domain expert. Once the annotated text is available, the training algorithm is run, and statistics or rules are derived automatically from the training corpus, which is then used to analyse new texts.

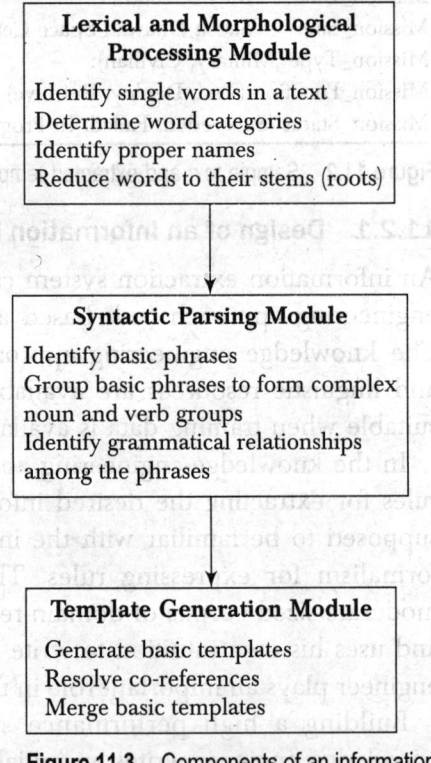

Figure 11.3 Components of an information extraction system

Usually, an information extraction system is composed of a series of modules that process text by applying rules (Hobbs 1994). Figure 11.3 shows the basic modules of an information extraction system, namely a morphological and lexical processing module, some sort of syntactic parsing and a module for domain specific processing (template generation).

Although there will be variations among systems, the functions for the modules will generally be performed somewhere in the processing.

As with an IR system, an information extraction system includes a tokenizer, which divides the text into sentences and tokens. The lexical and morphological processing also handles inflectional variants of a word. In languages with simple inflectional morphology such as English, Hindi, and Urdu, this module is relatively simple. The next task, performed by the lexical and morphological processing module, uses a lexicon to associate properties such as part-of-speech and meaning, with each word. Yet another important function of this module is to assign lexical features to items that have internal structures such as date, time, and various types of proper names. Proper names are particularly important for extraction systems, because typically, information is sought about the properties and relations of a particular object, and that object is usually identified by its name. Because of its importance in identifying objects of interest, almost every information extraction system includes a named-entity recognizer.

Names often have internal structures. For example, a company name can usually be identified by its final token. Patterns like, <word> <word> Company (e.g., Coca-Cola Company), <word> <word> Ltd. (e.g., Tata Tea Ltd), <word> <word> Inc., or <word> Pvt. Ltd., almost certainly designate a company name. After scanning a corpus of texts with several hundred names, it is possible to write rules for the different types of examples. However, things are not as simple as they appear. For example, McDonald can be tagged as a person, but it may also be a restaurant. Similarly, General Motors can be tagged as a person with a military title.

Detailed parsing is not required in an information extraction system. During syntactic analysis, sentence fragments (the basic phrases—noun groups, verb groups, etc.) are identified and exact roles are assigned to individual words. Usually, finite state grammars are used to identify these segments. Figure 11.5 shows the basic phrases identified from the sample text in Figure 11.4. These basic phrases are then combined to give complex noun and verb groups. Entities and events in the complex groups are recognized and later inserted into the proper templates. This recognition is usually facilitated with the help of hand-coded finite state automata. For example, using the regular expression,

NG (Company) VG (has launched) in India (Location),

we can identify the following semantic pattern:

Activity   LAUNCH
Company   Seagate technology
Location   India
Product   FreeAgent data mover

Seagate technology has launched in India three models of its FreeAgent data movers—FreeAgent Go, FreeAgent Desktop, and FreeAgent Pro—targeting the consumer segment

---

Figure 11.4   Sample text

In many situations, information about an event may be spread out in parts over many sentences, leading to a number of basic templates. These basic templates can be merged to produce a single hierarchical template. The template generation and merging involves discourse level and domain specific processing. For example, it must perform co-reference analysis, which, besides resolving anaphoric references by pronouns, also identifies the event or entities in individual sentences that refer to the same entity or event in the real world.

Seagate technology	Company name
has launched	Verb group
in	Preposition
India	Location
three models	Noun group
of	Preposition
its	Noun group
FreeAgent data movers	Product name
FreeAgent Go	Product name
FreeAgent Desktop	Product name
and	Conjunction
FreeAgent Pro	Product name
Targeting	Verb group
the consumer segment	Noun group

Figure 11.5   The basic phrases extracted from the text shown in Figure 11.4

11.2.2   Evaluating Information Extraction Systems

The first ambitious evaluation of information extraction systems occurred at the Message Understanding Conference-3 (MUC-3). The corpus used in the evaluation contained 1,300 documents describing terrorist activities in Latin America. For each text in the evaluation corpus, hand-coded template instantiations were designed to capture all the relevant information present in the source text. The template instantiations generated automatically were compared against those generated manually. The two basic measure used in comparison were precision (P) and recall (R). A more comprehensive overview can be found in Lehnert and Sundheim (1991).

The MUC-4 systems were evaluated using a single measure known as F-measure. Recall that all these measures have been used in evaluating information retrieval systems (Section 9.4.2, Chapter 9). We now discuss them in the context of information extraction systems.

Precision is a measure of accuracy. It is defined as the ratio of correct fillers extracted by the system to the total number of fillers extracted.

$$P = \frac{\text{Number of slots filled correct by system}}{\text{Total number of slots extracted by the system}}$$

Recall measures the amount of relevant information that a system is able to extract from the text. It is thus a measure of coverage. Recall can be defined as follows:

$$R = \frac{\text{Number of slots filled correct}}{\text{Total number of correct slots present in the text}}$$

There is a trade-off between recall and precision. Trying to achieve high coverage compromises on accuracy. Similarly, an attempt to improve accuracy lowers the recall. Due to this, a single scoring measure that combines both precision and recall, known as the F-measure, is often preferred. F-measure (F) is defined as follows:

$$F = \frac{(\beta^2 + 1) PR}{\beta^2 P + R}$$

When β is one, precision and recall are given equal weight. A β value greater than one favours precision and less than one favours recall.

11.3 AUTOMATIC TEXT SUMMARIZATION

Automatic text summarization is not new. Work in this area originated in the 1950s, when the first breakthroughs were achieved by Luhn (1958). Despite this, most of the significant research in this area has been carried out in the last few years. This recent resurgence is mainly due to the remarkable growth of the Internet, which has led to an enormous increase in the amount and availability of online documents. It is now difficult to absorb this huge amount of information. Automatic text summarization is, therefore, a vital tool in dispensing with information overload. For example, consider the number of results returned by a search engine in response to a query. Usually, there are several thousands. Even if we restrict ourselves to the first few pages of each document, it is difficult to decide which documents to read. It is also quite time consuming to open a document to see whether it is relevant or not. A summary of the retrieved documents can help us decide the relevance. This is just one of many possible scenarios where summaries can be useful.

In everyday life, we come across various forms of summaries without consciously recognizing them as such. Consider for example, the news headlines, which summarize the materials present in the news. The scoreboard showing various statistics of a cricket match summarizes what is happening on the ground. The trailer of a movie is very much a summary. The demo of a software is a summary of what will appear in the actual software. The temperature charts that doctors record when you have a fever is yet another example of a summary. The abstract in scholarly articles is a summary written by the authors. A book review is also a summary.

The preceding examples are summaries of documents of many different media forms, such as texts, audio segments, pictures and movies. However, in this section, we restrict our discussion to summaries of textual documents only. Regardless of the media, type and objective, there are few things common to all the summaries, and it is that all summaries are condensed representations of their source. Only information that captures the central meaning or theme of the input, or is deemed relevant to its user, are retained in the summary. Radev et al. (2002) defined a summary as follows:

> *A summary can be loosely defined as a text that is produced from one or more tools, that conveys important information in the original text(s) and that is no longer than half of the original text(s) and usually significantly less than that.*

The goal of automatic summarization is to *take an information source, extract content from it and provide the most important content to the user in a condensed form, in a manner sensible to user's or application's needs* (Mani 2001).

Before we discuss computational approaches to text summarization, let us first examine how humans perform this task.

11.3.1 Human Summarization Process

The general process that humans use to summarize written or spoken text can be described as a three-step process (Brandow 1995):

1. Understanding the content of the document
2. Identifying most important pieces of information
3. Rewriting this information

We use operations such as deletion, generalization, and compaction in this process. We identify important information, delete non-essential information and then, rewrite the remaining information to make it more general and more compact. Let us illustrate this with an example.

Example 11.2   Consider the sample text shown in Figure 11.6.

Yesterday morning, my friend asked me to her house. When I reached, she was preparing coffee. Her father was cleaning dishes. Her mother was busy writing her new book.

Figure 11.6   Sample text to be summarized

We can summarize the description of sample text as follows:

Yesterday, when I visited my friend, the whole family was busy.

Let us analyse how deletion, generalization, and compaction are involved in this process. First, we delete extraneous information—that my friend called me in the morning to visit her house, that the book mother was working on was new, and so forth. Then, we make generalizations: mother, father and daughter are all members of a family; washing, writing and preparing are all things that one can be busy with; dishes, book and coffee are all objects that can be replaced by some other legitimate object. Finally, we compact the three sentences into one, as follows:

One member of the family is busy doing something with an object
+ one member of the family is busy doing something with an object
+ one member of the family is busy doing something with an object
= the whole family is busy (with an object).

The phrase, *with an object,* can then be deleted (because it is extraneous information), leading to the final summary:

Yesterday, when I visited my friend, the whole family was busy.

However, this summarization does not work when the paragraph is expanded to include additional information, as shown in Figure 11.7.

> Yesterday morning, my friend called and asked me to visit her. When I reached her house, my friend was preparing coffee. Her father was cleaning dishes. Her mother was busy writing her new book. My friend offered me a cup of coffee. She told me that her mother had to submit the manuscript 10 days earlier than scheduled and asked me to help her in proofreading it.

Figure 11.7 Expanded text of Example 11.2.

In this case, the point of the story is not that the whole family is busy, but that my friend wants my help in proofreading her mother's manuscript, so that her mother finishes the book on time. This example illustrates why it is important to understand the content of the document to get the central theme and summarize it.

Endres-Niggemeyer (1998) described the human summarization process using the three stages—document exploration, relevance assessment, and summary production.

Document Exploration

This phase is concerned with document familiarization. The document title, its layout and formatting, table of contents and the overall structure are examined at this stage. The genre of the document is also established. Based on the document genre, the summarizer also applies prior knowledge of document type and structure to generate summary.

Relevance Assessment

In this phase, the abstractor constructs a structured mental representation of what the document is about. Niggemeyer called this representation, the theme. The theme forms the basis for the content of the summary.

Summary Production

This phase re-integrates important concepts of a document in a compact form to produce a summary. This process roughly corresponds to a cut-and-paste activity in which relevant text items are copied from the original document and reorganized to fit into a new structure, often with the help of standard sentence patterns (Endres-Niggemeyer 1998).

The items that are considered important by humans while creating an abstract for a document include statements of fact, items relating to the topic, items that discuss purpose, items that are stated positively (rather than negatively), items that contrast with each other and items that are stressed. The items that are usually discarded include reasons for an argument, comments about a topic and examples illustrating a point. Endress-Niggemeyer pointed out that abstractors use cue phrases in the text, e.g., expressions such as 'This paper is concerned with' and 'We conclude by', as well as in-text summary expressions, such as 'in summary', which cover the main theme of the document. Location and title cues are also used. Information at the beginning of a document, or at the beginning or end of a textual unit (e.g., section, subsection, and paragraph), is considered important as it usually carries theme-relevant information. Similarly, title and headings reflect the main theme of the text and are therefore considered useful in recognizing important passages from the text.

Studies conducted on the human summarization process suggest that even with one person, the process of summarizing the same document varies at different points in time. The difference becomes more pronounced in summaries created by two different individuals. However, all summaries created by human subjects have been judged as adequate. This is not the case with computer-generated summaries.

Some interesting characteristics of human summarizers, as observed by Niggemeyer, are as follows:

1. They never try to read the entire document, preferring instead to skim over the text to get a discourse level representation (the theme).
2. The abstractors follow a top-down approach. Paragraphs and sentences are examined before individual sentences.
3. During skimmed reading, shallow features such as cue phrases, location and structure information and information from titles and headings are utilized.
4. Finally, abstractors apply the cut-and-paste method using standard sentence patterns to produce a summary. This also involves specialization and generalization of information content in the article.

Automatic summarization systems use many of these aspects, though it is hard to find all of them in a single system. Some systems use shallow features while others focus on discourse level representation. Still others simply use cut and paste to create a summary.

There are a number of factors that are to be considered while designing a summarization system. Karen Sparck Jones (1999) offered a detailed characterization of these factors.

11.3.2 Types of Summaries

There are many different types of summaries. The distinction is mainly based on their objective (indicative vs informative), the information within the document that is considered in creating summaries (generic vs user-oriented), the number of documents considered (single document vs multiple documents) and the relationship the summaries bear with the source document (extracts vs abstract).

Indicative vs Informative Summary

According to the function that a summary serves (when presented to the reader), it can be classified as indicative or informative. An indicative summary merely points out what topics are addressed in the original document without giving any content. Thus, it can be used to alert the user to the source document so that the user can decide which of the original documents to read. The indicative summary does not, in any way, substitute the source document. The informative summary, on the other hand, provides a shorthand version of the content of the document. Its objective is to substitute the original document as far as coverage of information is concerned.

Generic vs User-Oriented Summary

A generic summary presents the author's viewpoint on the document. It considers all the information in the document to create a summary. A user-oriented summary on the other hand, considers only that information which is deemed relevant to a user query.

Many languages have changed and developed because of outside influences (1). For example, English as we know it today, has many words adapted from other cultures (2). It has some Latin words from the days when it was a part of the Roman Empire (3). English has a large number of words derived from French, the language of England's ruling class following the Norman invasion of 1066 (4). Spanish, Italian, French, Portuguese, and Romanian languages all have many similar words (5). This is because they are descendent from Latin (6). Latin is the language of Roman Empire, of which Spain, Italy, France, Portugal, and Romania were once part (7).

Extract

Spanish, Italian, French, Portuguese, and Romanian languages all have many similar words (5). Latin is the language of Roman Empire, of which Spain, Italy, France, Portugal, and Romania were once part (7).

Abstract

Many languages have changed and devolved because of outside influence including English which has many words from Latin and Roman. Spanish, Italian, French, Portuguese, and Romanian languages all have many similar words as they were once part of Roman Empire.

Figure 11.8 Sample text and its extract and abstract

Extracts vs Abstracts

Considering the relationship that a summary has to the source document, a summary can be either an abstract or extract. An extract involves the selection and verbatim inclusion of 'material' from the source document in the summary. This 'material' usually comprises sentences, paragraphs or verb phrases. The excerpted text units can be included in the summary verbatim, or they can be processed further to smooth the text flow. An abstract, on the other hand, involves the identification of salient concepts in the source document and their fusion and appropriate presentation, usually through natural language generation. Figure 11.8 shows a sample text and one possible extract and abstract for it. The numbers in parentheses are sentence numbers. The extract shown in the figure has been generated by extracting sentences 5 and 7 from the original text.

Single Document vs Multi-document

Based on the number of documents processed during summarization, a summary can be a single document or a multiple document summary. A single document summary is produced using information from a single document whereas a multiple document summary is produced by combining information from multiple documents.

11.3.3 Text Summarization Approaches

There are two basic approaches to text summarization. These are the *shallow* or *knowledge-poor* approach and *deep* or *knowledge-rich* approach.

Shallow or Knowledge-poor Approach

This approach usually involves only syntactic level processing. Though words may sometimes be analysed to a semantic level, sentences are not analysed beyond the syntactic level. This approach typically generates extracts by extracting sentences directly from the source text using statistical analysis at a surface level. Some post-processing is required to smooth out incoherence caused by extraction and to make it more compact. The main advantage of the shallow approach is its robustness.

Deep or Knowledge-rich Approach

This approach involves semantic level and discourse level analysis, usually to produce abstracts. Summary generation here consists of identifying the most important information in the document, encoding it appropriately and then, feeding it to a natural language generation (NLG) system which generates the summary (abstract). The encoding here refers to semantic or discourse level representation. As the semantic analysis and synthesis components can be knowledge intensive, these approaches usually require some coding for domain specific knowledge, since general-purpose knowledge bases are not usually available or feasible (Mani 2001). As the summaries are being generated, the problem of incoherence can be avoided by enforcing rules on how to link discourse segments. Deep approaches are computationally expensive but have the potential to offer more informative summaries.

We now discuss extractive and abstractive summary generation. In order to create an extractive summary, we need to extract important sentences from the document. We extract such sentences using the tf-idf style of scoring of sentences. Machine learning approaches can also be used to extract sentences (Kupiec et al. 1995). A number of other approaches have been discussed by Mani (2001). However, we confine our discussion to the basic issues involved in extractive and abstractive summary generation.

11.3.4 Extractive Summary

At this point, we turn our attention to what is actually involved in building an extractive summary, how the importance of sentences is determined and what factors contribute to sentences weights. A number of sentence extraction methods exist in literature. We first discuss Edmundson's work, which still forms the basis for most extractive methods. This discussion is followed by a graph-based sentence extraction method.

The Edmundsonian Paradigm

Edmundson (1969) provided the basic framework for extraction. He assigned a score to each sentence and extracted the highest-scoring sentences to create a summary. Edmundson considered four features to assign score to sentences:

- Cue words
- Keywords
- Location features
- Title words

Important sentences contain *indicative phrases* or *cue words* such as, 'The main aim of this paper is to describe...' , 'Our investigation has shown that...' , 'The purpose of this article is to review...' , etc. In Edmundson's work, cue words were extracted from a training corpus and divided into bonus words and stigma words. Bonus words are used as evidence for selection and have a frequency above an upper corpus frequency threshold. Stigma words have a frequency below a lower corpus frequency cut-off and are used as evidence for non-selection.

Keywords are extracted from the document. Only words with frequencies above a cut-off are considered. Each keyword is assigned a weight that equals its document frequency.

The *location features* are incorporated in two ways. First, sentences occurring under section headings, such as 'introduction' and 'conclusion', are assigned a positive weight for location. Second, sentences in the first and last paragraphs, or sentences that form the first or last sentence in a paragraph, are assigned a positive weight.

Title words are words that appear in titles. Edmundson assigned a positive weight to words appearing in title, subtitles, and headings. The underlying assumption here is that authors tend to use informative titles.

The weights of each of these four features are combined to yield a single score for a sentence, as follows.

$$W(s) = \alpha C(s) + \beta K(s) + \gamma L(s) + \delta T(s)$$

Edmundson adjusted the tuning parameters α, β, γ, and δ manually. The main problems in extraction are as follows.

Dangling anaphors   If an extracted sentence contains an anaphor whose referent is not included in the extract, then the extract may not be readable.

Gaps   The ideas in a text are usually connected. Extracting sentences breaks the connection and introduces gaps, leading to incoherent text.

Structured environments   Itemized lists and tables create problems in extraction. The structured integrity of these items needs to be preserved; they cannot be arbitrarily divided. For example, if a source sentence says, 'The three important crops of northern India are listed below', and lists all three crops as bulleted items, and only the first and third bullets are included in the summary, the summary will appear quite absurd to a user.

Usually, a post-processing 'repair' step is performed to fix these problems. Identifying the referent of anaphors requires linguistic and domain knowledge. Some systems solve this problem by simply disallowing all sentences containing specific anaphors. Others include some of the previous sentences as context to improve readability. Structured environment needs to be recognized to prevent arbitrary division. We can either exclude them from a summary or attempt to summarize them. Gaps can be handled by including lower-ranked sentences between two selected sentences or by including the first sentence of a paragraph if second and third sentences are part of the summary.

Graph-based Extraction

Graph-based methods map text into graphs. Nodes of the graph are textual units, which can be sentences or paragraphs. Two nodes are connected if they have a vocabulary overlap over a specified threshold. Highly connected nodes, called bushy nodes by Salton et al. (1997), are likely to discuss topics covered in many other textual units. Textual units represented by bushy nodes are good candidates for extraction. Figure 11.9 shows the sentence relationship map for text in Figure 11.8. Nodes in this graph represent sentences. Two nodes are connected if they have a similarity value above a cut-off value of 0.25. Node 1 is not linked with node 2, 3 and 5 in the graph, as it does not have a vocabulary overlap with these sentences, resulting in a zero similarity value.

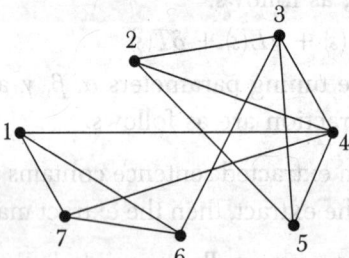

Figure 11.9   Sentence relationship map

Salton et al. (1997) use this approach to generate text summaries by passage extraction. Nodes in their work represent paragraphs. A vector-based representation of paragraphs was used to find the similarity between them. They used a paragraph similarity threshold of 0.20 to generate the graph, i.e., two paragraphs P_i and P_j are not to be linked if their similarity is less than 0.20. Important paragraphs are extracted to create the summary. Highly bushy nodes were considered important. They extracted the n most bushy nodes and arranged them in chronological order to create the summary. Some articles may contain short segments on specialized topics. The paragraphs in these segments are well connected to each other but not to other paragraphs. It is quite likely that bushy paths will miss them. To avoid this at least one paragraph was selected from each segment.

11.3.5   Abstractive Summary

An abstract is a type of summary where some of the content of the summary is not present in the source document. In general, creating abstracts is difficult, as it involves an understanding of the content and may also refer to background concepts not present in the text. Many different approaches have been proposed to generate abstracts. Here, we discuss the template filling approach for creating abstracts. In this approach, information extraction methods are used to fill the slots of a template from the source text. These filled templates produce a semantic level representation of the document which can be used to create abstracts. Slots of the template themselves are background concepts. The simplest approach to creating an abstract is to present filled templates in a tabular form to produce the abstract. This is inexpensive but ineffective from the point of view of readability. An alternative is to feed template-based representation to a natural language module. Generating abstracts is computationally expensive but leads to more readable, compact and coherent summaries. Figure 11.10 gives a modular representation of natural language generation.

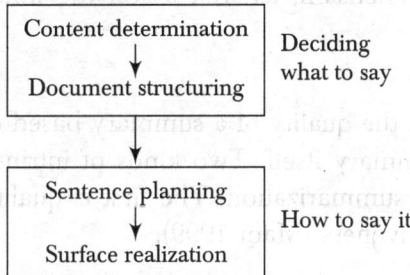

Figure 11.10   Language generation for abstraction

As discussed in Chapter 6, natural language generation involves two major tasks:

- Deciding what to say
- How to say it

Out of these two, the first task is trivial for text summarization, as the earlier stages of summarization already identify and build a semantic representation of what is to be communicated. What remains is to decide on how to say it, i.e., to decide the overall structure of the abstract, its layout, section and sub-section heading, segments to be presented as tables and captions and titles of figures. Sentence planning for abstraction involves message integration, i.e., deciding whether each message is to be combined in a single sentence or be broken across multiple sentences, or whether multiple sentences are to be combined into a single sentence. Lexical choice involves deciding on the types of the expressions to be used, e.g., a proper noun, a definite NP, and a pronoun. The result of lexical choice is a discourse level and syntactic level annotation on the semantic representation, which is used by the surface realization component to generate sentences.

11.3.6 Automatic Text Summarization Evaluation

Various methods that have been proposed need to be compared, so that their relative advantages and disadvantages can be understood. Much of the research done in the field of automatic text summarization is theoretical in nature. Though most of these works involve some sort of evaluation, there isn't a great deal of consensus on evaluation issues. The evaluation methods proposed for summarization can be broadly categorized as *intrinsic* and *extrinsic* (Radev et al. 2002). Intrinsic approaches assess the quality of the summary. Extrinsic evaluation measures the summary based on how it affects the completion of certain tasks. There have been a number of extrinsic evaluations, including impact of summarization on question-answering and comprehension, as well as on the information retrieval task.

Intrinsic Measures

These measures assess the quality of a summary based on the analysis of the content of the summary itself. Two kinds of intrinsic evaluation are usually carried out in summarization. The first is quality evaluation and the second is informativeness (Mani 1999).

Quality Evaluation   One way to assess the quality of a summary is to ask human judges to grade summaries for its readability or acceptability. *Readability* can be measured in terms of the absence of dangling anaphors, preservation of the integrity of structured environment, such as lists or tables. *Acceptability* can be defined in terms of good spelling and grammar, clear indication of the topic of the source document, conciseness, readability and understandability, presence of acronyms with their expansions, etc. Nevertheless, any judgement of quality can be made only by humans. Human judges may disagree in rating a summary. If they disagree a lot, then the judgments may not be a useful standard for evaluation. An alternative to human subjects is to automatically assess the quality of summaries using a grammar or style checker. Readability measures based on word and sentence length have been used to assess compaction of summaries (Mani 1999). However, compression alone cannot be a good indicator of quality, and word and sentence length cannot tell us much about the quality of a summary. Further, most quality factors focus on the well-formedness, which itself may not be important for determining whether the summary is good or not, for example, when the summary is to be used for assimilation of information. Hence, an information evaluation is often preferred for evaluating a summary.

Informativeness How informative a summary is, is measured in terms of the amount of information preserved from the source text at different levels of compression or from the gold or ideal summary—in case comparison is being made between machine summary and ideal summary—at different levels of compression. This approach requires a human-created abstract that cab be used as the gold (ideal or reference) summary. The problem however, is that there can be a number of summaries for a document. The same subject may generate different summaries at different times. Even if multiple summaries are used in evaluation, there is a possibility that the system may generate a different summary, though still a good one. This is especially true for abstracts.

Instead of asking judges to create a summary, we can give them a compression rate and ask them to pick a sufficient number of sentences in the document, or to rank all or 25% of the sentences in the document. Early studies suggest that the inter-judge agreement on good summary sentences is low. The first encouraging result, in terms of percent agreement, was reported by Jing et al. (1998). They found an average agreement of 96% when subjects were asked to create 10% extractive summaries of news articles drawn from TREC data, and 90% for 20% extractive summaries.

Salton et al. (1997) proposed an evaluation method based on the amount of overlap between the automatic and manual extracts. In their work, a summary was generated by extracting the most important paragraphs. Four different evaluations were made.

1. *Optimistic evaluation:* The optimistic evaluation is done by selecting the manual extract with which the automatic extract has a higher overlap, and measuring this overlap.
2. *Pessimistic evaluation:* A pessimistic evaluation is done by selecting the manual extract with which the automatic extract has a lower overlap.
3. *Intersection:* In this, the intersection of the two manually constructed summaries is first computed. Then, the percentage of these paragraphs included in the automatic extract, is computed. The fact that the paragraphs in this intersection were deemed important by both the judge, suggests that they may, in fact, be the most important paragraphs in the source text.
4. *Union:* Here the percentage of automatically selected paragraphs that are selected by at least one of the two users is calculated.

Sentence Recall and Sentence Precision The most widely known measures of IR, namely recall and precision, have also been used for summary evaluation. Sentence recall measures the fraction of sentences in the ideal summary that have been recalled in the automatically generated summary. Sentence precision measures the fraction of the summary that contains sentences matched with the model summary. Let m be the number of sentences in an ideal summary and n the number of sentences in a machine generated summary, of which k sentences also appear in the ideal summary. Then sentence recall (SR) and precision (SP) are:

$$\text{SR} = \frac{k}{m} \quad \text{and} \quad \text{SP} = \frac{k}{n}$$

Even though the precision/recall measures can give us an idea about the quality of the summary, they are not the best metrics for evaluating a system's quality. One problem with these measures is that they fail to discriminate many different summaries and may end up assigning the same score to summaries which are quite different. For example, consider the text shown in Figure 11.8. Figure 11.11 shows an ideal summary and two possible extracts of it. The ideal summary is constructed by extracting sentences 1, 2, and 4. The summary generated by System 1 contains sentences 1, 2, and 5, while the summary generated by System 2 contains sentences 1, 5, and 7. As both the summaries have one sentence in common with the ideal summary, they will both have the same recall value of 1/3. However, it is obvious that Summary 1 is more like the ideal summary than Summary 2. But this similarity cannot be captured by these measures. Further, these measures are applicable only to extracts, not to abstracts.

Ideal Summary

Many languages have changed and developed because of outside influences. English as we know it today, for example, has many words adapted from other cultures. English has a large number of words derived from French, the language of England's ruling classes following the Norman invasion of 1066.

Extract 1

Many languages have changed and developed because of outside influences. English as we know it today, for example, has many words adapted from other cultures. Spanish, Italian, French, Portuguese, and Romanian languages all have many similar words.

Extract 2

Many languages have changed and developed because of outside influences. Spanish, Italian, French, Portuguese, and Romanian languages all have many similar words. Latin is the language of Roman Empire, of which Spain, Italy, France, Portugal, and Romania were once part.

Figure 11.11   The ideal summary and two possible extracts for the text in Figure 11.8

Utility-based Measure   Radev et al. (2000) uses a fine-grained approach to judge summary worthiness of sentences. Instead of making Boolean judgements on whether or not a sentence is to be included in a reference summary, the judges are asked to assign a score in between 1 to 10 to the sentence. This score is called a utility point. A ten-point scale is used because all sentences are not equally important, so if a less important sentence is missed by a summary, it should not be penalized much. When evaluating a summary against a reference summary, the utility points of all sentences in the automatically generated summary that happen to be common with the ideal summary are added up. The major problem in this approach is that judges may disagree more widely when rating a sentence on a ten-point scale than when making a Boolean decision on summary worthiness.

Content-based Measures   All the measures discussed so far, focus on evaluating extracts and cannot be used for evaluating abstracts. Content-based measures attempt to measure content similarity between a summary and its source, or between a summary and the gold summary. Usually, content similarity is measured in terms of vocabulary overlap. As similarity is computed at the vocabulary level (instead of the sentence level), the content-based measures can be used to evaluate extracts as well as abstracts. To compute vocabulary overlap, we use Dice's coefficient or the cosine similarity measure. We have already used these measures to calculate the similarity between query and document vectors, to perform retrieval in Chapter 9. Vocabulary overlap is computed in a similar fashion. We first filter out stop words from the gold summary (or source text, if the comparison is to be made against it) and system summary, and then calculate the similarity between them using Dice's coefficient or the cosine similarity measure. In order to account for vocabulary differences, particularly in case of abstracts, we use synonyms from a thesaurus or from latent semantic indexing schemes.

Extrinsic Summary Evaluation

Extrinsic summary evaluations assess the quality of a summary in terms of how it affects the performance of the task for which it has been generated. For example, if a summary is being generated for a question-answering task, then an extrinsic evaluation will try to assess the quality of the summary in terms of performance of the question-answering system when summaries are used to find answers instead of source text. Similarly, if we generate summaries to help the user identify relevant document in an information retrieval task, then we can assess the quality by measuring the effectiveness of the summary in making the user's relevance decision. The impact of summarization on determining the relevance of a document has been investigated in Hobson et al. 2007, Siddiqui and Tiwary 2006, and Tombros and Sanderson 1998.

11.4   THE QUESTION-ANSWERING SYSTEM

Question-answering offers a more intuitive approach to information processing. Given a collection of documents and a natural language question posed by user, a question-answering system attempts to find the precise answer or at least the precise portion of text in which the answer appears. This is unlike the IR system in which the entire document that seems to contain information relevant to the user is returned. A question-answering system is different from an information extraction system in that the type of information to be extracted is unknown. In general, a question-answering system can use an information extraction system to identify entities in the text. A question-answering system also requires more NLP. It needs a precise analysis of questions and portions of texts and also semantic as well as background knowledge.

Question-answering systems are not new to this field. The use of natural language front ends to database and expert systems can be traced back to the 60s and 70s in systems such as BASEBALL (Green et al. 1961), LUNAR (Woods et al. 1972) and SHRDLU (Winogard 1973). The LUNAR system provides a natural language interface to the moon-rocks database. LUNAR uses an ATN parser and procedural semantics to answer questions. CHAT-8 is a logic-based system that answers questions about world geography. SHRDLU has knowledge of a small 'world' consisting of different coloured, different sized objects (cube, pyramids, etc.); it can understand instructions on how to modify the world. It consists of sub-systems that parse, interpret and construct sentences, carry out dictionary searches and semantic analyses, and make logical deductions. All these (early) systems are domain-specific in which syntactic analysis of a user's questions are intertwined with the semantic interpretation process. An important characteristic of these systems is the integration of domain-specific information.

MURAX (Kupiec 2001) is a more recent example of a domain-specific question-answering system; it provides a natural language interface to encyclopedias. Ask Jeeves and START (syntactic analysis using reversible transformations) are examples of open domain question-answering systems. The START natural language system (Katz 1990, 1997) is an example of a question-answering system that uses natural language annotations. The current trend in question-answering is towards the processing of large volumes of open domain text. This trend is partly motivated by the creation of the question-answering track in TREC, and the recent increase of systems that use the Web to extract answers.

An open domain question-answering system is supposed to answer questions on any possible topic. Such systems cannot rely on hand-coded domain-specific knowledge to provide correct answers. START is perhaps the first Web-based question-answering system. It has been developed by Boris Katz and his associates of the InfoLab Group at the MIT Computer Science and Artificial Intelligence Laboratory. We focus our discussion to open-domain question-answering systems, which use IR systems to get a set of documents in which answers can be found.

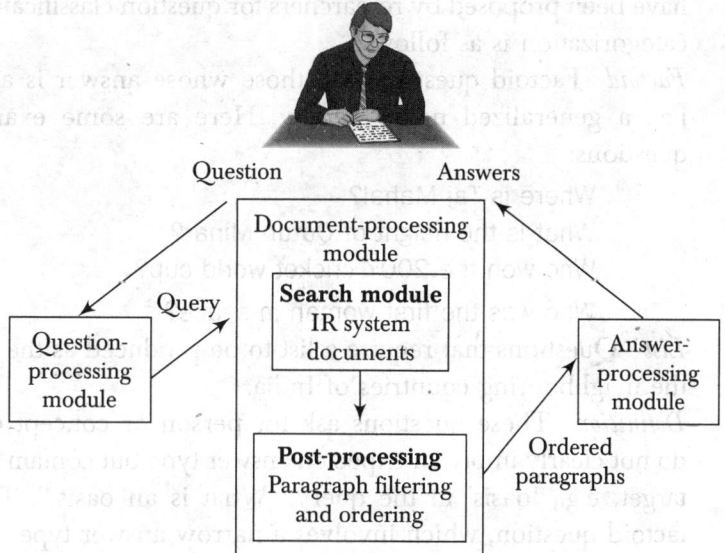

Figure 11.12 Architecture of an open-domain question-answering system

11.4.1 Architecture of an Open-Domain Question-Answering System

The general architecture of an open domain question-answering system is shown in Figure 11.12. It consists of three main modules: question analysis, document processing, and answer processing. The question-analysis module extracts clues from the question, such as question category, expected-answer type, interesting terms, focus, and semantically related words. The extracted terms are used to construct a query representation of the question for document retrieval.

Question-Analysis Module

The tasks performed by this module are as follows:

- To determine question type
- To determine answer type
- To identify topic and focus of the query
- To transform question into an IR query

Identifying Question Type   The first step in this analysis is to identify question type, i.e., time, location, person, place, size, number, etc. The questions are usually categorized based on the answer type. This categorization involves morpho-syntacitc analysis of the question. Knowing the question type defines what constitutes relevant data, which helps other modules to correctly locate the answer. No universally admitted classification of possible question type exist. Various different taxanomies have been proposed by researchers for question classification. One possible categorization is as follows.

Factoid   Factoid questions are those whose answer is a name or figure, i.e., a generalized named entity. Here are some examples of factoid questions:

> Where is Taj Mahal?
> What is the height of Qutub Minar?
> Who won the 2007 cricket world cup?
> Who was the first woman in space?

List   Questions that require a list to be produced as the answer, e.g., List the neighbouring countries of India.

Definition   These questions ask for person or concept definitions. They do not clearly imply an expected answer type but contain only the question target, e.g., 'oasis' in the query, 'What is an oasis?'. This is unlike the factoid question, which involves a narrow answer type.

Another classification used is as follows.

Relationship   Questions like 'What is the relationship between the cold wave and the western disturbances?'

Yes-No   Where the answer is a yes or no. For example, 'Is school closed today?'

Opinion   Questions like,'What do you think on multi-party electoral?'.

Cause-effect   These questions involve two components—the cause and its effect. For example: 'The driver has to wait for some faculty member for half an hour. That's why I was late in the Class.'

In TREC, question-answering tracks pattern-matching based on wh-words and simple part-of-speech information, combined with the use of semantic information provided by WordNet. Miller (1995) has been used to determine question types. Moldvan et al. (2000) proposed another classification of questions along with answer types, based on 200 questions used in TREC 8. The distinction between question classes is not always clear. A question may be ambiguous. To handle this ambiguity, certain questions are allowed to have multiple class labels.

Identifying Answer Type   The next step in question analysis is to identify the answer type. Answer types are often, but not always, tied to the classes recognized by named entity recognizers, e.g., method, explanation and recommendation. The expected answer type (EAT) is defined based on the question terms, especially the interrogative, using rules such as

> Who X ? → Answer type = Person
> Where X ? → Answer type = Place
> How many X ? → Answer type = Number
> When X→ Answer type = Date
> How X → Answer type = Method
> Why X → Answer type = Explanation

Identifying Question Topic and Focus   The topic of the question is the object or event that the question is about. The property of the topic that is the target answer for the question being asked is the focus. For example, the question, '*What is the height of Qutub Minar?*', has the topic, Qutub Minar, and the focus, height. Moldovan et al. (2000) defined the focus of a question as a word or sequence of words that indicate what information is being asked for in the question. If both the topic and focus of the question are known, then the type of question can be determined more easily. The focus of the question can be identified using either a pattern-based approach or a statistical approach. However, the statistical approach to focus detection requires a training corpus with known question focuses. Developing such a corpus may be expensive in terms of time and effort.

The question processing module stores the information acquired from the input question (e.g., the set of content words, the noun phrases, and focus of the question) for later use in ordering the retrieved document set and in answer selection.

Transforming Question into IR Query   The last step in question analysis is to transform the question to an IR query, which is used to retrieve candidate documents. The form of the query depends on the search engine used and its retrieval model. In a vector space model, the query may be a set of query terms, but a Boolean query consists of terms and logical operators. Similarly, the syntax and phrasal construction may vary from search engine to search engine. However, the main objective is to construct a query, and for that query, keywords have to be extracted. The standard techniques, such as stop word elimination, part-of-speech tagging and named entity recognition, are used to extract keywords from the question. Depending on the question type, different words can be extracted as keywords. For example, for questions beginning with 'what', 'which', and 'who', the first noun phrases that appear next to these words can be extracted as keywords. Moldovan (2000) described a set of ordered heuristics to detect keywords. These are listed here:

1. For each quoted expression in a question, extract all non-stop words in the quotation.
2. The proper nouns appearing in the question.
3. Complex nominals and their adjectival modifiers.
4. All other complex nominals, i.e., those that lack adjectival modifiers.
5. Nouns and their adjectival modifiers.
6. All other nouns.
7. Verbs.
8. Question focus.

All or a subset of these heuristics can be used to build keyword representation of the question. Application of each of the heuristics returns a set of keywords that are added to the set of question keywords. The order of the keywords in the set defines their priorities. So, if too many words are extracted, only the first N keywords from this set will be used to build the query representation of the question. Alternatively, a tf-idf style weighing can also be used to select keywords. We can expand the query with additional terms, such as synonyms of original question terms. These expanding terms can be obtained from lexical resources such as WordNet. The synonymy set in WordNet gives semantically related words that might occur in documents containing the correct answer. This expansion may follow an initial retrieval, which may fail to provide a sufficient number of candidate documents.

Document-Processing Module

The query generated by the question-analysis module is fed to a search engine to retrieve a set of candidate documents. However, a cosine similarity measure between document and query is not sufficient for IR in the question-answering system. This is because the keywords are selected to be the most representative words of a question. Therefore, a question-answering system will retrieve a document only when all the keywords are present in the document. By using the cosine similarity measure, a document can be retrieved even if all the keywords do not appear in it.

While performing retrieval for a question-answering system, recall is more important.

The results from the retrieval system are documents. For question-answering system, a whole document is too large a unit. Hence, they are filtered to remove paragraphs that do not contain all the keywords. This filtering reduces the amount of text that is to be analysed in detail. If too many or too few paragraphs survive the filter, then we revise the query representation by dropping or adding few of the keywords and perform the retrieval again. This ensures that a reasonable number of paragraphs are supplied to answer the processing module. The underlying assumption behind paragraph filtering is that the answer will appear in a few neighbouring paragraphs, rather than spread over the entire document. Using this heuristics, a set of consecutive paragraphs containing all the question keywords will survive, and the rest will be filtered out. Instead of extracting paragraphs, arbitrary-size window may also be extracted. Using shorter text segments reduces answer granularity, and increases the amount of information per unit length (Han et al. 2007).

Paragraph Ranking

The aim of paragraph ranking is to sort candidate passages according to the likelihood of their containing the answer. We can use several criteria to rank passages, such as redundancy, term statistics in passages, keyword sequence, separation between most distant keywords in the text segment, etc. We can also consider the source of the document while scoring paragraphs. For example, we can give more weights to local and known data sources than to unknown sources. Similarly, if the source is domain specific, and the domain matches the topic of the question, we can give it more weight. For example, if the question seeks the author of a book and the candidate document comes from Amazon, then we give it more weight. The ranked paragraphs (or segments) are forwarded to the answer-processing module.

Answer-processing Module

The answer-processing module extracts answers from ranked text segments passed to the answer-processing module by the document-processing module. As input, the module also receives the expected answer type, query terms, and other question-related information extracted in the question-processing phase (such as focus of the question). Based on this information, the exact answer should be carefully selected.

First, the paragraphs containing the correct answer type are identified. The answer type and the focus supplied by question-processing module, guide the search for paragraphs containing the correct answer type. Shallow parsing, such as named entity recognizer and part-of-speech tagger, is commonly used in this process.

If no match is found, then best-ranked paragraph is usually returned. Multiple answers are also possible. In TREC, usually the top five answers are returned.

Answer Extraction   Once the potential paragraphs are selected, answer candidates are extracted and ranked. Answer candidates are entities (relevant words or phrases) of the correct type. Shallow parsing is required to extract answer candidates. The ranking is mainly based on the context, i.e., the textual surroundings of the candidate. Various criteria are used for scoring candidate answers, such as proximity of question keywords and capitalization of words.

The candidate answers are passed to a validation algorithm, which selects the best answers, if any. Content-based answer validation (Magnini et al. 2002) can also be used. Magnini et al. exploited the redundancy of the Web for answer validation and found it quite effective. Their method consists in querying the Web with the question keywords and candidate answer, and assigning a score to each candidate answer based on co-occurrence statistics in the results. Figure 11.13 shows the basic steps in the Web validation algorithm as discussed by Tanev et al. (2004).

1. Query the web with a candidate answer CA and query keywords QK.
2. Consider the top k documents returned. For each text fragment in each of these k documents where the candidate answer CA co-occurs with some question keyword, we calculate a score based on the distance of CA and question keywords as:

$$\text{score (text fragment)} = \prod_{k \in \text{text fragment} \cap QK} 2^{1+|CA\,k|}$$

   where |CA k| is the distance n tokens between CA and k.
3. The score gained from different text fragments are summed up for each candidate answer.
4. The candidate answer that obtains the highest score is selected.

Figure 11.13   Web search-based answer-validation algorithm

The validation can be done with the help of WordNet.

11.4.2   Question-Answering Evaluation

Evaluating question-answering systems is a complex task. Nevertheless, to understand the effect of various processing steps on final results, and to compare the performances of various systems, the systems need to be evaluated. A number of evaluation measures have been developed as a result of large-scale campaigns in question-answering track, such as TREC, NTCIR, and CLEF. Commonly used measures of question-answering systems in these campaigns are recall, precision, mean reciprocal rank, confidence weighting score, and *k*-measure. We briefly discuss these measures here. To provide a framework for automatic evaluation, a set of correct answers, which we call *R*-set, is built. In TREC evaluation, *R*-set is constructed by taking the sum of all correct answers provided by all participants.

Recall and Precision

As with the other applications discussed in this chapter, the widely-known IR measures can be used to evaluate question-answering systems. Precision is a measure of accuracy, whereas recall is a measure of coverage. As discussed earlier, precision and recall can be combined in a weighted harmonic mean, called *F*-measure. However, calculating recall is a difficult task in question-answering, as it requires that all the correct answer for each question in the collection must be known in advance. This creates a problem, as there can be many correct answers to a question. Usually, this is approximated by the R-set.

Mean Reciprocal Rank

Mean reciprocal rank (MRR) is mean of the reciprocal rank (RR) of individual questions. *Reciprocal rank* is defined as inverse of the rank of the first correct answer, e.g., for an answer at Rank 1, RR will be 1; for an answer at Rank 5, RR will be 1/5. If no answer is returned for a question, its score (i.e., RR) will be 0. In TREC conferences, five answers are usually returned. The following equation is a more formal description of MRR:

$$MRR = \frac{\sum_{i=1}^{NQ} RR(q_i)}{NQ}$$

$$RR(q_i) = \begin{cases} \dfrac{1}{\text{Rank of the first correct answer for } q_i}, & \text{if an answer is returned} \\ 0, \text{otherwise} \end{cases}$$

where   NQ = Number of questions
  $RR(q_i)$ = Reciprocal rank of q_i (*i*th question)

The drawback with this measure is that systems do not receive credit for returning multiple correct answers, nor do they receive credit for realizing that they have not answered the query.

Confidence Weighted Score (CWS)

This score is also known as the average precision score and is inspired from the non-interpolated average precision of information retrieval. CWS is the average of N different precision measures. It is given by

$$CWS = \frac{1}{NQ} \sum_{i=1}^{NQ} \frac{\text{Number correct up to question } i}{i}$$

It is clear from the expression that CWS gives more weight to the initial terms in the sum. The first score participates in each subsequent term; the second score participates in all but one. This means that the order matters. Two systems that return the same number of correct answers may get different scores if the order of questions answered correctly is different. Let us elaborate on this with the help of an example.

Example 11.3 Consider a set of five questions. System A returns the correct answer for Questions 1, 3, 4, and 5 and System B returns correct answers for Questions 1, 2, 3, and 4. Assuming that the ordering in both the case is 1, 2, 3, 4, 5, the CWS for System A will be

$$\frac{1}{5}\left(\frac{1}{1}+\frac{1}{2}+\frac{2}{3}+\frac{3}{4}+\frac{4}{5}\right) = 0.74$$

and for System B will be

$$\frac{1}{5}\left(\frac{1}{1}+\frac{2}{2}+\frac{3}{3}+\frac{4}{4}+\frac{4}{5}\right) = 0.96$$

The main drawback of CWS is that it can reward the answer ordering strategies of the system rather than the correctness of the answers.

K-measure

K-measure was designed to reward systems that returned as many different correct answers for each question as possible, and to punish incorrect answers. The system requires a self-confidence score for each candidate answer. A human assessor judges each candidate answer and gives them a value 1, 0, or –1 depending on whether the answer is judged correct, already judged, or judged incorrect, respectively. K-measure is given by

$$K = \frac{1}{NQ} \sum_{i=1}^{NQ} \frac{\sum_{a \in AC_i} \text{confidence}(a) \times \text{judgement}(a)}{\max\{|R_i|, AC_i\}}$$

where for Question i,
  R_i is the set of correct answers,
  a is an answer candidate,
  AC_i is the set of answer candidates

Confidence(a) is the confidence score given by question-answering system to a

Judgement(a) is the judgement given by human assessor to a

The main difficulty in using K-measure is that it calculates an exhaustive R_i for each question.

SUMMARY

- Availability of a huge amount of information on the Web has necessitated the need of more effective means of information access.
- A number of information processing systems are already in place. This includes information extraction, text summarization, and question-answering systems.
- An information extraction system captures and outputs factual information contained within a document. Information need is expressed in the form of a template and the information extraction system fills the slots in the template.
- A text summarization system produces a summary (condensed form) of input text.
- The summary can be informative or indicative. It can be generic or user-specific. It may be based on a number of documents or on a single document.
- An extractive summary consists of 'in-text' material only.
- Edmundson (1969) provides the basic framework for most of the work on extraction. He considered four features, namely cue words, keywords, title words, and location, to assign scores to sentences.
- A question-answering system attempts to find the precise answer to a question, or at least the precise portion of text in which the answer appears.
- A question-answering system requires more NLP than both information retrieval systems and information extraction systems.
- Mean reciprocal rank, confidence weighted score or K-measure can be used to evaluate a question-answering system.

REFERENCES

Brandow, R., K. Mitze, and L. Rau, 1995, 'Automatic condensation of electronic publications by sentence selection,' *Information Processing and Management*, 3(5), Elsevier, UK, pp. 675–85.

Edmundson, H.P., 1969, 'New methods in automatic extracting,' *Journal of the Association for Computing Machinery*, 16(2), pp. 264–85.

Endres-Niggemeyer, B., 1998, *Summarizing Information*, Springer-Verlag, New York.

Hobbs, Jerry, 1994 'The Generic Information Extraction System,' *Proceedings of the Fifth Message Understanding Conference, (MUC-5)*, Morgan Kaufmann Publishers, San Mateo, CA.

Hobson, S.P., Bonnie J. Dorr, Christof Monz, and Richard Schwartz, 2007, 'Task-based evaluation of text summarization using relevance prediction,' *Information Processing and Management,* Elsevier, UK.

Jing, H., R. Barzilay, K. McKeown, and M Elhadad, 1998, 'Summarization evaluation methods experiments and analysis,' AAAI Intelligent Text Summarization Workshop, Stanford, CA, pp. 60–68.

Katz, Boris, 1990, 'Using English for indexing and retrieving,' *Artificial Intelligence at MIT: Expanding Frontiers*, 1, MIT Press, Cambridge, MA.

_____1997, 'Annotating the World Wide Web using natural language,' *Proceedings of the 5th RIAO Conference on Computer Assisted Information Searching on the Internet (RIAO' 97)*, McGill University, Montreal, Canada.

Kupiec, Julian, 1993, 'A robust linguistic approach for question answering using an online encyclopedia,' *SIGIR-93*, Philadelphia, PA.

Kyoung-Soo Han, Young-In Song, Sang-Bum Kim, and Hae-Chang Rim, 2007, 'Answer extraction and ranking strategies for definitional question answering using linguistic features and definition terminology,' *Information Processing and Management*, 43, Elsevier, UK, pp. 353–64.

Lehnert, W.G. and B. Sundheim, 1991, 'A performance evaluation of text analysis technologies,' *AI Magazine*, AAAI Press, Mento Park, pp. 81–94.

Luhn, H.P., 1958, 'The automatic creation of literature abstracts,' *IBM Journal of Research and Development*, 2(2), pp. 159–65.

Magnini, B., M. Negri, R. Prevete, and H. Tanev, 2002, 'Is it the right answer? Exploiting web redundancy for answer validation,' *Proceedings of 40th Annual Meeting of the Association for Computational Linguistics (ACL -2002)*, Philadelphia.

Mani, I. and M. Maybury, 1999, *Advances in Automatic Text Summarization*, MIT Press, Cambridge, MA.

Moldovan, D., Sanda Harabagju, Marius Pasca, Rada Mihalcea, Roxana Girji, Richard Goodrum, and Vasile RuS, 2000, 'The structure and performance of an open domain question answering system,' *Proceedings of the Conference of the Association for Computational Linguistics (ACL-2000)*, pp. 563–70.

Radev, D., H. Jing, and M. Budzikowska, 2000, 'Centroid-based summarization of multiple documents: sentence extraction, utility-based evaluation and user studies,' ANLP/NAACL Workshop on Automatic Summarization, pp. 21–29.

Radev, D., E. Hovy and K. McKeown, 2002, 'Introduction to special issue on summarization,' *Computational Linguistics*, 28(4), pp. 399–408.

Salton, Gerard, Amit Singhal, Mandar Mitra, and Chris Buckley, 1997, 'Automatic text structuring and summarization,' *Information Processing and Management*, 33(2), Elsevier, UK, pp. 193–207.

Siddiqui, Tanveer and U.S. Tiwary, 2006, 'Query based indicative summary for assessing document relevance,' *Proceedings of First IEEE International Conference on Digital Information Management*, Bangalore.

Sparck-Jones, K., 1999, 'Automatic summarizing: factors and directions,' *Advances in Automatic text summarization*.

Tanev, Hristo, Matteo Negri, Bernardo Magnini, and Milen Kouylikov, 2004, 'The DIALOG question answering system,' *Proceedings of CLEF-2004, UK, Lecture Notes in Computer Science*, 3491/2005, also available at *www.cles.iei.pi.cnr.it*.

Tombros, A. and M. Sanderson, 1998, 'The advantages of query-based summaries in information retrieval,' *Proceedings of the 21st Annual International SIGIR Conference*, August 24–28, Melbourne, Australia, ACM, NY, pp. 2–10.

Winogard, T., 1972, *Understanding Natural Language*, Academic Press, New York.

Woods, W.A., R.M. Kaplan, and B.L. Nash-Webber, 1972, 'The lunar sciences natural language information system: final report,' *Technical Report 2378*, BBN.

EXERCISES

1. You wish to extract information from news articles in order to maintain a database of all the matches being played in the Grand Slam tournaments. You have to specify the information output format to an information extraction system. Design a template for this problem.
2. Take a newspaper and go to financial news section. Find out the names of ten companies and write a set of patterns that cover all the names.
3. What are the advantages and disadvantages of using the knowledge-based approach to design an information extraction system?
4. Distinguish between indicative and informative summary.
5. What is the difference between an extract and an abstract?

6. What do you mean by 'gaps' in a summary? How is this problem handled?
7. Why are structured items a problem during summary generation?
8. Discuss content-based measure for summary evaluation.
9. Differentiate between intrinsic and extrinsic measures for summary evaluation.
10. How does question-answering system differ from an information retrieval system and an information extraction system?
11. The rank of the first correct answer returned by a question-answering system for a set of five questions is 2, 0, 4, 5, and 1 respectively. Compute MRR using this data.

LAB EXERCISES

1. Write a program that reads a text file and outputs another file containing sentence numbers and the actual sentences. Each sentence should appear in a new line and the output should contain the total number of sentences in the input file.
2. Write a program to extract all nouns and verbs from a sentence.
3. Write a program that takes a text file as input and assigns a score to each sentence based on the following simple scheme:
   Score (sentence) = $\alpha \times (1/\text{position} + \beta \times \text{number of words in the sentence}$
4. Write a program to identify the topic of simple wh-type questions.
5. Create a small document collection consisting of articles related to finance. Write a program to extract the names of all organizations from this collection. Use patterns identified in Exercise 2.

CHAPTER 12

LEXICAL RESOURCES

CHAPTER OVERVIEW

This chapter introduces various tools and lexical resources used in text processing applications and provides a ready reference of these. In particular, it introduces the reader with tools such as stemmers and taggers, lexical resources such as WordNet and FrameNet, and test collections (corpuses) that are freely available for research purpose.

12.1 INTRODUCTION

A whole range of tools and lexical resources have been developed to ease the task of researchers working with natural language processing (NLP). Many of these are open sources, i.e., readers can download them off the Internet. This chapter introduces some of the freely available resources. The motivation behind including this chapter comes from the belief that *knowing where the information is, is half of the information.*

We hope that providing a ready reference of what is available where, will save a lot of time and effort, especially for young researchers and those who are new to the field. All the material presented in this chapter is already available, at the links provided with the discussion or in the form of scholarly articles published on that resource. We bring these resources together and offer a brief discussion on them. In particular, we discuss lexical resources such as WordNet and FrameNet, and tools such as stemmers, taggers, and parsers, and freely available test corpuses for various text-processing applications. We begin our discussion with WordNet in Section 12.2. Section 12.3 discusses FrameNet. Stemmers are discussed in Section 12.4. We present a list of available part-of-speech taggers in Section 12.5. The next section presents a list of document collections. And finally, relevant journals and conferences are listed in Section 12.7.

12.2 WORDNET

WordNet[1] (Miller 1990, 1995) is a large lexical database for the English language. Inspired by psycholinguistic theories, it was developed and is being maintained at the Cognitive Science Laboratory, Princeton University, under the direction of George A. Miller. WordNet consists of three databases—one for nouns, one for verbs, and one for both adjectives and adverbs. Information is organized into sets of synonymous words called *synsets*, each representing one base concept. The synsets are linked to each other by means of lexical and semantic relations. Lexical relations occur between word-forms (i.e., senses) and semantic relations between word meanings. These relations include synonymy, hypernymy/hyponymy, antonymy, meronymy/holonymy, troponymy, etc. A word may appear in more than one synset and in more than one part-of-speech. The meaning of a word is called sense. WordNet lists all senses of a word, each sense belonging to a different synset. WordNet's sense-entries consist of a set of synonyms and a gloss. A gloss consists of a dictionary-style definition and examples demonstrating the use of a synset in a sentence, as shown in Figure 12.1. The figure shows the entries for

---

12.1.1   Noun

1. **read** (something that is read) *"the article was a very good read"*

12.1.2   Verb

1. **read** (interpret something that is written or printed) *"read the advertisement"; "Have you read Salman Rushdie?"*
2. **read**, say (have or contain a certain wording or form) *"The passage reads as follows"; "What does the law say?"*
3. **read** (look at, interpret, and say out loud something that is written or printed) *"The King will read the proclamation at noon"*
4. **read**, scan (obtain data from magnetic tapes) *"This dictionary can be read by the computer"*
5. **read** (interpret the significance of, as of palms, tea leaves, intestines, the sky; also of human behaviour) *"She read the sky and predicted rain"; "I can't read his strange behavior"; "The fortune teller read his fate in the crystal ball"*
6. take, **read** (interpret something in a certain way; convey a particular meaning or impression) *"I read this address as a satire"; "How should I take this message?"; "You can't take credit for this!"*
7. learn, study, **read**, take (be a student of a certain subject) *"She is reading for the bar exam"*
8. **read**, register, show, record (indicate a certain reading; of gauges and instruments) *"The thermometer showed thirteen degrees below zero"; "The gauge read `empty'"*
9. **read** (audition for a stage role by reading parts of a role) *"He is auditioning for `Julius Caesar' at Stratford this year"*
10. **read** (to hear and understand) *"I read you loud and clear!"*
11. understand, **read**, interpret, translate (make sense of a language) *"She understands French"; "Can you read Greek?"*

Figure 12.1   WordNet 2.0 entry for 'read'

---

[1]http://wordnet.princeton.edu/

the word 'read'. 'Read' has one sense as a noun and 11 senses as a verb. Glosses help differentiate meanings. Figures 12.2, 12.3, and 12.4 show some of the relationships that hold between nouns, verbs, and adjectives and adverbs. Nouns and verbs are organized into hierarchies based on the hypernymy/hyponymy relation, whereas adjectives are organized into clusters based on antonym pairs (or triplets). Figure 12.5 shows a hypernym chain for 'river' extracted from WordNet. Figure 12.6 shows the troponym relations for the verb 'laugh'.

Relation	Definition	Example
Hypernym	From concepts to super-ordinates	oak→ tree
Hyponym	From concepts to subtypes	oak→ white oak
Meronym	From wholes to parts	tree → trunk
Holonym	From parts to wholes	trunk → tree
Antonym	Opposites	victory → defeat

Figure 12.2   Noun relations in WordNet

Relation	Definition	Example
Hypernym	From events to super-ordinate events	wander→ travel
Troponym	From events to their subtypes	walk → stroll
Entails	From events to the events they entail	snore → sleep
Antonym	Opposites	increase → decrease

Figure 12.3   Verb relations in WordNet

Relation	Definition	Example
Antonym (adjective)	Opposite	heavy → light
Antonym (adverb)	Opposite	quickly → slowly

Figure 12.4   Adjective and adverb relations in WordNet

```
1 sense of 'river'
Sense 1
river — (a large natural stream of water (larger than a creek); 'the
river was navigable for 50 miles')
     => stream, watercourse — (a natural body of running water
flowing on or under the earth)
     => body of water, water — (the part of the earth's surface
covered with water (such as a river or lake or ocean); 'they invaded
our territorial waters'; 'they were sitting by the water's edge')
     => thing — (a separate and self-contained entity)
     => entity — (that which is perceived or known or inferred to
have its own distinct existence (living or nonliving))
```

Figure 12.5   Hypernym chain for 'river'

WordNet is freely and publicly available for download from *http:// wordnet.princeton.edu/obtain.*

WordNets for other languages have also been developed, e.g., EuroWordNet and Hindi WordNet. EuroWordNet covers European languages, including English, Dutch, Spanish, Italian, German, French, Czech, and Estonian. Other than language internal relations, it also contains multilingual relations from each WordNet to English meanings.

Hindi WordNet has been developed by CFILT (Resource Center for Indian Language Technology Solutions), IIT Bombay.[2] Its database consists of more than 26,208 synsets and 56,928 Hindi words.[3] It is organized using the same principles as English WordNet but includes some Hindi specific relations (e.g., causative relations). A total of 16 relations have been used in Hindi WordNet. Each entry consists of synset, gloss, and position of synset in ontology. Figure 12.7 shows the Hindi WordNet entry for the word 'आकांक्षा' (*aakanksha*).

Sense 1
laugh, express joy, express mirth — (produce laughter)
   => bray — (laugh loudly and harshly)
   => bellylaugh — (laugh a deep, hearty laugh)
   => roar, howl — (laugh unrestrainedly and heartily)
   => snicker, snigger — (laugh quietly)
   => giggle, titter — (laugh nervously; 'The girls giggled when the
        rock star came into the classroom')
   => break up, crack up — (laugh unrestrainedly)
   => cackle — (emit a loud, unpleasant kind of laughing)
   => guffaw, laugh loudly — (laugh boisterously)
   => chuckle, chortle, laugh softly — (laugh quietly or with
        restraint)
   => convulse — (be overcome with laughter)
   => cachinnate — (laugh loudly and in an unrestrained way)

Figure 12.6   Troponym relation for the word 'laugh'

Hindi WordNet can be obtained from the URL http://*www.cfilt.iitb.ac.in/ wordnet/webhwn/.*

CFLIT has also developed a Marathi WordNet. Figure 12.8 shows the Marathi WordNet (*http://www.cfilt.iitb.ac.in/wordnet/webmwn/wn.php*) entry for the word 'पाव' (*pau*).

12.2.1   Applications of WordNet

WordNet has found numerous applications in problems related with IR and NLP. Some of these are discussed here.

---

[2]http://www.cfilt.iitb.ac.in/
[3]Hindi WordNet Documentation http://www.cfilt.iitb.ac.in/hindi-wordnet-license01.pdf

Concept Identification in Natural Language

WordNet can be used to identify concepts pertaining to a term, to suit them to the full semantic richness and complexity of a given information need.

Word Sense Disambiguation

WordNet combines features of a number of the other resources commonly used in disambiguation work. It offers sense definitions of words, identifies synstes of synonyms, defines a number of semantic relations and is freely available. This makes it the (currently) best known and most utilized resource for word sense disambiguation. One of the earliest attempts to use WordNet for word sense disambiguation was in IR by Voorheese (1993). She used WordNet noun hierarchy (hypernym / hyponym) to achieve disambiguation. A number of other researchers have also used WordNet for the same purpose (Resnik 1995, 1997; Sussna 1993).

1. (R) अपेक्षा, आकांक्षा, अन्ववेक्षा – किसी पर भरोसा रखने की क्रिया कि अमुक कार्य उसके द्वारा हो जायेगा " हर पिता की अपने पुत्र से यह अपेक्षा रहती है कि वह अपने जीवन में सफल हो "

2. (R) इच्छा, अभिलाषा, आकांक्षा, ख्वाहिश, आरजू, तमन्ना, कामना, तलब, चेष्टा, हसरत, मुराद, पिपासा, प्यास, तृष्णा, मनोकामना, मनोकामना, मनोवांछा, मनोरथ, मनोभावना, मरजी, रजा, मर्जी, मन, रजा, मंशा, लिप्सा, लालसा, तुषा, चाह, अरमान, क्षुधा, भूख, भूक, छुधा, हवस, स्पृहा, अभीप्सा, अनु, अपेक्षिता, अभिकांक्षा, अभिकाम, वांछा, वाछना, वाञ्छा, अभिध्या, अभिलाष, अभिप्रीति, अभिमत, अभिमतता, अभिमति, अभि_लास, अभिलासा, अभिलाख्या, अभिलाख, अभिलाखना — वह मनोवृत्ति जो किसी बात या वस्तु की प्राप्ति की ओर ध्यान ले जाती है " इंसान की हर इच्छा पूरी नही होती / उसकी ज्ञान पिपासा बढ़ती जा रही है / मेरा आज खाने का मन नही है "

Figure 12.7 WordNet entry for the Hindi word आकांक्षा (*aakanksha*)

1. (R) पाव एक चतुर्थांशं-चौथा भाग "मी बाजारातून एक पाव चुरमुरे आणले"
2. (R) पाव, पावरोटी-गव्हाचे पीठ आंबवून केलेल एक खाद्यविशेष "मुंबईत बरेच लोक पाव खाऊन गुजराण करतात"
3. (R) पाव-पावाचे वजन "दुकानदार चहाच्या पूडीचे वजन करण्यासाठी पाव शोधत आहे"
4. (R) पाव-एक चतुर्थांशं बाटली दारू "एक पाव प्यायल्यावर तो वटवट करू लागला"

Figure 12.8 WordNet entry for the Marathi word पाव (*pau*)

Automatic Query Expansion

WordNet semantic relations can be used to expand queries so that the search for a document is not confined to the pattern-matching of query terms, but also covers synonyms. The work performed by Voorhees (1994) is based on the use of WordNet relations, such as synonyms, hypernyms, and hyponyms, to expand queries.

Document Structuring and Categorization

The semantic information extracted from WordNet, and WordNet conceptual representation of knowledge, have been used for text categorization (Scott and Matwin 1998).

Document Summarization

WordNet has found useful application in text summarization. The approach presented by Barzilay and Elhadad (1997) utilizes information from WordNet to compute lexical chains.

12.3   FRAMENET

FrameNet[4] is a large database of semantically annotated English sentences. It is based on principles of frame semantics. It defines a tagset of semantic roles called the frame element. Sentences from the British National Corpus are tagged with these frame elements. The basic philosophy involved is that each word evokes a particular situation with particular participants. FrameNet aims at capturing these situations through case-frame representation of words (verbs, adjectives, and nouns). The word that invokes a frame is called *target word* or *predicate*, and the participant entities are defined using semantic roles, which are called *frame elements*. The FrameNet ontology can be viewed as a semantic level representation of predicate argument structure.

Each frame contains a main lexical item as predicate and associated frame-specific semantic roles, such as AUTHORITIES, TIME, and SUSPECT in the ARREST frame, called frame elements. As an example, consider sentence (12.1) annotated with the semantic roles AUTHORITIES and SUSPECT. The target word in sentence (12.1) is 'nab' which is a verb in the ARREST frame.

[Authorities The police] nabbed [suspect the snatcher].   (12.1)

A COMMUNICATON frame has the semantic roles ADDRESSEE, COMMUNICATOR, TOPIC, and MEDIUM. Figure 12.9 shows the core and non-core frame elements of the COMMUNICATION frame, along with other details. A JUDGEMENT frame contains roles such as JUDGE, EVALUEE, and REASON. A frame may inherit roles from another frame. For example, a STATEMENT frame may inherit from a COMMUNICATION frame; it contains roles such as SPEAKER, ADDRESSEE, and MESSAGE. The following sentences show some of these roles:

[Judge She] [Evaluee blames the police] [Reason for failing to provide enough protection].   (12.2)

[Speaker She] told [Addressee me] [Message 'I'll return by 7:00 pm today'].
(12.3)

---

[4]http://framenet.icsi.berkeley.edu/

12.3.1   FrameNet Applications

Gildea and Jurafsky (2002) and Kwon et al. (2004) used FrameNet data for automatic semantic parsing. The shallow semantic role obtained from FrameNet can play an important role in information extraction. For example, a semantic role makes it possible to identify that the theme role played by 'match' is same in sentences (12.4) and (12.5) though the syntactic role is different.

The umpire stopped the match.   (12.4)
The match stopped due to bad weather.   (12.5)

In sentence (12.4), the word 'match' is the object, while it is the subject in sentence (12.5).

Semantic roles may help in the question-answering system. For example, the verb 'send' and 'receive' would share the semantic roles SENDER, RECIPIENT, GOODS, etc., (Gildea and Jurafsky 2002) when defined with respect to a common TRANSFER frame. Such common frames allow a question-answering system to answer a question such as 'Who sent packet to Khushbu?' using sentence (12.6).

Khushbu received a packet from the examination cell.   (12.6)

Other applications include IR (Mohit and Narayanan 2003), interlingua for machine translation, text summarization, and word sense disambiguation.

	Communication
Frame Elements	
Core:	
Addressee [Add]	Receiver of Message from the Communicator.
Communicator [Com]	The person conveying (written or spoken) a message to another person.
Message [Msg]	A proposition or set of propositions that the Communicator wants the Addressee to convey
Topic [Top]	The entity that the proposition(s) are about.
Non-core:	
Amount_of_information [Amo]	The amount of information exchanged when communication occurs.
Depictive [Dep-Act]	The Depictive describes the state of the Communicator.
Duration []	The length of time during which the communication takes place.
Manner [Manr]	The Manner in which the Communicator communicates.
Means [Mns]	The Means by which the Communicator communicates.
Medium [Medium]	The physical or abstract setting in which the Message is conveyed.
Time []	The time at which the communication takes place.

Inherits From:
Is Inherited By: Communication_noise, Statement
Subframe of:
Has Subframes:
Uses: Topic
Is Used By: Claim_ownership, Communication_response, Contacting, Deny_permission, Discussion, Hear, Questioning, Reasoning, Reporting, Request, etc
Is Inchoative of:
Is Causative of:
See Also:
Sample Predicates
communicate, indicate, signal, speech

Figure 12.9   Frame elements of communication frame

12.4   STEMMERS

As discussed in Chapter 3, stemming, often called conflation, is the process of reducing inflected (or sometimes derived) words to their base or root form. The stem need not be identical to the morphological base of the word; it is usually sufficient that related words map to the same stem, even if this stem is not in itself a valid root. Stemming is useful in search engines for query expansion or indexing and other NLP problems. Stemming programs are commonly referred to as stemmers. The most common algorithm for stemming English is Porter's algorithm[5] (Porter 1980). Other existing stemmers include Lovins[6] stemmer (Lovins 1968) and a more recent one called the Paice/Husk stemmer[7] (Paice 1990). Figure 12.10 shows a sample text and output produced using these stemmers.

Input Text:
Such an analysis can reveal features that are not easily visible from the variations in the individual genes and can lead to a picture of expression that is more biologically transparent and accessible to interpretation.

Output:
Lovins stemmer: such an analys can rev feature that ar not eas vis from the vari in th individu gen and can lead to a picture of expres that is mor biolog transpar and acces to interpres

Porter's stemmer: such an analysi can reveal feature that ar not easily visible from the variat in the individu gene and can lead to a picture of express that is more biolog transpar and access to interpret

Paice stemmer: Such an analys can rev feat that are not easy vis from the vary in the individ gen and can lead to a pict of express that is mor biolog transp and access to interpret

---

Figure 12.10   Stemmed text using different stemmers

12.4.1   Stemmers for European Languages

There are many stemmers available for English and other languages. Snowball[8] presents stemmers for English, Russian, and a number of other European languages, including French, Spanish, Portuguese, Hungarian, Italian, German, Dutch, Swedish, Norwegian, Danish, and Finnish. The links for stemming algorithms for these languages can be found at *http://snowball.tartarus.org/texts/stemmersoverview.html.*

---

[5]http://tartarus.org/~martin/PorterStemmer/
[6]http://sourceforge.net/project/showfiles.php?group_id=24260
[7]http://www.comp.lancs.ac.uk/computing/research/stemming/Links/implementations.htm
[8]http://snowball.tartarus.org/

12.4.2 Stemmers for Indian Languages

Standard stemmers are not yet available for Hindi and other Indian languages. The major research on Hindi stemming has been accomplished by Ramanathan and Rao (2003) and Majumder et al. (2007). Ramanathan and Rao (2003) based their work on the use of handcrafted suffix lists. Majumder et al. (2007) used a cluster-based approach to find classes of root words and their morphological variants. They used a task-based evaluation of their approach and reported that stemming improves recall for Indian languages. Their observation on Indian languages was based on a Bengali data set. The Resource Centre of Indian Language Technology (CFILT), IIT Bombay has also developed stemmers for Indian languages, which are available at *http://www.cfilt.iitb.ac.in.*

12.4.3 Stemming Applications

Stemmers are common elements in search and retrieval systems such as Web search engines. Stemming reduces the variants of a word to same stem. This reduces the size of the index and also helps retrieve documents that contain variants of a query terms. For example, a user issuing a query for documents on 'astronauts' would like documents on 'astronaut' as well. Stemming permits this by reducing both versions of the word to the same stem. However, the effectiveness of stemming for English query systems is not too great, and in some cases may even reduce precision.

Text summarization and text categorization also involve term frequency analysis to find features. In this analysis, stemming is used to transform various morphological forms of words into their stems.

12.5 PART-OF-SPEECH TAGGER

Part-of-speech tagging is used at an early stage of text processing in many NLP applications such as speech synthesis, machine translation, IR, and information extraction. In IR, part-of-speech tagging can be used in indexing (for identifying useful tokens like nouns), extracting phrases and for disambiguating word senses. The rest of this section presents a number of part-of-speech taggers that are already in place.

12.5.1 Stanford Log-linear Part-of-Speech (POS) Tagger

This POS Tagger is based on maximum entropy Markov models. The key features of the tagger are as follows:
(i) It makes explicit use of both the preceding and following tag contexts via a dependency network representation.

(ii) It uses a broad range of lexical features.

(iii) It utilizes priors in conditional log-linear models.

The reported accuracy of this tagger on the Penn Treebank WSJ is 97.24%, which amounts to an error reduction of 4.4% on the best previous single automatically learned tagging result (Tuotanova et al. 2003). Details on the tagger can be found at the link *http://nlp.stanford.edu/software/tagger.shtml*.

12.5.2   A Part-of-Speech Tagger for English[9]

This tagger uses a bi-directional inference algorithm for part-of-speech tagging. It is based on maximum entropy Markov models (MEMM). The algorithm can enumerate all possible decomposition structures and find the highest probability sequence together with the corresponding decomposition structure in polynomial time. Experimental results of this part-of-speech tagger show that the proposed bi-directional inference methods consistently outperform unidirectional inference methods and bi-directional MEMMs give comparable performance to that achieved by state-of-the-art learning algorithms, including kernel support vector machines (Tsuruoka and Tsujii 2005).

12.5.3   TnT tagger[10]

Trigrams'n'Tags or TnT (Brants 2000) is an efficient statistical part-of-speech tagger. This tagger is based on hidden Markov models (HMM) and uses some optimization techniques for smoothing and handling unknown words. It performs at least as well as other current approaches, including the maximum entropy framework. Table 12.1 shows tagged text of document #93 of the CACM collection.

Table 12.1   Doc #93 of CACM collection tagged using TnT tagger

A	DT	simple	JJ
technique	NN	algebraic	JJ
is	VBZ	formulas	NNS
shown	VBN	into	IN
for	IN	a	DT
enabling	VBG	three	CD
a	DT	address	NN
computer	NN	computer	NN
to	TO	code	NN
translate	VB		

---

[9]http://www.tsujii.is.s.u-tokyo.ac.jp/~tsuruoka/postagger/
[10]http://www.coli.uni-saarland.de/~thorsten/tnt/

12.5.4 Brill Tagger

Brill (1992) described a trainable rule-based tagger that obtained performance comparable to that of stochastic taggers. It uses transformation-based learning to automatically induce rules. A number of extensions to this rule-based tagger have been proposed by Brill (1994). He describes a method for expressing lexical relations in tagging that stochastic taggers are currently unable to express. It implements a rule-based approach to tagging unknown words. It demonstrates how the tagger can be extended into a k-best tagger, where multiple tags can be assigned to words in some cases of uncertainty. Brill tagger is available for download at the link *http://www.cs.jhu.edu/~brill/RBT1_14.tar.Z*.

12.5.5 CLAWS Part-of-Speech Tagger for English

Constituent likelihood automatic word-tagging system (CLAWS) is one of the earliest probabilistic taggers for English. It was developed at the University of Lancaster (*http://ucrel.lancs.ac.uk/claws*). The latest version of the tagger, CLAWS4, can be considered a hybrid tagger as it involves both probabilistic and rule-based elements. It has been designed so that it can be easily adapted to different types of text in different input formats. CLAWS has achieved 96–97% accuracy. The precise degree of accuracy varies according to the type of text. For more information on the CLAWS tagger, see Garside (1987), Leech, Garside, and Bryant (1994), Garside (1996), and Garside and Smith (1997).

12.5.6 Tree-Tagger

Tree-Tagger (Schmidt 1994) is a probabilistic tagging method. It avoids problems faced by the Markov model methods when estimating transition probabilities from sparse data, by using a decision tree to estimate transition probabilities. The decision tree automatically determines the appropriate size of the context to be used in estimation. The reported accuracy for the tagger is above 96% on the Penn-Treebank WSJ corpus. The tagger is available at the link *http://www.ims.uni-stuttgart.de/projekte/corplex/TreeTagger/DecisionTreeTagger.html*.

12.5.7 ACOPOST: A Collection of POS Taggers[11]

ACOPOST is a set of freely available POS taggers. The taggers in the set are based on different frameworks. The programs are written in C. ACOPOST currently consists of the following four taggers.

---

[11] http://acopost.sourceforge.net/

Maximum Entropy Tagger (MET)

This tagger is based on a framework suggested by Ratnaparkhi (1997). It uses an iterative procedure to successively improve parameters for a set of features that help to distinguish between relevant contexts.

Trigram Tagger (T3)

This tagger is based on HMM. The states in the model are tag pairs that emit words. The technique has been suggested by Rabiner (1990) and the implementation is influenced by Brants (2000).

Error-driven Transformation-based Tagger (TBT)

This tagger is based on the transformation-based tagging approach proposed by Brill (1993). It uses annotated corpuses to learn transformation rules, which are then used to change the assigned tag using contextual information.

Example-based Tagger (ET)

The underlying assumption of example-based models (also called memory-based, instance-based or distance-based models) is that cognitive behaviour can be achieved by looking at past experiences that match the current problem, instead of learning and applying abstract rules. This framework has been suggested for NLP by Daelemans et al. (1996).

12.5.7   POS Tagger for Indian Languages

The automatic text processing of Hindi and other Indian languages is constrained heavily due to lack of basic tools and large annotated corpuses. Research groups are now focusing on removing these bottlenecks. The work on the development of tools, techniques, and corpora is going on at several places such as CDAC, IIT Bombay, IIIT Hyderabad, University of Hyderabad, CIIL Mysore, and University of Lancaster. IIT Bombay is involved in the development of morphology analysers and part-of-speech taggers for Hindi and Marathi. Both these languages have rich morphological structures. Their approach is based on *bootstrapping on a small corpus tagged* by a rule-based tagger and then applying statistical techniques to train a machine. More information can be found at *http:// ltrc.iiit.net* and *www.cse.iitb.ac.in.* Work on Urdu part-of-speech taggers has been reported by Hardie (2003) and Baker et al. (2004).

12.6 RESEARCH CORPORA

Research corpora have been developed for a number of NLP-related tasks. In the following section, we point out few of the available standard document collections for a variety of NLP-related tasks, along with their Internet links.

12.6.1 IR Test Collection

We have already provided a list of IR test document collection in Chapter 9. Glasgow University, UK, maintains a list of freely available IR test collections. Table 12.2 lists the sources of those and few more IR test collections.

LETOR (learning to rank) is a package of benchmark data sets released by Microsoft Research Asia. It consists of two datasets OHSUMED and TREC (TD2003 and TD2004). LETOR is packaged with extracted features for each query-document pair in the collection, baseline results of several state-of-the-art learning-to-rank algorithms on the data and evaluation tools. The data set is aimed at supporting future research in the area of learning ranking function for information retrieval.

Table 12.2 IR test collection

LETOR	*http://research.microsoft.com/users/tyliu/LETOR/*
LISA	
CACM	
CISI	
MEDLINE	*http://www.dcs.gla.ac.uk/idom/ir_resources/test_collections/*
Cranfield	
TIME	
ADI	

12.6.2 Summarization Data

Evaluating a text summarizing system requires existence of 'gold summaries'. DUC provides document collections with known extracts and abstracts, which are used for evaluating performance of summarization systems submitted at TREC conferences. Figure12.11 shows a sample document and its extract from DUC 2002 summarization data.

```
<DOC>
<DOCNO> AP880911-0016 </DOCNO>
<FILEID>AP-NR-09-11-88 0423EDT</FILEID>
<FIRST>r i BC-HurricaneGilbert 09-11 0339</FIRST>
<SECOND>BC-Hurricane Gilbert,0348</SECOND>
<HEAD>Hurricane Gilbert Heads Toward Dominican Coast</HEAD>
<BYLINE>By RUDDY GONZALEZ</BYLINE>
<BYLINE>Associated Press Writer</BYLINE>
<DATELINE>SANTO DOMINGO, Dominican Republic (AP) </DATELINE>
<TEXT>
```

   Hurricane Gilbert swept toward the Dominican Republic Sunday, and the Civil Defense alerted its heavily populated south coast to prepare for high winds, heavy rains, and high seas.

   The storm was approaching from the southeast with sustained winds of 75 mph gusting to 92 mph.

   "There is no need for alarm," Civil Defense Director Eugenio Cabral said in a television alert shortly before midnight Saturday.

   Cabral said residents of the province of Barahona should closely follow Gilbert's movement. An estimated 100,000 people live in the province, including 70,000 in the city of Barahona, about 125 miles west of Santo Domingo.

   Tropical Storm Gilbert formed in the eastern Caribbean and strengthened into a hurricane Saturday night. The National Hurricane Center in Miami reported its position at 2 a.m. Sunday at latitude 16.1 north, longitude 67.5 west, about 140 miles south of Ponce, Puerto Rico, and 200 miles southeast of Santo Domingo.

   The National Weather Service in San Juan, Puerto Rico, said Gilbert was moving westward at 15 mph with a "broad area of cloudiness and heavy weather" rotating around the center of the storm.

   The weather service issued a flash flood watch for Puerto Rico and the Virgin Islands until at least 6 p.m. Sunday.

   Strong winds associated with the Gilbert brought coastal flooding, strong southeast winds and up to 12 feet to Puerto Rico's south coast. There were no reports of casualties.

   San Juan, on the north coast, had heavy rains and gusts Saturday, but they subsided during the night.

   On Saturday, Hurricane Florence was downgraded to a tropical storm and its remnants pushed inland from the U.S. Gulf Coast. Residents returned home, happy to find little damage from 80 mph winds and sheets of rain.

   Florence, the sixth named storm of the 1988 Atlantic storm season, was the second hurricane. The first, Debby, reached minimal hurricane strength briefly before hitting the Mexican coast last month.

```
</TEXT>
</DOC>
```

Extract

Tropical Storm Gilbert in the eastern Caribbean strengthened into a hurricane Saturday night. The National Hurricane Center in Miami reported its position at 2 a.m. Sunday to be about 140 miles south of Puerto Rico and 200 miles southeast of Santo Domingo. It is moving westward at 15mph with a broad area of cloudiness and heavy weather with sustained winds of 75mph gusting to 92mph. The Dominican Republic's Civil Defense alerted that country's heavily populated south coast and the National Weather Service in San Juan, Puerto Rico issued a flood watch for Puerto Rico and the Virgin Islands until at least 6 p.m. Sunday.

Figure 12.11   Sample document from DUC 2002 and its extract

12.6.3   Word Sense Disambiguation

SEMCOR[12] is a sense-tagged corpus used in disambiguation. It is a subset of the Brown corpus, sense-tagged with WordNet synsets. Open Mind Word Expert[13] attempts to create a very large sense-tagged corpus. It collects word sense tagging from the general public over the Web.

12.6.4   Asian Language Corpora

The multilingual EMILLE corpus is the result of the enabling minority language engineering (EMILLE) project at Lancaster University, UK. The project focuses on generation of data, software resources and basic language engineering tools for the NLP of south Asian languages. Central Institute for Indian Languages (CIIL), the Indian partner in the project, extended the set of target languages to include a number of Indian languages. CIIL provides a wider range of data in these languages from a wide range of genres. The data sources that EMILLE made available include monolingual written and spoken corpuses, parallel and annotated corpuses. The full EMILLE/CIIL corpus is available for free, but for research use only, at the link *http://www.elda.org/catalogue/en/text/W0037.html*. Further details about the corpus can be found in the manual at the site *http://www.emille.lancs.ac.uk/manual.pdf*.

Corpus building in these languages is constrained by the scarcity of repositories of electronic text. The monolingual corpus includes written data for 14 South Asian languages and spoken data for five languages (Hindi, Bengali, Gujrati, Punjabi, and Urdu). The spoken corpus was constructed from radio broadcasts on the BBC Asia network. The parallel corpus contains English text and its translation in five languages. The text includes UK government advice leaflets which are published in multiple languages. The corpus is aligned at sentence level. The parallel corpus provided by EMILLE corpus is a valuable resource for statistical machine translation research. The annotated component includes Urdu data annotated for part-of-speech tagging, and a Hindi corpus annotated to show nature of demonstrative use.

12.7   JOURNALS AND CONFERENCES IN THE AREA

A wide number of conference proceedings and journals report research in the various areas of NLP. Most notable among them are those associated with Association for Computing Machinery (ACM), Association for

---

[12] http://www.cs.unt.edu/~rada/downloads.html#semcor
[13] http://teach-computers.org.

Computational Linguistics (ACL), its European counterpart EACL, Recherche d'Information Assistie par Ordinateur (RIAO) and the International Conferences on Computational Linguistics (COLING). The ACM SIGIR Conference is one of the major conferences held on research and development in information retrieval. It provides the international forum for dissemination of research and demonstration of new systems and techniques. The 30th Annual International ACM SIGIR Conference[14] was held on 23–27 July 2007 at Amsterdam. The Proceedings of Text Retrieval Conferences (TRECs)[15] are another important source of information. These proceedings report results from standardized evaluations organized by the US government. The TRECs have been organized regularly since 1992 as a part of the TIPSTER text retrieval. They were earlier known as the Document Understanding Conference or Message Understanding Conferences. The conference series is sponsored by the National Institute of Standards and Technology (NIST) with additional support from other US government agencies. The ACM Special Interest Group on Information Retrieval (ACM-SIGIR) focuses on IR related tasks, and ECIR is its European counterpart. The NTCIR (NII test collection for IR) focuses on information retrieval with Japanese and other Asian languages.

KES[16] International Conferences in Knowledge-Based and Intelligent Engineering & Information Systems have been a regular feature since 1997. The conference mainly focuses on applications of intelligent systems. The topics covered by KES includes general intelligent topics like neural networks, fuzzy techniques, genetic algorithms, knowledge representation and management, applications using intelligent techniques (e.g., speech processing and synthesis and NLP) and emerging intelligent technologies like intelligent information retrieval, intelligent web mining and applications, intelligent user interfaces, etc.

HLT-NACCL is sponsored by the North American chapter of the Association for Computational Linguistics.

The *Journal of Computational Linguistics* is a leading premier publication focussing on theoretical and linguistics aspects. More practical applications are covered in the *Natural Language Engineering Journal*. *Information Retrieval* by Kluwer, *Information Processing and Management* by Elsevier, ACM's *Transactions on Information Systems* (TOIS), *Journal of American Society for Information Sciences* are major journals covering a wide range of information

---

[14] http://www.sigir2007.org/
[15] http://trec.nist.gov/
[16] http://www.kesinternational.org/conferences.php

Research, International Journal of Information Technology and Decision Making (World Scientific), and *Journal of Digital Information Management and Information System.*

A few AI publications also report work on language processing. Among these are *Artificial Intelligence, Computational Intelligence,* IEEE's *Transaction on Intelligent Systems,* and *Journal of AI Research.*

SUMMARY

- Lexical resources such as WordNet and FrameNet can be used in a number of NLP-related tasks.
- Stemmers are useful in a number of information processing tasks such as information retrieval, text summarization, and text categorization.
- Widely known stemmers include Porter's and Lovins stemmers.
- Part-of-speech tagger is used to assign a part-of-speech, such as noun, verb, pronoun, preposition, adverb, and adjective, to each word in a sentence (or text).
- Taggers include stanford log-linear part-of-speech tagger, TnT, CLAWS, and Brill's tagger.
- TREC and SIGIR conferences offer useful resources for a number of information processing-related tasks.

REFERENCES

Ananthkrishnan, R. and Durgesh Rao, 2003, 'A lightweight stemmer for Hindi,' Workshop on Computational Linguistics for South Asian Languages, *The 10th Conference of the European Chapter of the Association for Computational Linguistics (EACL'03),* ACL, Morristown, NJ.

Baker, P., A. Hardie, A.M. McEnery, and B.D. Jayaram, 2004, 'Corpus linguistics and South Asian languages: corpus creation and tool development,' *Literary and Linguistic Computing,* 19(4), 509–24.

Barzilay, Regina and Michael Elhadad, 1997, 'Using lexical chains for text summarization,' *Proceedings of the Intelligent Scalable Text Summarization Workshop (ISTS'97),* ACL, Madrid.

Brants, Thosrten, 2000, 'TnT—as statistical part-of-speech tagger,' *Proceedings of the Sixth Applied Natural Language Processing Conference (ANLP-2000),* Seattle, WA.

Brill E., 1992, 'A simple rule-based part-of-speech tagger,' *Proceedings of the Third Conference on Applied Natural Language Processing,* ACL, Budapest, Hungary.

_____1994, 'Some advances in rule-based part-of-speech tagging,' *Proceedings of the Twelfth National Conference on Artificial Intelligence (AAAI-94)*, Seattle.

Daelemans, Walter, Jakub Zavrel, Peter Berck, and Steven Gillis, 1996, 'MBT: A memory-based part-of-speech tagger generator,' *Proceedings of the Fourth Workshop on Very Large Corpora*, pp. 14–27, Kopenhagen, Denmark.

Garside, R., 1987, 'The CLAWS Word-tagging System,' *The Computational Analysis of English: A Corpus-based Approach*, Longman, London.

_____1996, 'The robust tagging of unrestricted text: the BNC experience,' *Using Corpora for Language Research: Studies in the honour of Geoffrey Leech*, Longman, London, pp. 167–80.

Garside, R. and N. Smith, 1997, 'A hybrid grammatical tagger: CLAWS4,' *Corpus Annotation: Linguistic Information from Computer Text Corpora*, Longman, London, pp. 102–21.

Gildea, D. and D. Jurafsky, 2002, 'Automatic labelling of semantic roles,' *Computational Linguistics*, 28(3), pp. 245–88.

Hardie, A., 2003, 'Developing a tagset for automated part-of-speech tagging in Urdu,' *Proceedings of the Corpus Linguistics 2003 Conference*, UCREL Technical Papers, 16, Department of Linguistics, Lancaster University, UK.

Kwon, Namhee, Michael Fleischman, and Eduard Hovy, 2004, 'FrameNet-based semantic parsing using maximum entropy models,' *Proceedings of the 20th International Conference on Computational Linguistics*, Geneva, Switzerland.

Leech, G., R. Garside, and M. Bryant, 1994, 'CLAWS4: The tagging of the British National Corpus,' *Proceedings of the 15th International Conference on Computational Linguistics (COLING 94)*, Kyoto, Japan, pp. 622–28.

Majumder, Prasenjit, Mandar Mitra, Swapan K. Parul, and Gobinda Kole, 2007, 'YASS: Yet Another Suffix Stripper,' *ACM Transactions on Information Systems*, 25(4).

Miller, G., 1990, 'WordNet: an online lexical database,' *International Journal of Lexicography*, 24, pp. 513–23.

_____1995, 'WordNet: a lexical database for English,' *Communication of the ACM*, 38(11).

Narayanan, Srinivas, Charles J. Fillmore, Collin F. Baker, and Miriam R.L. Petruck, 2002, 'FrameNet meets the semantic web: A DAML+OIL frame representation,' *Proceedings of the 18th National Conference on Artificial Intelligence*, AAAI Edmonton, Alberta.

Ramanathan, A. and D. Rao, 2003, 'Lightweight stemmer for Hindi,' *Proceedings of the 10th Conference of the European Chapter of the Association for Computational Linguistics (EACL) on Computational Linguistics for South Asian Languages*, Budapest.

Ratnaparkhi, Adwait, 1998, 'Maximum entropy models for natural language ambiguity resolution,' *PhD Thesis*, University of Pennsylvania.

Resnik, P., 1995, 'Disambiguating noun groupings with respect to WordNet senses,' *Proceedings of the Third Workshopon Very Large Corpora*, Cambridge, Massachusetts, pp. 54–68.

_____1997, 'Selectional preference and sense disambiguation,' *Proceedings of the ACL SIGLEX Workshop on Tagging Text with Lexical Semantics: Why, What, and How?*, ACL, Somerset, New Jersey, pp. 52–57.

Schmid, H., 1994, 'Probabilistic part-of-speech tagging using decision trees,' *Proceedings of International Conference on New Methods in Language Processing*.

Scott S. and S. Matwin, 1998, 'Text classification using WordNet hypernyms,' *Proceedings of the COLING/ACL Workshop on Usage of WordNet in Natural Language Processing Systems*, Montreal.

Sussna, Michael, 1993, 'Word sense disambiguation for free-text indexing using a massive semantic network,' *Proceedings of the Second International Conference on Information and Knowledge Base Management, CIKM'93*, Arlington, Virginia, pp. 67–74.

Toutanova, Kristina, Dan Klein, Christopher D. Manning, and Yoram Singer, 2003, 'Feature-rich part-of-speech tagging with a cyclic dependency network,' *Proceedings of HLT-NAACL* (available at http://acl.ldc.upenn.edu/N/N03-1033.pdf), pp. 252–59.

Voorhees, E.M., 1993, 'Using WordNet to disambiguate word senses for text retrieval,' *Proceedings of the 16th Annual International ACM SIGIR Conference on Research and Development in Information Retrieval*, Pittsburgh, Pennsylvania, pp.171–80.

_____1994, 'Query expansion using lexical-semantic relations,' *Proceedings of the 17th Annual International ACM-SIGIR Conference on Research and Development in Information Retrieval*, Springer-Verlag, London, pp. 61–69.

Yoshimasa Tsuruoka and Jun'ichi Tsujii, 2005, 'Bi-directional inference with the easiest-first strategy for tagging sequence data,' *Proceedings of HLT/EMNLP*, pp. 467–74.

APPENDIX A

PENN TREEBANK TAGSET

CC	Coordinating conjunction—for example and, but, and or
CD	Cardinal number
DT	Determiner
EX	Existential there
FW	Foreign word
IN	Preposition or subordinating conjunction
JJ	Adjective
JJR	Adjective, comparative
JJS	Adjective, superlative
LS	List item marker
MD	Modal—for example can, could, might, and may
NN	Noun, singular or mass
NNP	Proper noun, singular
NNPS	Proper noun, plural
NNS	Noun, plural
PDT	Pre-determiner—for example all and both when they precede an article
POS	Possessive ending—for example nouns ending in 's'
PRP	Personal pronoun—for example I, me, you, and he
PRP$	Possessive pronoun—for example my, your, mine, and yours
RB	Adverb—most words that end in -ly as well as degree words such as quite, too, and very
RBR	Adverb, comparative—adverbs with the comparative ending -er, with a strictly comparative meaning

RBS	Adverb, superlative
RP	Particle
SYM	Symbol—should be used for mathematical, scientific, or technical symbols
TO	to
UH	Interjection—for example uh, well, yes, and my
VB	Verb, base form—subsumes imperatives, infinitives, and subjunctives
VBD	Verb, past tense—includes the conditional form of the verb to be
VBG	Verb, gerund or present participle
VBN	Verb, past participle
VBP	Verb, non-third person singular present
VBZ	Verb, third person singular present
WDT	Wh-determiner—for example which and that when it is used as a relative pronoun
WP	Wh-pronoun—for example what, who, and whom
WP$	Possessive wh-pronoun
WRB	Wh-adverb—for example how, where, and why

Punctuation tags
#
$
"
(
)
,
.
:
"

APPENDIX B

PORTER STEMMER

The Porter stemming algorithm consists of condition/action pairs. Actions are in the form of rewrite rules. The conditions may be on stems, suffix, or rules. The stem conditions take either of the following form:

(i) $m = 0$, 1, or 2.

(ii) Stem contains or ends with (pattern).

where m, the measure, is the number of vowel–consonant (VC) sequence. For example, for the word in 'tea' and 'coffee', $m = 0$ as the number of vowel–consonant sequence is 0, whereas for the word 'astronaut', $m = 3$ ('as', 'on', and 'ut').

The patterns are of the form:

*$<x>$ stem ends with a given letter x

v   stem contains a vowel

*d stem ends in a double consonant

*o stem ends in a consonant–vowel–consonant sequence, where the final consonant is not w, x, or y.

Suffix condition takes the form 'current_suffix = = pattern' and rule conditions take the form 'rule was used'. Action rules are of the form 'old_suffix → new_suffix'.

The steps involved in Porter stemming algorithm are given below:

Step 1: Deal with plurals and past participles.

Step 1a:

Action rules	Examples
SSES → SS	dresses → dress
IES → I	patties → patti
SS → SS	success → success
S → null	girls → girl

Step 1b:

Conditions	Action rules	Examples
$(m > 0)$	EED → EE	succeed → succee
		deed → deed
$(*v*)$	ED → null	walked → walk
		fled → fled
$(*v*)$	ING → null	playing → play
		sing → sing

Step 1b1:

Conditions	Action rules	Examples
Null	AT → ATE	rotat(ed) → rotate
Null	BL → BLE	struggl(ed) → struggle
Null	IZ → IZE	recogniz(ed) → recognize
(*d and not	double letter	letter scann(ed) → scan
(*L or *S or *Z))	→ single	call(ing) → call
		miss(ing) → miss
$(m = 1$ and *o$)$	null → E	trail(ing) → trail
		pil(ing) → pile

Step 1c:

$(*v*)$ Y → I   lazy → lazi

spy → spy

Step 2:

$(m > 0)$ ATIONAL → ATE	rotational → rotate
$(m > 0)$ TIONAL → TION	proportional → proportion
$(m > 0)$ ENCI → ENCE	valenci → valence
$(m > 0)$ ANCI → ANCE	hesitanci → hesitance
$(m > 0)$ IZER → IZE	recognizer → recognize
$(m > 0)$ ABLI → ABLE	stabli → stable
$(m > 0)$ ALLI → AL	practicalli → practical
$(m > 0)$ ENTLI → ENT	differentli → different
$(m > 0)$ ELI → E	vileli → vile
$(m > 0)$ OUSLI → OUS	previousli → previous
$(m > 0)$ IZATION → IZE	privatization → privatize
$(m > 0)$ ATION → ATE	predication → predicate
$(m > 0)$ ATOR → ATE	dictator → dictate
$(m > 0)$ ALISM → AL	socialism → social

($m > 0$) IVENESS → IVE	comprehensiveness → comprehensive	
($m > 0$) FULNESS → FUL	successfulness → successful	
($m > 0$) OUSNESS → OUS	obviousness → obvious	
($m > 0$) ALITI → AL	realiti → real	
($m > 0$) IVITI → IVE	transitiviti → transitive	
($m > 0$) BILITI → BLE	reliabiliti → reliable	

Step 3:

($m > 0$) ICATE → IC	replicate → replic
($m > 0$) ATIVE → null	formative → form
($m > 0$) ALIZE → AL	conceptualize → conceptual
($m > 0$) ICITI → IC	electriciti → electric
($m > 0$) ICAL → IC	aeronautical → aeronautic
($m > 0$) FUL →	hopeful → hope
($m > 0$) NESS →	goodness → good

Step 4:

($m > 1$) AL →	revival → reviv
($m > 1$) ANCE →	allowance → allow
($m > 1$) ENCE →	preference → prefer
($m > 1$) ER →	hardliner → hardlin
($m > 1$) IC →	endoscopic → endoscop
($m > 1$) ABLE →	adjustable → adjust
($m > 1$) IBLE →	defensible → defens
($m > 1$) ANT →	irritant → irrit
($m > 1$) EMENT →	replacement → replac
($m > 1$) MENT →	adjustment → adjust
($m > 1$) ENT →	dependent → depend
($m > 1$ and (*S or *T)) ION →	adoption → adopt
($m > 1$) OU →	homologou → homolog
($m > 1$) ISM →	communism → commun
($m > 1$) ATE →	activate → activ
($m > 1$) ITI →	singulariti → singular
($m > 1$) OUS →	anologous → homolog
($m > 1$) IVE →	defective → defect
($m > 1$) IZE →	equalize → equal

The suffixes are now removed. All that remains is a little tidying up.

Step 5a:

$(m > 1)$ E →	equate → equat
	date → date
$(m = 1$ and not *o$)$ E → null	cease → ceas

Step 5b:

$(m > 1$ and *d and *L$)$	null → single letter
	controll → control
	roll → roll

CONCEPTUAL RELATIONS (CONRELS)

AGNT (Agent) Links [ACT] to the concept representing actor of the act. Example: Tuba bit an apple.

[PERRSON: TUBA] ←(AGNT) ←[BITE] →(PTNT) → [APPLE]

ATTR (Attribute) Links an entity to its attribute. That is to say, (ATTR) links [ENTITY: *x] TO [ENTITY: *y], where y is an attribute of x. Example: The water is cool.

[WATER] → (ATTR) → [COOL]

BETW (between) Has three arcs, first two arcs are linked to the things on either side of the third.
Example: A person is located between a rock and a hard place.

[PERSON] ←(BETW) –
← [ROCK]
← [PLACE]→(ATTR)→[HARD]   [156]

CAUS (Cause) It is a second order relation, links [state:*x] to [state:*y] where *x has a cause *y.

CHRC (Characteristics) Links [ENTITY:*x] to {ENTITY:*y], where *x has characteristics *y.
Example: Water is cold.

[WATER]→(CHRC) →[COLD]

DSCR (Description) The relation between a situation and the proposition that describes it.
Example: The cat on mat.

[SITUATION]→(DSCR)→[PROPOSITION:[CAT]→(ON)→[MAT]]. [156]

EXPR (Experiencer) It links an state to an animate who is experiencing that state.

MNR (Manner)   It links an [ACT] to an [ATTRIBUTE].

Example: Shereen drives slowly.

[SHEREEN] ← (AGNT) ← [DRIVE] → (MNR) →[SLOWLY]

MEAS (Measure)   It links a [DIMENSION] to a [MEASURE] of that dimension.

Example: The road is 10 km long.

[ROAD]→(CHRC)→[LENGTH]→(MEAS)→[MEASURE:10 KM]

OBJ(Object)   It links an [ACT] to an [ENTITY], which is acted upon.

Example: The cat swallowed the canary.

[CAT:#]←(AGNT) ←[SWALLOW]→(OBJ)→[CANARY:#]   [154]

PART   It links an [ENTITY:*x] to an [ENTITY:*y], where *y is part of *x.

PAST   It is a monadic relation that links to a [PROPOSITION] that was true at some time preceding the present.

Example: Zuha left.

(PAST) →[PROPOSITION: [PERSON: ZUHA] ←(AGNT)←[LEAVE]]

PTNT (Patient)   It links an [ACT] to an [ANIMATE].

QTY   It links a set of [ENTITY: {*}] to a [NUMBER] that indicates the number of entities in that set.

Example: There are 10 students in the class.

[CLASS]←(LOC) ←[STUDENT: {*}] → (QTY) → [NUMBER: 50]

RCPT (Recipient)   It links an [ACT] to an [ANIMATE], which receives the object or result of an action.

RSLT(Result)   It links an [ACT] to an [ENTITY], which is generated by the act.

Example: Arjun built a house.

[PERSON: Arjun]←(AGNT)←[BUILD]→[RSLT]→[HOUSE]   [154]

SUBT(Subtype)   It links a type to subtype its subtype.

Example: Elephant is a subtype of animal.

[TYPE: ELEPPHANT] ←(SUBT)←[TYPE: ANIMAL]   [156]

APPENDIX D

KNOWLEDGE-REPRESENTATION FORMALISM

A number of knowledge-representation formalisms have been introduced to represent natural language sentences. The most important among them are:

- Logic-based representations, e.g. propositional logic and first-order predicate logic
- Network-based representations, e.g. semantic networks and conceptual graphs
- Structured representations, e.g. frames and scripts

Use of conceptual graphs for knowledge representation has already been discussed in Chapter 10. Here, we briefly discuss some of the knowledge representation formalisms: propositional logic, first-order predicate logic (FOPL), semantic network, and frames.

Prepositional logic represents elementary sentences such as 'It is raining' using propositional symbols such as P, Q, and R. Compound propositions are formed from atomic formulas using the logical connectives conjunction (), disjunction (), negation (), implication (→), equivalence ().

Let   P = She works hard.

and   Q = She will get good marks.

Then the sentence 'If she works hard then she will get good marks' will be represented as: $P \rightarrow Q$.

Propositional logic uses inference rules to perform logical proofs or deductions. The FOPL was developed to extend the expressive capacity of propositional logic. It uses predicates, variables, functions, constants, quantifiers and logical connectives. The sentence

      Suha eats an apple.

is represented in FOPL as

       eats (Suha, apple)

where 'eats' is a predicate and 'apple' and 'Suha' are its arguments.

The FOPL makes use of two variable quantifiers, namely existential and universal quantifier. Existential quantifier is used to represent the fact that the sentence is true for at least one value in the domain, whereas universal quantifier is used to represent generalized statements. For example, consider the sentence:

All children like apple.

It can be represented in FOPL as

$(\forall)(children\ (x) \rightarrow like\ (apple, x)$

where $\forall$ is universal quantifier and $\rightarrow$ is implication read as 'implies'.

The previous examples demonstrate how natural language sentences can be represented in FOPL. Once translated, these logical expressions can be used to infer new facts.

A semantic network represents knowledge as a graph with labelled nodes and arcs. Nodes in the graph represent facts or concepts and arcs represent relations between them. For example, 'Suha eats apple' will be represented as

$$\boxed{\text{Suha}} \xrightarrow{\text{eats}} \boxed{\text{Apple}}$$

Common relations used are OWNS, HAS, IS-A, MEMBER-OF, SUBSET-OF, AKO, HAS-PART, INSTANCE-OF, ATTRIBUTES, etc.
For example, the sentence

Suha is a girl.

can be represented using IS-A relation as

$$\boxed{\text{Suha}} \xrightarrow{\text{IS-A}} \boxed{\text{Girl}}$$

The form of inferencing used in semantic network is termed as property inheritance, wherein an entity inherits characteristics from its ancestor nodes as long it does not leads to contradiction. One of the main problems in using semantic networks is the lack of standard relations. It uses general relations that fail to capture deeper semantic aspects of knowledge. Some efforts to standardization have been made, most notably among them is using case structure of English verbs (Simmon). Case relationships include agents, objects, instrument, location, and time. Conceptual dependency theory (Schank & Reiger, 1974) attempts to model deeper semantic structure of natural language.

Frames can be considered as a structuring device using which we can organize our knowledge about world. They were introduced by Mervin

Minsky (1975)[1] as a data structure to represent a mental model of a stereotypical situation such as driving a car or eating in a restaurant (Patterson 1990).[2] Frames are record-like structures that use slots and values to represent general or specific characteristics of entity and their values. A slot may have another frame as its value. This permits us to organize frames in a hierarchically.

---

[1]Mervin Minsky, 1975, 'A framework for representing knowledge,' In *The Psychology of Computer Vision*, P.H. Winston (Ed.), McGraw Hill, New York.
[2]Patterson, D.W., 1990, *Introduction to Artificial Intelligence and Expert Systems*, Prentice-Hall of India.

INDEX